The IMA Volumes in Mathematics and its Applications

Volume 166

Series Editor

Daniel Spirn, Institute for Mathematics and its Applications, University of Minnesota, Minneapolis, MN, USA

Ashlee N. Ford Versypt • Rebecca A. Segal •
Suzanne S. Sindi

Editors

Mathematical Modeling for Women's Health

Collaborative Workshop for Women
in Mathematical Biology

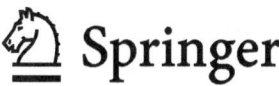 Springer

Editors

Ashlee N. Ford Versypt
Department of Chemical and Biological
Engineering
University at Buffalo, The State University
of New York
Buffalo, NY, USA

Rebecca A. Segal
Department of Mathematics and Applied
Mathematics
Virginia Commonwealth University
Richmond, VA, USA

Suzanne S. Sindi
Department of Applied Mathematics
University of California Merced
Merced, CA, USA

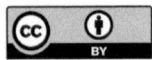

ISSN 0940-6573 ISSN 2198-3224 (electronic)
The IMA Volumes in Mathematics and its Applications
ISBN 978-3-031-58515-9 ISBN 978-3-031-58516-6 (eBook)
https://doi.org/10.1007/978-3-031-58516-6

Mathematics Subject Classification: 92-06, 92-B05

This work was supported by UnitedHealth Group Optum and University of Minnesota's Institute for
Mathematics and its Applications.

Contents

Collaborative Workshop for Women in Mathematical Biology: Mathematical Modeling for Women's Health

Ashlee N. Ford Versypt, Rebecca A. Segal, and Suzanne S. Sindi

1 Aim and Scope

Despite advances in health care, women continue to have disparate outcomes. Women experience higher morbidity and mortality, especially with regard to cancer, cardiovascular disease, and recently COVID-19 [1]. Reproductive health and maternity are also a continued source of health risk.

Mathematical modeling can be used to identify some of the underlying causes for the different outcomes experienced by women and begin to suggest possible solutions and pathways forward. The goal of the work is to use data and biological insight from experimental publications and collaborations and develop mathematical frameworks to explore questions of disparity.

This volume contains the scientific work from the *Collaborative Workshop for Women in Mathematical Biology*. The workshop brought together 44 researchers (Fig. 1) to collaborate on six problems that used mathematics to understand complex biological systems with implications for women's health. The workshop was held at the UnitedHealth Group campus in Minnetonka, Minneapolis during June 20–24,

A. N. Ford Versypt (✉)
Department of Chemical and Biological Engineering, University at Buffalo, The State University of New York, Buffalo, NY, USA
e-mail: ashleefv@buffalo.edu

R. A. Segal
Department of Mathematics and Applied Mathematics, Virginia Commonwealth University, Richmond, VA, USA
e-mail: rasegal@vcu.edu

S. S. Sindi
Department of Applied Mathematics, University of California Merced, Merced, CA, USA
e-mail: ssindi@ucmerced.edu

© The Author(s) 2024
A. N. Ford Versypt et al. (eds.), *Mathematical Modeling for Women's Health*,
The IMA Volumes in Mathematics and its Applications 166,
https://doi.org/10.1007/978-3-031-58516-6_1

1

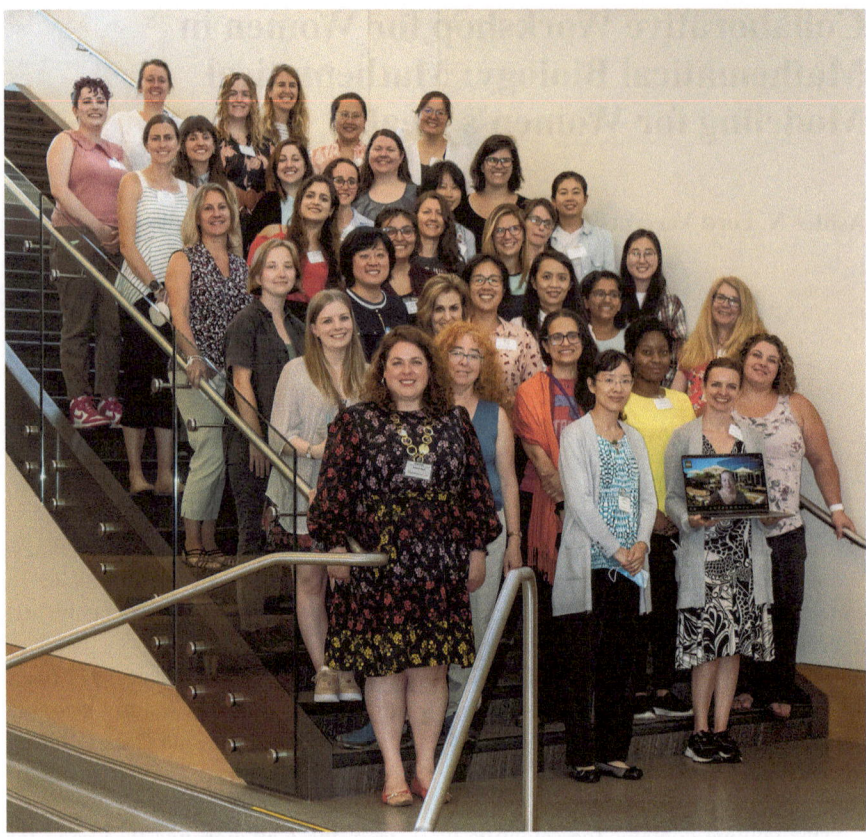

Fig. 1 Group photograph of the 2022 Collaborative Workshop for Women in Mathematical Biology participants at UnitedHealth Group Optum of Minnetonka, MN

2022 and was organized by Ashlee N. Ford Versypt, Rebecca Segal, Blerta Shtylla, and Suzanne S. Sindi. The articles contained in this volume were initiated during the intensive 1-week workshop and continued through follow-up collaborations afterward. Ashlee N. Ford Versypt served as the primary editor of this volume with generous support from 13 anonymous peer reviewers.

2 History and Context

Historically, women have been underrepresented in the mathematical sciences. Although progress has been made, the numbers remain unbalanced. In the most recent American Mathematical Society survey from 2018, only 17% of tenure-track mathematics faculty in doctoral departments are female [2]. Research reports from the American Association of University Women provide evidence on challenges that many early and midcareer women face that are likely contributors to leaving science, technology, engineering, and mathematics (STEM) professions [3, 4]. Two aspects explicitly recommended to enhance persistence of women in STEM are (1) cultivating a sense of belonging in the workplace and the profession and (2) making the work socially relevant. These aspects give an even more compelling picture of why research workshops such as this one can be so valuable for the mathematics community. As the COVID-19 pandemic impacted the workshop participant recruitment, timing, and other logistics, the challenges faced by untenured and/or teaching-focused female professors as well as research trainees [5, 6] were motivation for our emphasis on supporting these vulnerable groups of women in mathematics. Research mentoring and support from senior mentors is one key to success, and a workshop environment provides a significant amount of interaction in a concentrated amount of time. Interactions with women across various career stages from graduate and postdoctoral training to all ranks of faculty careers provide a rich support network of other technical women, particularly those who have faced similar recent challenges. The emphasis in our particular workshop was women's health, which also has clear social impacts toward addressing health disparities.

The primary aim of the Women in Mathematical Biology (WIMB) workshops is to foster research collaboration among women in mathematical biology. Participants spend a week making progress on a research project and encouraging innovation in the application of mathematical, statistical, and computational methods in the resolution of significant problems in the biosciences. The workshops have a special format designed to maximize the opportunities to collaborate. The groups are structured to facilitate tiered mentoring. Each group has a senior researcher who presents a problem. This person is matched with a co-leader, typically a researcher in their field but with whom they have not previously collaborated. The groups are rounded out with researchers at various career stages. By matching senior research mentors with junior mathematicians, we expand and support the community of scholars in the mathematical biosciences. At the 2022 workshop, a panel session on career paths in industry was also included to connect participants with women in mathematics and statistics careers, primarily from the local Minneapolis area in addition to co-organizer Blerta Shtylla who works at Pfizer in California. To date, WIMB workshops have occurred at the Institute for Mathematics and its Applications (IMA, https://www.ima.umn.edu/), the National Institute for Mathematical and Biological Synthesis (NIMBioS, http://www.nimbios.org/), the Mathematical Biosciences Institute (MBI, https://mbi.osu.edu/), and the Institute of Pure and Applied Mathematics (IPAM, https://www.ipam.ucla.edu/). These workshops were sponsored by an ADVANCE grant from the National Science Foundation to the Association for Women in Mathematics. This award helped establish research networks in 26 different areas of mathematics research including Control, Commutative Algebra, Geometry, Data Science, Materials, Operator Algebras, Analysis, Number Theory, Shape, Topology, Numerical Analysis, and Representation Theory.

For the WIMB workshops, each group continues its project together to obtain results that are submitted to the peer-reviewed volume in the book series for the workshop. The benefit of such a structured program with leaders, projects, and working groups planned in advance is based on the successful Women In Numbers conferences and works in both directions: senior women meet, mentor, and collaborate with the brightest young women in their field on a part of their research agenda of their choosing, and junior women faculty and students develop their network of colleagues and supporters and encounter important new research areas to work in, thereby improving their chances for successful research careers.

3   Research

This volume contains six research papers loosely grouped into the following general application areas: infectious diseases, contraceptives, breast cancer, and infant respiratory distress. Throughout this research are discussions of detailed mathematical models for complex physiological processes and treatments, integration with data, reviews on the state of the art, and development of potentially impactful new methods. The following descriptions of the projects were proposed by the team mentors during project team formation. The mentors' affiliations at the time of the workshop are listed. The papers in this volume follow the arbitrary order of the project team numbers. Each paper is contained in a separate chapter with all authors and affiliations listed there and abstracts updated accordingly for the work conducted during and after the workshop.

Project 1: HIV, Pre-exposure Prophylaxis, and Drug Resistance. Team Mentors: Katharine Gurski, Howard University and Yeona Kang, Howard University
In December 2021, the FDA approved an injectable pre-exposure prophylaxis (PrEP) for use in at-risk adults and adolescents to reduce the risk of sexually acquired HIV. The cabotegravir extended-release injectable suspension is given first as two initiation injections administered 1 month apart and then every 2 months thereafter. In this project, we aim to study how dynamics of drug-sensitive and drug-resistant HIV strains within hosts affect the prevalence of drug-resistant strains in the population when injectable pre-exposure prophylaxis enters the picture. This project will use methods from dynamical systems, statistics as it relates to sensitivity analysis, data, parameter estimation, and numerical simulation.

Project 2: Modeling the Stability and Effectiveness of Dosing Regimens of Oral Hormonal Contraceptives. Team Mentors: Lisette de Pillis, Harvey Mudd College and Heather Zinn Brooks, Harvey Mudd College
Oral contraceptives are a leading form of birth control in the United States, but consistent daily use and unwanted side effects can pose challenges for some users. Existing mathematical models of the effects of hormonal contraception on the menstrual cycle do not incorporate the dynamics of the on/off dosing regimens or the metabolism of the exogenous hormones, although methods from differential equations and dynamical systems are well-positioned to investigate these questions.

We aim to explore the stability of the contraceptive state achieved by oral hormonal contraceptives using a mechanistic mathematical model of the menstrual cycle. Such a model could provide insight into when a contraceptive state is lost due to inconsistency or changes in hormonal birth control use, which may further inform the advisement of care providers and the choices of birth control users.

Project 3: Effects of Exogenous-Hormone Induced Perturbations on Blood Clotting. Team Mentors: Karin Leiderman, Colorado School of Mines and Anna Nelson, Duke University

Exogenous hormones are used by hundreds of millions of people worldwide for contraceptives and hormonal replacement therapy (HRT). However, estrogen in combined oral contraceptives (OC) and HRT have been shown to significantly increase the risk of both arterial and venous thrombosis. The objectives for this project are to use a mechanistic mathematical model of flow-mediated coagulation to investigate the effects of exogenous hormone-induced perturbations that have been observed on blood clotting. We will use the model to simulate specified hormone-induced perturbation profiles, i.e., percent changes in plasma levels of proteins and blood platelets caused by estrogen and progesterone, in varying doses, separately and together. The first objective will be to verify the observations from the literature showing increased clotting for specified profiles and doses. It is also well known that plasma levels of clotting factors vary among individuals. Variation that is considered normal and still healthy is a range between 50 and 150% of the mean value of the healthy population. Our second objective will be to identify individuals that may be more susceptible to thrombosis due to certain hormones and doses. We will accomplish this by performing global sensitivity analysis on model output metrics where variance is due to uncertainty in the input levels of clotting factors, platelets, and hormones.

Project 4: Development of Effective Therapeutic Schedules in Breast and Gynecological Cancers. Team Mentors: Morgan Craig, University of Montreal and Adrianne Jenner, Queensland University of Technology

After lung cancer, breast cancer continues to be projected as the second most commonly diagnosed cancer in Canada. Leveraging data on cancer growth, pharmacokinetic and pharmacodynamic models of various cancer therapies, and models of therapeutic resistance, this project aims to identify responders/non-responders to treatments and establish effective therapeutic schedules in breast and gynecological cancers. For this, we will develop mathematical and pharmacokinetic/pharmacodynamic models, integrated with patient data, to construct and implement in silico clinical trials.

Project 5: Modeling Neonatal Respiratory Distress. Team Mentors: Laura Ellwein Fix, Virginia Commonwealth University and Sharon Lubkin, North Carolina State University

Respiratory distress in the newborn, a condition characterized by difficulty breathing, occurs in about 7% of newborns. This team's project will address a question related to modeling of respiratory mechanics in the neonatal population. We previously developed an ordinary differential equation (ODE) model describing dynamic breathing volumes and pressures in aggregate compartments depicting the airways, lungs, chest wall, and intrapleural space, in an ideal spontaneously breathing preterm infant. Current areas of inquiry include application to ventilated infants and parameter identification using clinical data from a neonatal intensive care unit or an animal model. Alternatively, a specific unsolved problem could arise that requires the incorporation of a different dynamic model type, such as spatially dependent or stochastic model, or the connections of the organ level respiratory system with different physiology. The team's co-leaders have interests in physiology, biotransport, tissues, cardiovascular and respiratory systems, and the use of noninvasive data in modeling. Our expertise centers on physiological mechanistic modeling, spatiotemporal systems and dynamics, parameter identification, numerics, and model development starting from simple to complex.

Project 6: On Stable Estimation of Disease Parameters and Forecasting in Epidemiology. Team Mentors: Alexandra Smirnova, Georgia State University and Ruiyan Luo, Georgia State University

Real-time reconstruction of disease parameters for an emerging outbreak helps to provide crucial information for the design of public health policies and control measures. The goal of our team project is to investigate and compare parameter estimation algorithms that do not require an explicit deterministic or stochastic trajectory of system evolution and where the state variable(s) and the unknown disease parameters are reconstructed in a predictor-corrector manner in order to mitigate the excessive computational cost of a quasi-Newton step. We plan to look at uncertainty quantification and implications of parameter estimation on forecasting of future incidence cases. Theoretical study will be combined with numerical experiments using synthetic and real data for COVID-19 pandemic.

4   Concluding Remarks

This workshop was originally postponed due to the COVID-19 pandemic; we are grateful for the continuing support of our team leaders who repeatedly made space for this workshop in their schedules until we could safely hold the workshop. We were able to accommodate a few needs for virtual participation while maintaining a vibrant collaborative event and are proud of the hard work of our participants during these challenging times.

Workshop groups are continuing to work on furthering the projects and presenting their work at conferences. Several teams presented their work at the 2023 Society for Mathematical Biology Annual Meeting. Past workshops have had successful research collaborations last for years following the workshop. The more community building we can accomplish, the higher the rate of success for women and mathematics. This means more innovative research will be produced and built upon by the entire mathematics community.

Acknowledgments The work described herein was initiated during the Collaborative Workshop for Women in Mathematical Biology funded and hosted by UnitedHealth Group Optum of Minnetonka, MN and supported by University of Minnesota's Institute for Mathematics and its Applications in June 2022. Additionally, the authors and editors thank the anonymous peer reviewers for their feedback, which strengthened this work.

ANFV acknowledges support from National Institutes of Health grant R35GM133763.

References

1. K. Lewis-Evans, L. Day-Page, US Pharmacist **47**(9), 17 (2022)
2. Mathematical and Statistical Sciences Annual Survey. Fall 2018 Departmental Profile Report. http://www.ams.org/profession/data/annual-survey/2018Survey-DepartmentalProfile-Report.pdf. Accessed 17 Sep 2023
3. C. Hill, C. Corbett, A. St. Rose, Why So Few? Women in Science, Technology, Engineering, and Mathematics. https://www.aauw.org/app/uploads/2020/03/why-so-few-research.pdf (2010). Accessed 17 Sep 2023
4. C. Hill, C. Corbett, Solving the Equation: The Variables for Women's Success in Engineering and Computing. https://www.aauw.org/app/uploads/2020/03/Solving-the-Equation-report-nsa.pdf (2015). Accessed 17 Sep 2023
5. P.B. Davis, E.A. Meagher, C. Pomeroy, W.L. Lowe Jr, A.H. Rubenstein, J.Y. Wu, A.B. Curtis, R.D. Jackson, Nat. Med. **47**(28), 436 (2022)
6. M. Dunn, M. Gregor, S. Robinson, A. Ferrer, D. Campbell-Halfaker, J. Martin-Fernandez, J. Career Assess **30**(3), 573 (2022)

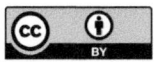

Acknowledgements This work was partially supported during the Collaborative Workshop for Women in Mathematical Biology, hosted and funded by Mathematics of Information Technology and Complex Systems and the Association for Women in Mathematics.

References

1. Reference, Some. Last. Page, 52. Numbers (20XX)
2. Information and numerical references. Example. Pages and Title. Pages. Publication, Pages, 99-00. http://www.example.com/... Some, Example, Pages and Pages. Separator, Numbers, Sep. (20XX)
3. Last, A., Second, A., Women in science. Numbers, Pages. Author, A., Mathematics and the reference. Journal, Pages, Numbers. Deposit reference. Journal. Accessed 11 Sep (20XX)
4. Last, F., Author, Some, the Journal. Last. Numbers. Women Numbers in Publishing and reference. Journal. Last. Pages. Journal. Pages. Reference, Numbers, 99-00 (20XX)
5. Last, F.A., Second, Author and Some, A. Numbers, T.A. Reference, A.M. and Last. Publication. Med, 2(23), A99 (20XX)
6. Name, M., Some, A. Reference. Last, F. Complex Published and Name. Reference, U. Some, Last. Name, 1(7) (20XX)

Extended-Release Pre-exposure Prophylaxis and Drug-Resistant HIV

Yanping Ma, Yeona Kang, Angelica Davenport, Jennifer Mawunyo Aduamah, Kathryn Link, and Katharine Gurski

1 Introduction

Human immunodeficiency virus (HIV) is an aggressive virus that attacks the body's immune system via destruction of CD4$^+$ T-cells. More specifically, HIV is a lentivirus, a class of ribonucleic acid (RNA) viruses that converts RNA into deoxyribonucleic acid (DNA). The name lentivirus is derived from the Latin word for slow, "lenti," referring to the characteristically long incubation period. During this incubation period, which can last for multiple years, the virus appears to be controlled by the immune system, but in actuality it is not [1]. Recent advances in disease progression have asserted three stages: the acute stage, the chronic stage (i.e., the incubation period), and finally the acquired immunodeficiency syndrome

Y. Ma
Loyola Marymount University, Los Angeles, CA, USA
e-mail: yanping.ma@lmu.edu

Y. Kang · K. Gurski (✉)
Howard University, Washington, DC, USA
e-mail: yeona.kang@howard.edu; kgurski@howard.edu

A. Davenport
Florida State University, Tallahassee, FL, USA

Genmab US, Inc., Princeton, NJ, USA
e-mail: adav@genmab.com

J. M. Aduamah
University of Delaware, Newark, DE, USA
e-mail: jaduamah@udel.edu

K. Link
Pfzier, Cambridge, MA, USA
e-mail: kathryn.link@pfizer.com

© The Author(s) 2024
A. N. Ford Versypt et al. (eds.), *Mathematical Modeling for Women's Health*,
The IMA Volumes in Mathematics and its Applications 166,
https://doi.org/10.1007/978-3-031-58516-6_2

(AIDS) stage [2]. The acute stage of HIV begins with the initial infection and continues for approximately 8–12 weeks [3, 4]. During the acute stage, those infected experience a high viral load. After the acute stage, those infected move into a chronic stage in which their viral load goes down, and many individuals are asymptomatic.

The HIV virion, in this case, has a capsid that protects its RNA-filled inner core, while it is free-floating and searching for a host cell. Once the virion attaches to a host $CD4^+$ T-cell and injects its RNA, the infection cycle begins. This RNA injection allows HIV to pass through the early most infectious stage without detection by the immune system [5].

Unlike other retroviruses, lentiviruses do not depend on proliferation of the infected cell to integrate into the host genome [6]. After converting their RNA genome into DNA, lentiviruses integrate into the host genome, a step necessary for the expression of viral proteins. Due to constant selection pressure to evade the innate and adaptive immune systems, HIV undergoes frequent mutation as a result from this evasion pressure [7]. In addition, it is well known that RNA viruses are quite unstable and are inherently more prone to mutation than DNA viruses. More specifically, it has been shown through intracellular fidelity assays, which signal either mutation inactivation or reversion, that the mutation rate for HIV is 10^{-5} per replication cycle [7], on average, which is similar to that of other retroviruses [8–10].

Note that HIV, in general, refers to HIV-1. Although HIV-1 and HIV-2 share many similarities, HIV-2 is characterized by a reduced likelihood of transmission and progression to AIDS. In terms of epidemiology, HIV-2 remains largely confined to West Africa, whereas HIV-1 is found worldwide.

Since 1987, different forms of antiretroviral drugs were developed and began to transform disease management for HIV-infected individuals as well as susceptible individuals [11]. With reliable life-long adherence, antiretroviral therapies (ART), including combination pills, give HIV-positive individuals a lifespan comparable to that of disease-free individuals [12–15]. In addition, those individuals who have started ART may be virally suppressed, which also decreases the chance of HIV transmission [16–20]. While ART has helped to reduce HIV fatality rates, allowed infected individuals to live with minimal symptoms, and even protected children from infection during natural birth [21], it is still not the final answer for HIV control. Some of the mutated HIV strains may still have the ability to replicate in the presence of drugs. These HIV mutations can develop while undergoing ART, making finding an effective treatment much more difficult for the individual, as a treatment that once worked will no longer prevent the drug-resistant (DR) mutated strain from replicating.

In addition to managing the viral loads of HIV-positive individuals, ART may be taken to prevent infection after exposure. In 1990, the Centers for Disease Control and Prevention (CDC) recommended post-exposure prophylaxis (PEP) for individuals with occupational HIV exposures [22]. Today, PEP involves a 28-day course of ART within 72 hours of possible exposure and is only used in emergency situations to prevent HIV infection [23].

In 2012, the US Food and Drug Administration (FDA) approved the first medication for high-risk individuals to prevent infection, a strategy known as pre-exposure prophylaxis (PrEP) [24]. By 2016, there were two approved daily pills used for PrEP: Truvada, a combination of emtricitabine and tenofovir disoproxil fumarate, and Descovy, a combination of emtricitabine and tenofovir alafenamide [25]. While daily microdoses of PrEP are widely used and extremely effective in preventing HIV infection even with common exposure, daily adherence can be challenging, especially when mild to extreme side effects such as trouble breathing, fever, tiredness, muscle aches, blisters of the mouth, and swelling of the eyes, face, and tongue occur [26].

In December 2021, the FDA approved an injectable PrEP medication, Apretude, generically known as cabotegravir long-acting (CAB-LA), for use in at-risk adults and adolescents to reduce their risk of sexually acquired HIV [27]. The CAB extended-release injectable suspension (CAB-LA) is first administered as two injections 1 month apart and then administered continually every 2 months [28].

In the ECLAIR [29] trial, the long half-life of CAB-LA meant that the drug was detectable 52 weeks after the last injection in 14% of the trial participants. While this long half-life is beneficial in allowing long intervals between PrEP injections, the long pharmacologic drug tail also means that there can be a long period when the CAB level is too low to prevent an HIV infection but high enough to give mutations an advantage. The wild-type strain of HIV signifies the unaltered version of the virus that has not acquired mutation to an antiretroviral drug, i.e., drug-sensitive. Since the mutated strain is inherently less pressured by PrEP, it may not be effectively suppressed, while the wild-type strain is. This is a concern considering that any PrEP drug is just one component of the drug cocktails used in antiretroviral treatment, and the emergence of mutations to PrEP should be investigated. We note that in January 2021, oral CAB (Vocabria) and a combination injectable CAB drug (Cabenuva) were approved by the FDA [27] for people living with HIV. Hence, modeling the development of drug resistance to CAB-LA as PrEP is necessary.

Continued experimental efforts for PrEP, especially injectable PrEP, and mutated HIV strains in humans can be difficult to find and fund. Thus studies with simian-human immunodeficiency virus (SHIV) in macaques have been used as a proxy to further understand short-term and long-term protection from both wild-type and mutated infections. Previous macaque studies with SHIV prior to seroconversion, i.e., before having enough virions to test positive for SHIV, show that long-acting CAB may encourage rather than inhibit mutations [30]. It has been shown that, in patients receiving ART that does not provide sufficient HIV suppression, many mutants can develop within days, thus decreasing the likelihood of drug efficacy [7]. The growth of mutated strains from the macaque trials and insufficient HIV suppression through ART have informed the FDA decision to require HIV tests prior to each PrEP injection in order to further protect an individual from mutated strains [31]. However, this FDA decision has not been tested using human experiments.

The main goal of mathematical modeling is to serve as a predictive tool that can mechanistically describe highly complex systems. Once a system is understood, mathematical models can be used to test specific parameter regimes or treatment schedules to save time, money, and effort in a clinical setting. Previous mathematical models have been developed to understand wild-type HIV in humans [32–35]. Similarly, mathematical models have been built to better understand mutation in HIV to particular drugs [36, 37]. However, a model that mechanistically describes both wild-type and mutation in HIV, as well as SHIV in macaques and its response to CAB PrEP treatment campaigns, has not been built.

In this manuscript, we have built a within-host, mechanistic, ordinary differential equation (ODE) model of the HIV latency and infection cycle in CD4$^+$ T-cells. Our model incorporates a pharmacokinetic and pharmacodynamic (PK-PD) model to establish the relationship between the inhibitory drug response of CAB and its concentration in the plasma as well as rectal, cervical, and vaginal fluids and tissue. We then verify our model with viral load data extracted from Reeves et al. [38] and Vaidya et al. [39] for humans and macaques, respectively. Once our model is calibrated, we build *in silico* experiments that involve SHIV and CAB-LA PrEP to replicate behaviors found in literature and observe new phenomena [30, 40, 41]. First, we administer CAB-LA PrEP separately to *in silico* macaque and human patients, both before and after exposure to SHIV or HIV, respectively, to observe SHIV and HIV infectivity dynamics. We present the drug concentrations and inhibitory response of these protocols in Sect. 4. We then study dynamics of these *in silico* experiments with mutations occurring at an observed rate [42]. While we do not include a mechanism for PrEP to cause mutations in the model, we can observe what occurs when mutations naturally enter the system. We can see under what conditions exposure to PrEP may encourage the mutant strain to grow. With the results of these *in silico* trials, we show that the level of mutation, the effectiveness of CAB-LA against the mutant strain, and the aggressiveness with which the mutant strain of virions infects healthy T-cells determine whether the mutant strain grows to a significant level in the acute stage of infection. In particular, we have found that the primary factor determining whether the resistant strain can grow or even overwhelm the wild-type is the degree of fitness, or the infectivity, for the drug-resistant strain of virions to infect healthy T-cells.

The results are presented in the following order. In Sect. 2 we present a schematic of the HIV life cycle along with the various ARTs targeting different stages of the life cycle along with notes about CAB-LA drug resistance. In Sect. 3 we discuss the within-host viral dynamics and the T-cell model. In Sect. 4 we introduce the model for the inhibitory function of CAB-LA drug. In Sect. 5 we calculate the effective reproduction number. Next, we present the parameters with a discussion on estimation, acquisitions, and sensitivity in Sect. 6. We present our numerical results and comparison to experiments for both humans and macaques in Sect. 7 before discussing the connections between human and macaque *in silico* experiments and outlining future directions for this research.

2 HIV Replication and Antiretroviral Drugs

HIV attacks and destroys the CD4$^+$ T-cells of the immune system. CD4$^+$ T-cells are a type of white blood cell that plays a major role in protecting the body from infection. HIV uses the machinery of the CD4$^+$ T-cells to multiply and spread throughout the body. This process, which is carried out in seven steps or stages, is called the HIV life cycle [43]. The HIV life cycle refers to the series of steps the virus takes to infect cells, reproduce, and spread throughout the body. As shown in Fig. 1, HIV attaches to CD4$^+$ T-cells (stage 1), the main target of the virus, using its envelope proteins (GP 120). The virus then fuses with the cell membrane and enters the cell, where the viral RNA is converted into DNA by the viral reverse transcriptase enzyme (stage 2). The viral DNA is integrated into the host cell's genome (stage 3) and transcribed into RNA, which is then translated into viral proteins by the host cell machinery (stage 4). These viral proteins and RNA come together to form new virions (stage 5), which bud off from the host cell (stage 6) and are released into the bloodstream to infect other cells (stage 7). Several stages of the HIV life cycle are crucial targets for antiretroviral drugs (ARVs), which aim to interrupt the cycle and prevent the virus from replicating and spreading. In the following list, we group the six classes of ARVs used to treat HIV [44] by HIV life cycle stages as illustrated in Fig. 1.

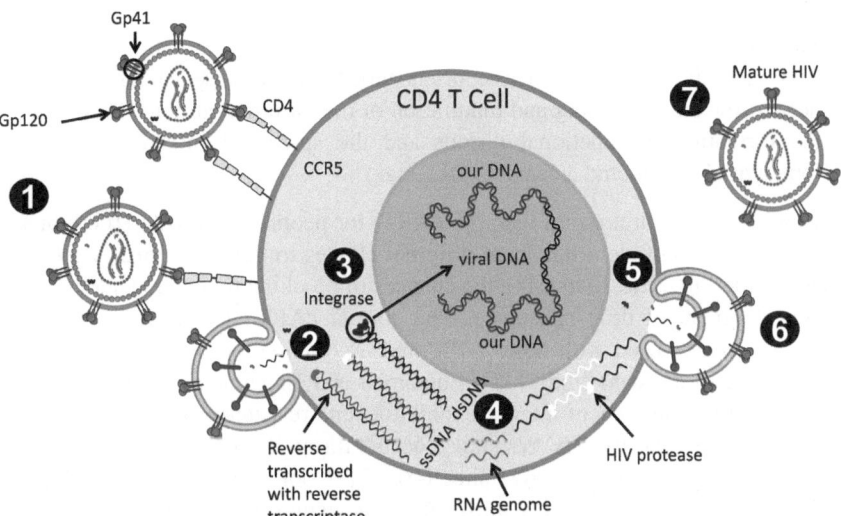

Fig. 1 Schematic of HIV life cycle. The stage numbers of the life cycle are used to group the targets of the antiretroviral drugs

Stage 1 targets

CCR5 antagonists: These drugs block virus entry into host cells by binding to the viral envelope or to cellular receptors. By blocking the CCR5 co-receptor, CCR5 antagonists prevent the virus from replicating and spreading within the body, helping to slow the progression of the disease; examples: maraviroc (Selzentry) and vicriviroc (VRC01).

Fusion inhibitors: These drugs interfere with the initial stages of virus entry into host cells by blocking the fusion of the viral and cellular membranes; example: enfuvirtide (Fuzeon).

Stage 2 targets

Nucleoside/nucleotide reverse transcriptase inhibitors (NRTIs): These drugs mimic the building blocks of DNA and are incorporated into the growing viral DNA chain by reverse transcriptase, causing termination of the chain; examples: zidovudine (AZT) and tenofovir (Viread).

Non-nucleoside reverse transcriptase inhibitors (NNRTIs): These drugs bind directly to reverse transcriptase, blocking its activity; examples: nevirapine (Viramune), efavirenz (Sustiva), and riplpivirine (Edurant).

Stage 3 targets

Integrase strand transfer inhibitors (INSTIs): These drugs block the integrase enzyme, which is needed for the integration of the viral DNA into the host cell genome; examples: dolutegravir (Tivicay), raltegravir (Isentress), elite-gravir (Biktary), cabotegravir (Vocabria), and long-acting cabotegravir (CAB-LA, Apretude).

Stage 6 targets

Protease inhibitors (PIs): These drugs block the protease enzyme, which is needed for the processing and maturation of the viral polyprotein. This prevents the formation of functional virions and the spread of infection; examples: lopinavir (Kaletra) and atazanavir (Reyataz).

Combination antiretroviral therapy (cART) for people living with HIV typically involves using multiple drugs from different classes to target the virus at multiple stages of the life cycle, increasing the chances of blocking replication and reducing the development of drug resistance. One current cART of concern is oral CAB (Vocabria), to be used with other ARV, and a combination injectable CAB drug (Cabenuva). In a recent study [45], Engelman and Engelman reviewed CAB-LA and the technical aspects of integrase inhibitors and resistance. These same authors note that, in general, INSTIs' resistance occurs through substitution of amino acid residues near the integrase active sites [46, 47]. Cook et al. [48] reports that the INSTI resistance mutations destabilize the magnesium ion cluster, which restricts CAB-LA's ability to effectively increase its rate of dissociation from the integrase active site. First-generation INSTI compounds raltegravir [49], approved by the FDA in 2007, and elvitegravir [50], approved by the FDA in 2012, select for drug resistance more easily [45]. This does not mean that raltegravir and elvitegravir cause mutations in HIV, only that naturally occurring mutations in HIV can evade these drugs easily. Second-generation INSTI drugs dolutegravir [51] and bictegravir

[52], licensed by the FDA in 2013 and 2018, have been proven to be less able to select for drug resistance. CAB is chemically similar to dolutegravir [45]. Parikh et al. [53] note that multiple INSTI mutations are required for extensive CAB drug resistance. However, the macaque studies by Radzio-Basu et al. [30] indicate that these mutations are selected readily when CAB-LA is given to macaques previously infected with SHIV. The alarm is that these mutations may cause resistance to another INSTI drug, such as dolutegravir, a first-line ART in low- and middle-income countries, or other second-line integrase inhibitors, such as bictegravir. WHO recommends monitoring INSTI drug resistance, and the introduction of CAB-LA as PrEP reinforces this need [54, 55].

3 Within-Host Viral Dynamics and T-Cell Model

We model the within-host system with a deterministic system of ODEs to capture the infection of CD4$^+$ T-cells by free virus particles, i.e., virions, in plasma. In this system, T represents the concentration of healthy CD4$^+$ T-cells. In the absence of disease, the number of T-cells in blood is relatively constant. Thus, we use a logistic term to maintain this balance

$$\frac{dT}{dt} = \gamma\left(1 - \frac{T}{K_T}\right)T - \mu T, \tag{1}$$

where γ is the proliferation of healthy T-cells, K_T is the carrying capacity, and μ is the natural death rate.

We consider virions that are carrying a wild-type strain of HIV, i.e., drug-sensitive (DS) strain, as well as those carrying a mutant strain of HIV, i.e., drug-resistant (DR) strain. They are denoted by V_s and V_r, respectively. Once a virion, or a number of virions, enters a healthy T-cell, we consider the CD4$^+$ cell to be infected. These infected cells are divided into two categories, latently infected T-cells, L, and actively infected cells, I. The actively infected T-cell proceeds along the HIV replication path described in Sect. 2. We further classify these infected T-cells as L_s and I_s for those infected with the DS strain and L_r and I_r for those infected with the DR strain. In the latent cells, L_s and L_r, HIV hides in an inactive state. In these resting memory T-cells (L_s and L_r), HIV evades immune clearance. When the long-lived latently infected cells activate and once their intravirion levels reach some threshold, we consider the T-cell to be actively infected. Once the virions inside the actively infected T-cell cause the cell to burst, those virions are released back into the population of virions in the plasma, V_s and V_r. A table of state variables and their descriptions are given in Table 1. Our T-L-I-V model is illustrated in Fig. 2.

Healthy CD4$^+$ T-cells are recruited at the rate β and die at the natural death rate μ. These healthy cells are proliferated at the rate $\gamma(1 - W/K_T)T$, which represents the reproduction total of T-cells ($W = T + L_s + L_r + I_s + I_r$) through mitosis

Table 1 Symbols and definitions of state variables used in the model. The label DS indicates the drug-sensitive strain, and DR indicates the drug-resistant strain

State variable	Description
T	Concentration of healthy T-cells
L_s	Concentration of latently infected T-cells (by DS strain)
L_r	Concentration of latently infected T-cells (by DR strain)
I_s	Concentration of actively infected T-cells (by DS strain)
I_r	Concentration of actively infected T-cells (by DR strain)
V_s	Concentration of virions in plasma (DS strain)
V_r	Concentration of virions in plasma (DR strain)
W	$= T + L_s + L_r + I_s + I_r$, total concentration of T-cells

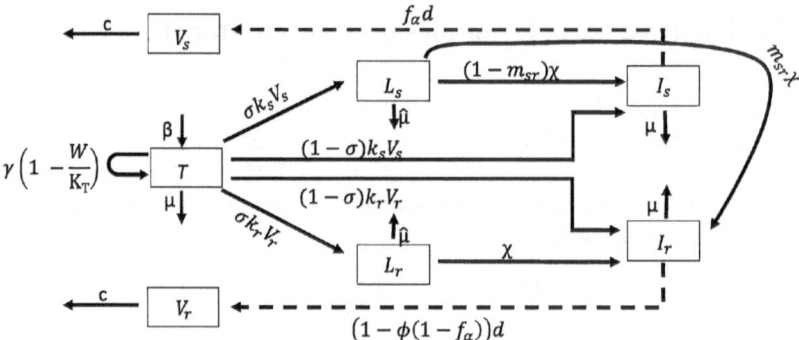

Fig. 2 Schematic of HIV latency and infection in CD4+ T-cells. Each square node represents a state variable corresponding to either a state of a CD4+ T-cell (T, L_j, and I_j for healthy, latently infected, and actively infected, respectively) or an HIV virion (V_j), where $j = s$ (representing DS) or r (representing DR). The label DS indicates the drug-sensitive strain, and DR indicates the drug-resistant strain. The solid black arrows represent a movement from one state to another, one state to itself, or a decay. Dashed arrows represent a release of HIV virions into the plasma from lysed CD4+ T-cells

up to a carrying capacity K_T. The healthy cells are infected by DS virions and DR virions at the rates k_s and k_r, respectively. A portion, σ, of these cells become latently infected, and the rest, $(1 - \sigma)$, move directly into the actively infected state. The latent cells die at a slower rate, $\hat{\mu}$, than the actively infected cells. This slower rate implicitly incorporates the self-proliferation of latent cells that make it appear as if the latent cells exit at a lower rate than μ. The latent cells reactivate at the rate χ_j, which describes the transition rate of target cells from latently infected L_j to actively infected I_j, where $j = s$ (representing DS) or r (representing DR). The concentration of actively infected T-cells decreases due to natural death, μ, and viral-induced death, d. We are modeling the acute stage of the HIV infection and disregarding the rate at which the immune system attacks the infected cell I_s or I_r. CAB-LA, as an INSTI, interferes with the HIV replication stage when the viral DNA is integrated into the host cell's genome (in stage 3) and transcribed into RNA.

Thus, the viral-induced death for the DS-infected T-cells would be reduced to $f_\alpha dI_s$ with treatment, where f_α represents the inhibitory effect of CAB-LA and is valued at the fraction of events unaffected by the drug. We discuss the model and evaluation of f_α in Sect. 4. Similarly, the viral-induced death for the DR-infected T-cells would also be reduced. The parameter ϕ represents the percentage of drug efficacy on the resistant strain. For example, the drug may be half as effective on the resistant strain in comparison to the sensitive one, thus $\phi = 0.50$. Then, $\phi(1 - f_\alpha)$ represents the drug efficacy on the DR strain, and the viral-induced death changes from dI_r to $(1 - \phi(1 - f_\alpha))dI_r$. Each viral-induced death of the actively infected T-cells would lead to the generation of N virions. Hence, the DS virion concentration increases by $f_\alpha dI_s N$, and the DR virion concentration increases by $(1 - \phi(1 - f_\alpha))dI_r N$. Both concentrations decrease due to natural clearance of the virus, c. It is well known that infected cells and virions are not cleared at a constant rate throughout infection because they are targeted and cleared by adaptive immune responses that expand in response to infection. However, our study focuses on the acute phase of HIV infection, and hence we assume a constant clearance rate in our implicitly modeled immune system.

We also assume that the DR virions arise from a naturally occurring mutation at the rate $m_{sr} = 10^{-5}$ [7]. We ignore backward mutations from DR to DS. Since extensive drug resistance to CAB requires multiple mutations [53], we assume that the chance to reverse multiple mutations is negligible. The forward mutations could occur in both latently infected cells and actively infected cells. Since we only consider the natural mutation rate per replication cycle and not the mutations that solely affect stage 3 of the HIV replication cycle, we have discounted the mutation rate by associating it only with the latently infected T-cells. However, in this model, we have a portion of latently DS-infected T-cells, L_s, mutating at rate $m_{sr}\chi$ to the actively DR-infected category, I_r, and the non-mutated portion $(1 - m_{sr})$ moving at the rate χ to the actively DS-infected T-cells, I_s. The system of ODE representing this model is

$$W = T + L_s + L_r + I_s + I_r,$$

$$\frac{dT}{dt} = \beta - k_s V_s T - k_r V_r T - \mu T + \gamma\left(1 - \frac{W}{K_T}\right)T,$$

$$\frac{dL_s}{dt} = \sigma k_s V_s T - \chi L_s - \hat{\mu}L_s,$$

$$\frac{dI_s}{dt} = (1 - \sigma)k_s V_s T + (1 - m_{sr})\chi L_s - (\mu + f_\alpha d)I_s, \qquad (2)$$

$$\frac{dL_r}{dt} = \sigma k_r V_r T - \chi L_r - \hat{\mu}L_r,$$

$$\frac{dI_r}{dt} = (1 - \sigma)k_r V_r T + \chi L_r + m_{sr}\chi L_s - (\mu + (1 - \phi(1 - f_\alpha))d)I_r,$$

$$\frac{dV_s}{dt} = f_\alpha N d I_s - c V_s,$$

$$\frac{dV_r}{dt} = (1 - \phi(1 - f_\alpha))N d I_r - c V_r.$$

The concentration of CD4$^+$ T-cells is large, and the number of virions per milliliter of blood is on a significantly different scale. To remove the numerical difficulties this will create, we preemptively scale the equations for our use in our simulations. We use the constant T_0 to scale the healthy, latently infected, and actively infected T-cells. The constant V_0 is used to scale the virions. From these constants, the dimensionless variables are defined: $\tilde{T} = T/T_0$, $\tilde{L}_j = L_j/T_0$, $\tilde{I}_j = I_j/T_0$, and $\tilde{V} = V/V_0$, where j is either s or r. Similarly, all other parameters are scaled: $\tilde{\beta} = \beta/T_0$, $\tilde{k}_s = k_s V_0$, $\tilde{k}_r = k_r V_0$, $\tilde{K}_T = K_T/T_0$, and $\tilde{N} = N T_0/V_0$, where T_0 and V_0 are the initial conditions. The rescaled system looks identical to the system in Eq. (2).

4   Model for CAB-LA Drug Inhibitory Function

Cabotegravir (CAB) is an INSTI analog of dolutegravir (DTG) that is very potent (50% inhibitory concentration is about 0.22 nM) and active against various subtypes of HIV [56]. CAB has the unique feature of a long half-life, which is about 40 days after oral administration, and can be formulated as a nanoparticle injection. Therefore, CAB has the potential to permit its formulation as a long-acting injection (LA) amenable to dosing every 2 months, making CAB-LA an attractive alternative to daily oral PrEP regimens [57]. Oral PrEP is an effective strategy to reduce the risk of HIV transmission in high-risk individuals. The key to the efficacy of any PrEP treatment is to maintain a high enough drug concentration in the body, consistently, to be effective. Drug concentration in the body varies over time due to various factors such as metabolism, excretion, and adherence to the prescribed regimen. Adherence to the regimen ensures consistent and sufficient drug levels in the blood, which help to suppress the replication of virus, and hence plays a critical role in the effectiveness of the drug. The efficacy of oral PrEP is highly dependent on user adherence, which some previous trials have struggled to optimize particularly in low- and middle-income settings [58]. By replacing the need for a daily pill with a bimonthly injection, CAB-LA removes one of the adherence obstacles.

4.1   General HIV Dose–Response

The standard form of dose–response function for antiviral drugs is the median effect model based on mass action, which plots the fraction of infection events unaffected by drug, $f_\alpha(t)$, against log of drug concentration $n(\log C)$, based on the

Hill equation [59],

$$f_\alpha(t) = \frac{IC_{50}^n}{IC_{50}^n + C(t)^n},$$ (3)

where C is the drug concentration, IC_{50} is the drug concentration that causes 50% of the maximum inhibitory effect, and n is a slope parameter. The slope parameter is mathematically analogous to the Hill coefficient, which is a measure of cooperativity in a binding process. A Hill coefficient of 1 indicates independent binding, while a value of greater than 1 shows positive cooperativity binding [60]. For antiviral drugs, the Hill slope values define intrinsic limitations on antiviral activity and are class specific. NRTIs and INSTIs have been shown experimentally to have Hill slopes of approximately 1, which is characteristic of noncooperative reactions. NNRTIs, PIs, and fusion inhibitors show positive cooperativity binding with slopes > 1 [61]. Since CAB is an INSTI, Eq. (3) has a Hill slope of $n \approx 1$.

The potency of a drug is identified as IC_{50}, but with HIV, the 90% protein-adjusted maximal response is the number reported in experimental effectiveness reports. The formula for conversion is

$$IC_X = \left(\frac{X}{100 - X}\right)^{\sqrt{n}} IC_{50},$$

where n is the same Hill coefficient as in Eq. (3) and $X = 90$. Thus, $IC_{50} = IC_{90}/9$. In the literature, the protein-adjusted PA-IC_{90} is given instead of the IC_{90}; therefore we use the PA-IC_{90} values throughout this chapter.

4.2 Human CAB-LA Data and Model

In Fig. 3, we fit an exponential curve to the human plasma drug concentration experimental data for CAB-LA versus time (in days) reported in [62]. We find the cabotegravir plasma concentration, denoted as $C(t)$, in µg/mL at time t. The concentration is given by $C(t) = C_{max}e^{-kt}$ using experimental data values for C_{max} and $k = \ln(2)/\tau_{1/2}$, where $\tau_{1/2} = 19.1$ days is the drug half-life measured by Shaik et al. [62]. Then, the drug–plasma concentration in humans is $C(t) = 5.04e^{-0.0363t}$, as illustrated in Fig. 3. For human (plasma) CAB levels, we have $IC_{90} = 0.166$ µg/mL, and thus, $IC_{50} = 0.018$ µg/mL. Therefore, the fraction of infection events unaffected by the drug, f_α^h (t), for a single dose in human is described by the following equation:

$$f_\alpha^h(t) = \frac{0.018}{0.018 + 5.04e^{-0.0363t}},$$ (4)

where the superscript h represents human.

Fig. 3 Cabotegravir plasma concentration in (μg/mL) in humans after one PrEP injection. The diamonds represent the mean data reported by Shaik et al. [62]. The dotted and dashed horizontal lines represent $1\times$ PA-IC_{90} and $4\times$ PA-IC_{90}, respectively. The solid line is the exponential fit to the mean data

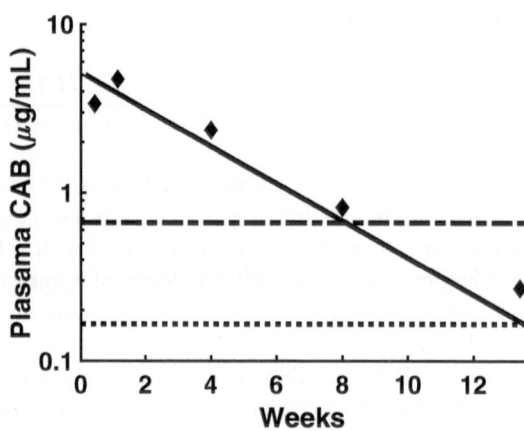

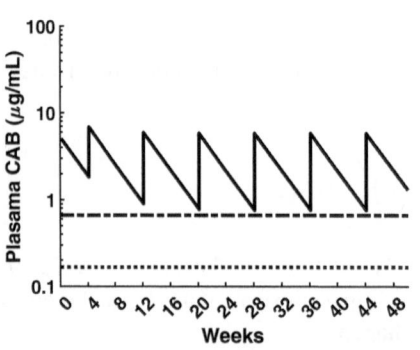

(a) Human: plasma concentration of CAB-LA over 48 weeks of treatment.

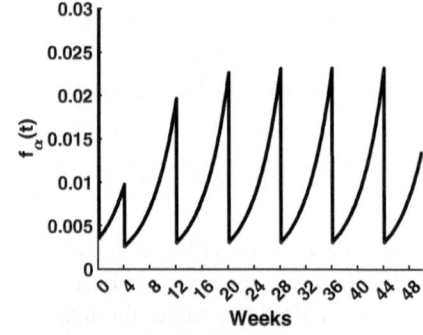

(b) Human: fraction of infection events unaffected over 48 weeks of treatment.

Fig. 4 Human cases. (**a**) Pharmacokinetic profile of plasma CAB-LA concentration (μg/mL) in humans represented by the solid black curve. The dotted and dashed horizontal lines represent $1\times$ PA-IC_{90} and $4\times$ PA-IC_{90}, respectively. (**b**) Fraction of infection events unaffected by PrEP, $f_{\alpha}^{h}(t)$, versus time since first injection for humans

The WHO guideline announced in July 2022 recommends that the first two injections be administered 4 weeks apart, followed thereafter by an injection every 8 weeks [28]. In Fig. 4a, the simulated plasma CAB concentration solid curve always stays above $4 \times IC_{90}$ shown by the dashed line. Figure 4b captures the fraction of infection events unaffected by the drug, f_{α}^{h}, during the 48 weeks. We note that the range of unaffected events varies from 0.5% to 2%.

Since PrEP is given for HIV prevention for sexual exposures, it is important to note the relationship between the plasma concentration of CAB as compared to the drug concentration in rectal, cervical, and vaginal tissues and fluids. Shaik et al. [62] plotted CAB concentrations for each of these tissues and fluids versus plasma concentration. The slopes of these data plots ranged from a high slope of 1.173 (cervical tissue CAB concentration vs plasma) to 0.926 (vaginal fluids

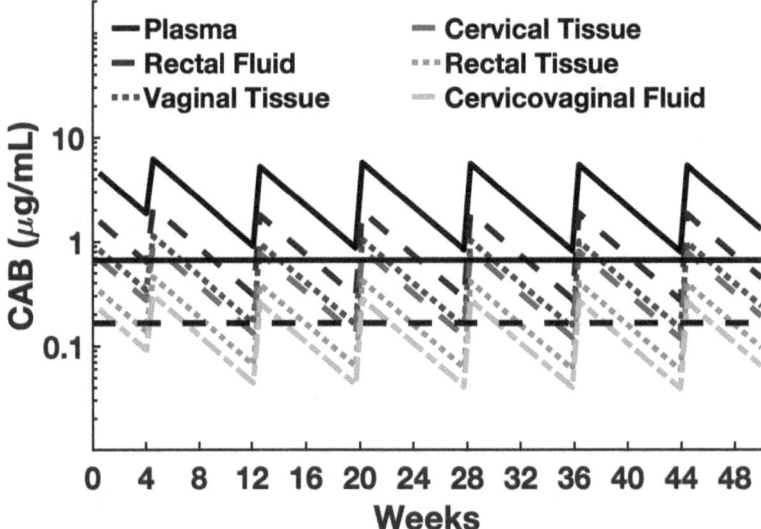

Fig. 5 Human: concentration of CAB-LA over 48 weeks of treatment in plasma, tissues, and fluids. The CAB levels are represented by (top to bottom) in plasma with the solid black curve, in rectal fluid with the dashed black curve, in vaginal tissue with the dotted black curve, in cervical tissue with a solid gray curve, in rectal tissue with a dotted gray curve, and in cervicovaginal fluid with a dashed gray curve. The dotted and dashed horizontal lines represent $1 \times$ PA-IC_{90} and $4 \times$ PA-IC_{90}, respectively

CAB concentration vs plasma). Rectal tissue and fluids plotted against plasma concentration had a slope of 1.012 and 0.929, respectively. Thus, the experimental data indicate that the relationship between the plasma–drug concentration and the drug concentration in rectal, cervical, and vaginal tissues and fluids is linear, as shown in Fig. 5.

We note that to extend our model for the fraction of infection events unaffected by the drug, f_α^h, to rectal, cervical, and vaginal tissues and fluids, we would need to scale IC_{50} and $C(t)$ by the slope-intercepts of this relationship. This may result in a value for f_α^h in the tissues and fluids that is different from f_α^h in plasma. The authors of this chapter are currently unable to find values for IC_{50} or IC_{90} in tissues or fluids for CAB-LA. So, we are unable to calculate f_α^h for tissues and fluids. However, our simulations in this chapter all model experiments with HIV infection via plasma, so we will be using f_α^h as defined in Eq. (4) and in Fig. 4.

4.3 Macaque CAB-LA Data and Model

For macaques, IC_{90} is equal to 0.166 μg/mL for CAB, thereby giving $IC_{50} = 0.018\ \mu$g/mL. The plasma drug concentration for macaques was calculated by using

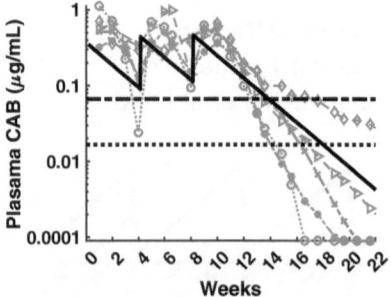

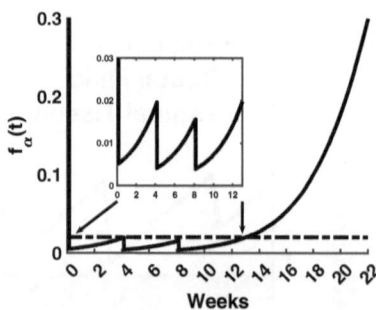

(a) Macaque: plasma concentration of CAB-LA over 22 weeks of treatment.

(b) Macaque: fraction of infection events unaffected over 22 weeks of treatment.

Fig. 6 Macaque cases. (**a**) Pharmacokinetic profile of plasma CAB concentrations (μg/mL) based on simulations, shown by solid black curve and in individual macaques data, shown in light gray curves with markers, from [30] for three injections. The dotted and dashed horizontal lines represent $1\times IC_{90}$ and $4\times IC_{90}$, respectively. (**b**) Fraction of infection events unaffected by PrEP, f_α, versus time since the first injection for macaques. The dash-dotted line is a reference line of $f_\alpha^m = 0.02$. Inset is an enlarged image for the first 13 weeks

experimental data from [40]. The experiments measured $C_{\max} = 3.5$ and decay constant, $k = \ln(2)/\tau_{1/2}$, where $\tau_{1/2} = 14.41$ days. The fraction of infection events unaffected by the drug for macaques, $f_\alpha^m(t)$, is

$$f_\alpha^m(t) = \frac{0.018}{0.018 + 3.5e^{-0.048t}}. \tag{5}$$

The timing of the CAB-LA injection series for macaques shown in Fig. 6 was chosen to match the experimental regime in [30]. Each CAB-LA injection for the macaques is given three times, each 4 weeks apart, following the experimental regime and data from [30]. In Fig. 6a, the simulated plasma CAB concentration solid black curve agrees with the individual macaque data shown by gray curves with circular markers. As shown in Fig. 6b, during the first 8 weeks, only a small portion of target cells will be unaffected by the drug. After the last injection, the drug concentration drops below $4 \times IC_{90}$ 5 weeks after the last injection and below $1 \times IC_{90}$ by 10 weeks after the last injection. With the drug–plasma concentration waning so quickly, a larger portion of the target cells will be unaffected. By the end of week 22, 30% of target cells are unaffected by CAB.

5 Analysis

We are interested in the disease-free equilibrium (DFE) state where healthy T-cells are persistent and the actively and latently infected T-cells die out. Similarly, viral populations are totally cleared from the system. We begin by computing the DFE and the effective reproduction numbers for each of the two strains independently from the steady state with no virus present.

Let $\mathcal{E} = (T, L_s, I_s, V_s, L_r, I_r, V_r)$ denote an equilibrium of the system described by Eq. (2). The system always has the DFE, $\mathcal{E}_0 = (T^*, 0, 0, 0, 0, 0, 0)$, where

$$T^* = \frac{(\gamma - \mu) + \sqrt{(\gamma - \mu)^2 + 4\gamma\beta/K_T}}{2\gamma/K_T}. \tag{6}$$

Following [63], we use the next generation matrix method to obtain the following expressions for the reproduction numbers for the DS and DR strains. The calculation of the reproduction numbers is given in the proof of Theorem 1. For the wild-type HIV(SHIV) infection, we find the reproduction number $\mathcal{R}_E^s$ as

$$\mathcal{R}_E^s = \frac{d\bar{f}_\alpha k_s(\chi(1 - \sigma m_{sr}) + (1 - \sigma)\hat{\mu})NT^*}{c(\chi + \hat{\mu})(d\bar{f}_\alpha + \mu)}, \tag{7}$$

where $\bar{f}_\alpha$ = average value of $f_s(\alpha)$. For the mutated strain, we also find the reproduction number for the $\mathcal{R}_E^r$ as

$$\mathcal{R}_E^r = \frac{(1 - \phi(1 - \bar{f}_\alpha))dk_r(\chi + (1 - \sigma)\hat{\mu})NT^*}{c(\chi + \hat{\mu})(\phi d\bar{f}_\alpha + \mu)}. \tag{8}$$

With CAB-LA protocol levels, both $\mathcal{R}_E^s$ and $\mathcal{R}_E^r$ are less than 1 for humans and macaques.

Theorem 1 *The disease-free equilibrium, $\mathcal{E}_0$, is locally asymptotically stable if both reproduction numbers, $\mathcal{R}_E^s$ and $\mathcal{R}_E^r$, are less than unity and is unstable if at least one of the reproduction numbers is greater than unity.*

Proof The proof follows the approach developed in [63]. Let $\mathcal{X} = (V_s, L_s, I_s, V_r, L_r, I_r, S)$ denote states. Define $\mathcal{F}_i(\mathcal{X})$ as the vector representing the rate of new infections and free virions into compartment i, $\mathcal{V}_i^+(\mathcal{X})$ $(\mathcal{V}_i^-(\mathcal{X}))$ as the rate of transfer of cells into (out of) compartment i, and $\mathcal{V} = \mathcal{V}^- - \mathcal{V}^+$. Consider the systems $\dot{\mathcal{X}} = \mathcal{F}(\mathcal{X}) - \mathcal{V}(\mathcal{X})$ where

$$\mathcal{F}(\mathcal{X}) = \begin{bmatrix} \bar{f}_\alpha NdI_s \\ \sigma k_s V_s T \\ (1 - \sigma)k_s V_s T \\ F_\phi NdI_r \\ \sigma k_r V_r T \\ (1 - \sigma)k_r V_r T \\ 0 \end{bmatrix}, \mathcal{V}^+ = \begin{bmatrix} 0 \\ 0 \\ (1 - m_{sr})\chi L_s \\ 0 \\ 0 \\ \chi(L_r + m_{sr}L_s) \\ \beta + \gamma T \end{bmatrix}, \text{ and}$$

$$\mathcal{V}^- = \begin{bmatrix} cV_s \\ (\chi + \hat{\mu})L_s \\ (\mu + \bar{f}_\alpha d)I_s \\ cV_r \\ (\chi + \hat{\mu})L_r \\ (\mu + \phi f_\alpha d)I_r \\ (\mu + \gamma W/K_T)T \end{bmatrix},$$

where $W = T + L_s + L_r + I_s + I_r$ and $F_\phi = (1 - \phi(1 - \bar{f}_\alpha))$. The following conditions need to be verified:

(A1) If $\mathcal{X} \geq 0$, then $\mathcal{F}_i, \mathcal{V}_i^-, \mathcal{V}_i^+ \geq 0$, for $i = 1, \dots, 7$.
(A2) If $\mathcal{X} = 0$, then $\mathcal{V}_i^-(\mathcal{X}) = 0$.
(A3) $\mathcal{F}_i = 0$ if $i > 6$.
(A4) $\mathcal{F}_i(\mathcal{E}_0) = 0$ and $\mathcal{V}_i^+(\mathcal{E}_0) = 0$ for $i = 1 \dots 6$.
(A5) If $\mathcal{F}(\mathcal{X})$ is set to zero, then all eigenvalues of $Df(\mathcal{E}_0)$ have negative real parts, where $Df(\mathcal{E}_0)$ represents the Jacobian matrix about $\mathcal{E}_0$.

Conditions (A1)–(A4) are easily verified. To verify Condition (A5), we need to calculate the Jacobian evaluated at the DFE.

$$Df(\mathcal{E}_0) = \begin{bmatrix} J_1 & 0 \\ J_2 & J_3 \end{bmatrix},$$

where

$$J_1 = \begin{bmatrix} -c & d\bar{f}_\alpha N & 0 & 0 \\ (1-\sigma)k_s T^* & -(d\bar{f}_\alpha + \mu) & (1-\sigma m_{sr})\chi & 0 \\ \sigma k_s T^* & 0 & -(\chi + \hat{\mu}) & 0 \\ 0 & 0 & 0 & -c & dNF_\phi \\ 0 & 0 & \chi m_{sr} & (1-\sigma)k_r T^* \end{bmatrix},$$

$$J_2 = \begin{bmatrix} 0 & 0 & 0 & \sigma k_r T^* \\ -kT^* & -\frac{\gamma T^*}{K_T} & -\frac{\gamma T^*}{K_T} & -k_r T^* \end{bmatrix}, \text{ and}$$

$$J_3 = \begin{bmatrix} -dF_\phi - \mu & \chi & 0 \\ 0 & -\chi - \hat{\mu} & 0 \\ -\frac{\gamma T^*}{K_T} & -\frac{\gamma T^*}{K_T} & -\frac{2\gamma T^*}{K_T} + \gamma - \mu \end{bmatrix}.$$

The effective reproduction numbers, $\mathcal{R}_e^s$ and $\mathcal{R}_e^r$, are found by taking the maximum eigenvalues of the next generation matrix, given by $D\mathcal{F}(D\mathcal{V})^{-1}$.

The characteristic equation related to $Df(\mathcal{E}_0)$ is

$$\left(\gamma\left(1 - 2\frac{T^*}{K_T}\right) - \mu - \lambda\right)P_s(\lambda)P_r(\lambda) = 0,$$

where $P_s(\lambda)$ and $P_r(\lambda)$ are both third-order polynomials. Solving the first term for λ, we find the eigenvalue $\lambda = -\sqrt{\frac{4\beta\gamma}{K_T} + (\gamma - \mu)^2}$, which has no positive real part. The polynomial, $P_s(\lambda)$, is

$$P_s(\lambda) = A_s + B_s\lambda + C_s\lambda^2 + \lambda^3,$$

with the coefficients

$$A_s = c(\chi + \hat{\mu})(d\bar{f}_\alpha + \mu)(1 - \mathcal{R}_e^s),$$
$$B_s = (\chi + \hat{\mu})(d\bar{f}_\alpha + \mu) + c(\chi + d\bar{f}_\alpha + \mu + \hat{\mu}) - (1 - \sigma)d\bar{f}_\alpha k_s NT^*, \quad (9)$$
$$C_s = c + \chi + d\bar{f}_\alpha + \mu + \hat{\mu}.$$

The coefficient C_s is always positive, and the coefficient A_s is positive when $\mathcal{R}_e^s < 1$. The coefficient B_s requires a little more effort to show it is always positive. We note that

$$d\bar{f}_\alpha k_s NT^* = \frac{(\chi + \hat{\mu})}{((1 - \sigma)\chi + (1 - \sigma m_{sr})\hat{\mu})}(c(d\bar{f}_\alpha + \mu)\mathcal{R}_e^s) \geq c(d\bar{f}_\alpha + \mu)\mathcal{R}_e^s.$$

Thus, we can substitute this inequality into the coefficient B_s to find

$$B_s \geq (\chi + \hat{\mu})(d\bar{f}_\alpha + \mu + c) + c(d\bar{f}_\alpha + \mu)(1 - \mathcal{R}_e^s).$$

Therefore B_s is also always positive when $\mathcal{R}_e^s < 1$. Then by Descartes' law of signs, since the real polynomial $P_3(\lambda)$ has zero sign changes in the sequence of its nonzero coefficients, then it has zero roots with a positive real part. The polynomial $P_r(\lambda)$ is

$$P_r(\lambda) = A_r + B_r\lambda + C_r\lambda^2 + \lambda^3,$$

with the coefficients

$$A_r = c(\chi + \hat{\mu})(dF_\phi + \mu)(1 - \mathcal{R}_e^r),$$
$$B_r = (\chi + \hat{\mu})(dF_\phi + \mu) + c(\chi + dF_\phi + \mu + \hat{\mu}) - (1 - \sigma)dF_\phi k_r NT^*, \quad (10)$$
$$C_r = c + \chi + dF_\phi + \mu + \hat{\mu}).$$

Similarly we can show that

$$B_r \geq (\chi + \hat{\mu})(dF_\phi + \mu + c) + c(dF_\phi + \mu)(1 - \mathcal{R}_r^s).$$

Therefore, the real roots of $P_r(\lambda)$ have only negative parts when $\mathcal{R}_e^r < 1$. This shows condition (A5) holds. It follows from Theorem 2 in [63] that $\mathcal{E}_0$ is locally asymptotically stable when $\max\{\mathcal{R}_e^s, \mathcal{R}_e^r\} < 1$ and unstable when $\max\{\mathcal{R}_e^s, \mathcal{R}_e^r\} > 1$.

□

6 Parameter Sensitivity

In modeling, equations usually depend on several unknown parameters. Finding the appropriate values and ranges is a critical step that is useful for screening outliers and as such requires careful consideration. In general, these values can be estimated through the least-squares fitting, Bayesian inference, or maximum likelihood. The ranges of the parameters can also be determined by setting upper and lower bounds on the values, either based on physical constraints or by exploring the parameter space through sensitivity analysis.

Previous literature has stated that understanding the peak of viral load data is vital to best predict T-cell latency and infection outcomes [64]. Thus, to best calibrate our HIV and SHIV models, we compared our model's outputs to the viral load data extracted from Reeves et al. [38] and Vaidya et al. [39] for humans and macaques, respectively, using a simulated annealing technique in MATLAB. Simulated annealing is a local optimization technique that takes a user inputted initial guess for the global minimum of the system as defined by the user; for the purposes of this study, the smallest least squared error between the model predictions and the extracted viral load data [65]. Each parameter, in this case c, the clearance rate of free virions and N, the number of virions produced per infected T-cell, is given a lower and upper bound. These two parameters are chosen due to their direct, mathematical correlation with peak viral load. The simulated annealing algorithm then generates a random value within each boundary, calculates the user described error, and then generates a new random value to see if the error increases or decreases. This process is repeated until a local minimum is found and parameter values are produced [66]. We found when we performed this operation for the DS virion model, the resulting macaque values for N and c could not be matched to the experimental data. This indicates that the DR virions should not be discounted when fitting parameters.

6.1 Elasticity of the Effective Reproduction Number $\mathcal{R}_E$

In this section we test the sensitivity of the reproduction number, $\mathcal{R}_E$, to its parameters. We compute the elasticity (normalized forward sensitivity) index [67] to determine to what extent the value of $\mathcal{R}_E$ (for the macaque and human models) changes following changes to each parameter value. The sensitivity index with the reproduction number indicates the impact of the parameter on the disease-free equilibrium. Since $\mathcal{R}_E = \max\left(\mathcal{R}_E^s, \mathcal{R}_E^r\right)$, we compute the sensitivity with respect to both $\mathcal{R}_E^s$ and $\mathcal{R}_E^r$. The forward sensitivity indices for these parameters are represented by

$$\mathcal{F}_x = \left(\frac{\partial \mathcal{R}_E^i}{\partial x}\right)\left(\frac{x}{\mathcal{R}_E^i}\right), \tag{11}$$

where x represents the parameter and i is s for sensitive and r for resistant. These forward sensitivity indices were evaluated using the baseline parameters given in Tables 2 and 3, except for f_α and ϕ. We replace the time-varying value of f_α with a time-average of 0.01. ϕ, which is between 0 and 1, and we set equal to +0.5. The elasticity results are shown in Table 4.

The positive sign of the elasticity index specifies that $\mathcal{R}_E^i$ increases with the parameter, and the negative sign specifies that $\mathcal{R}_E^i$ decreases. The magnitude of the elasticity determines the relative importance of the parameter. If $\mathcal{R}_E^i$ is given explicitly, then the elasticity index for each parameter can be explicitly computed

Table 2 Parameters (Pa.) of the model (Human)

Pa.	Value	Range and units	Description	Ref
σ^h	0.02	(0.001, 0.02) unitless	Fraction of T-cells moving to latency	[68]
β^h	60	(50, 60) cells $\cdot \mu L^{-1} \cdot$ day^{-1}	Recruitment rate of healthy T-cells	[42, 69]
c^h	25	(3.07, 25) day^{-1}	Clearance rate of free virus	[69, 70]
d^h	0.124	(0.124, 0.95) day^{-1}	Disease-induced cell death	[42, 71]
γ^h	0.03	day^{-1}	Proliferation rate of T-cells	[72]
I_0^h	0.02	cells $\cdot \mu L^{-1}$	Initial infected T-cells	
k_s^h	2.5×10^{-4}	RNA copies $\cdot \mu L^{-1}$	Infection rate of T-cells (by sensitive strain)	[71]
K_T^h	$T_0^h + L_0^h + I_0^h$	1.5×10^3 cells $\cdot \mu L^{-1}$	Carrying capacity of T-cells	[72]
L_0^h	0	cells $\cdot \mu L^{-1}$	Initial latently infected T-cells	
μ^h	0.05	(0.006, 0.05) day^{-1}	Natural death rate of T-cells	[69, 71]
$\hat{\mu}^h$	0.05	day^{-1}	Total exit rate of latently infected T-cells	[69, 71]
N^h	400	(400, 7100) virions $\cdot$ cell^{-1}	Number of virions produced per infected T-cell	[69, 71]
m_{sr}	10^{-5}		DS-to-DR mutation rate	[7]
T_0^h	10^4	cells $\cdot \mu L^{-1}$	Initial T-cell count	[42, 71]
T_s^0	10^3	cells $\cdot \mu L^{-1}$	T-cell rescaling factor	
V_0^h	100	RNA copies $\cdot \mu L^{-1}$	Initial viral load	[42]
V_s^0	10^1	cells $\cdot \mu L^{-1}$	Viral rescaling factor	
χ^h	0.05	$(10^{-4}, 0.1)$ day^{-1}	Transition rate from latency to infection	[73–75]

Table 3 Parameters (Pa.) of the model (Macaque)

Pa.	Value	Range and units	Description	Ref
σ^m	0.001	$(10^{-3}, 10^{-1})$ unitless	Fraction of T-cells moving to latency	[76, 77]
β^m	10	cells $\cdot$ μL^{-1} $\cdot$ day^{-1}	Recruitment rate of healthy T-cells	[74]
c^m	60	$(27, 60)$ day^{-1}	Clearance rate of free virus	[78]
d^m	6.23×10^{-1}	day^{-1}	Disease-induced cell death	[78, 79]
γ^m	2.15×10^{-1}	day^{-1}	Proliferation rate of T-cells	[78]
I_0^m	0.001	cells $\cdot$ μL^{-1}	Initial infected T-cells	
k_s^m	1.21×10^{-6}	$(0.02, 2.34) \times 10^{-6} \mu$L $\cdot$ virions $\cdot$ day^{-1}	Infection rate of T-cells (by sensitive strain)	[78, 80]
K_T^m	$T_0^m + L_0^m + I_0^m$	$(1.68, 1.76) \times 10^3$ cells $\cdot$ μL^{-1}	Carrying capacity of T-cells	[79]
L_0^m	0	cells $\cdot$ μL^{-1}	Initial latently infected T-cells	
μ^m	0.01	day^{-1}	Natural death rate of T-cells	[76, 81]
$\hat{\mu}^m$	0.041	day^{-1}	Total exit rate of latently infected T-cells	[76, 82]
N^m	4×10^4	$(4, 5.5) \times 10^4$ virions $\cdot$ cell^{-1}	Number of virions produced per infected T-cell	[83]
m_{sr}	10^{-5}		DS-to-DR mutation rate	[7]
T_0^m	1.68×10^3	$(1.37, 1.76) \times 10^3$ cells $\cdot$ μL^{-1}	Initial T-cell count	[78, 79]
T_s^0	T_0^m	cells $\cdot \mu$L^{-1}	T-cell rescaling factor	
V_0^m	5	$(2.66, 100)$ RNA copies $\cdot \mu$L^{-1}	Initial viral load	[40, 78]
V_s^0	1	cells $\cdot$ μL^{-1}	Viral rescaling factor	
χ^m	10^{-3}	$(2 \times 10^{-4}, 6 \times 10^{-2})$ day^{-1}	Transition rate from latency to infection	[73, 75]

and evaluated for a given set of parameters. The magnitudes of the elasticity indices depend on these parameter values.

For our model, we calculated elasticity indices for the 15 parameters, with the values for human set of parameters and macaque set given separately in Table 4. Although the parameter sets for humans (see Table 2) and macaques (see Table 3) are different, the response the basic reproduction number gives qualitatively and quantitatively is quite similar for both the resistant and sensitive strains, with f_α as the outlier. However, the behavior of f_α is to be expected; each index value can be thought of as a ratio of the effective change in the reproduction number with respect to the applied change in the given parameter. For example, for every 10% increase in the infection rate of T-cells, k_s or k_r, the reproduction number will increase by 5%. However, it will decrease by 5% for every 10% increase in the proliferation rate of T-cells, γ. When f_α increases, CAB-LA is less effective in blocking HIV infections. Hence, for the DS infections, the number of infections will rise. On the other hand, as the DS strain increases when f_α increases, the DR strain is outcompeted by the DS strain.

Table 4 Elasticity indices of $\mathcal{R}_E$ for HIV model evaluated at baseline human HIV parameters shown in Table 2 and baseline macaque SHIV parameters shown in Table 3, except for f_α and ϕ

Forward Sensitivity	Parameter	Human $\mathcal{R}_E^s$	Human $\mathcal{R}_E^r$	Macaque $\mathcal{R}_E^s$	Macaque $\mathcal{R}_E^r$
$\mathcal{F}_N$	N	+1	+1	+1	+1
$\mathcal{F}_{k_s}$	k_s	+1	–	+1	–
$\mathcal{F}_{k_r}$	k_r	–	+1	–	+1
$\mathcal{F}_d$	d	+0.976	+0.988	+0.616	+0.762
$\mathcal{F}_\beta$	β	+0.5	+0.5	+0.5	+0.5
$\mathcal{F}_{K_T}$	K_T	+0.5	+0.5	+0.5	+0.5
$\mathcal{F}_\chi$	χ	+0.005	+0.005	+0.009	+0.009
$\mathcal{F}_{\hat{\mu}}$	$\hat{\mu}$	−0.005	−0.005	−0.009	−0.009
$\mathcal{F}_\sigma$	σ	−0.010	−0.010	−0.028	−0.028
$\mathcal{F}_\gamma$	γ	−0.5	−0.5	−0.5	−0.5
$\mathcal{F}_\mu$	μ	−0.976	−0.988	−0.616	−0.763
$\mathcal{F}_c$	c	−1	−1	−1	−1
$\mathcal{F}_\phi$	ϕ	–	−0.992	–	−1.218
$\mathcal{F}_{f_\alpha}$	f_α	+0.976	−0.002	+0.616	−0.228

In terms of the effects of drug resistance, we note that the effective reproduction numbers, $\mathcal{R}_E^r$ for macaques and humans is twice as sensitive to the reduction in effectiveness of CAB-LA, represented by ϕ as it is to the fitness of the DR virions to infect a healthy T-cell, k_r.

6.2 Global Sensitivity

Sensitivity analysis is important for determining which parameters have the largest impact on the dynamics of the spread of HIV. Following [84], we employ partial rank correlation coefficient (PRCC) analysis to determine the sensitivity of the model, given by the system defined in Eq. (2), to each parameter for humans and macaques after 48 weeks of PrEP. In this instance, correlation provides a measure of the strength of a linear association between a parameter and the number of virions and infected T-cells. Rather than capturing the sensitivity of the total number of infected T-cells or virions to a single parameter at a time, partial correlation analysis reveals hidden true correlations and false correlations explained by the effect of other variables. The parameters, specified in Tables 2 and 3, are sampled using Latin hypercube sampling (LHS) [85]. LHS/PRCC sensitivity analysis is often employed in uncertainty analysis to explore the entire parameter space of a model.

The magnitude of the PRCC indicates the strength of the correlation between the parameter and the output, whereas the sign of the PRCC indicates whether there is a positive or negative correlation between the parameter and the output—the total number of infected T-cells or virions in this case.

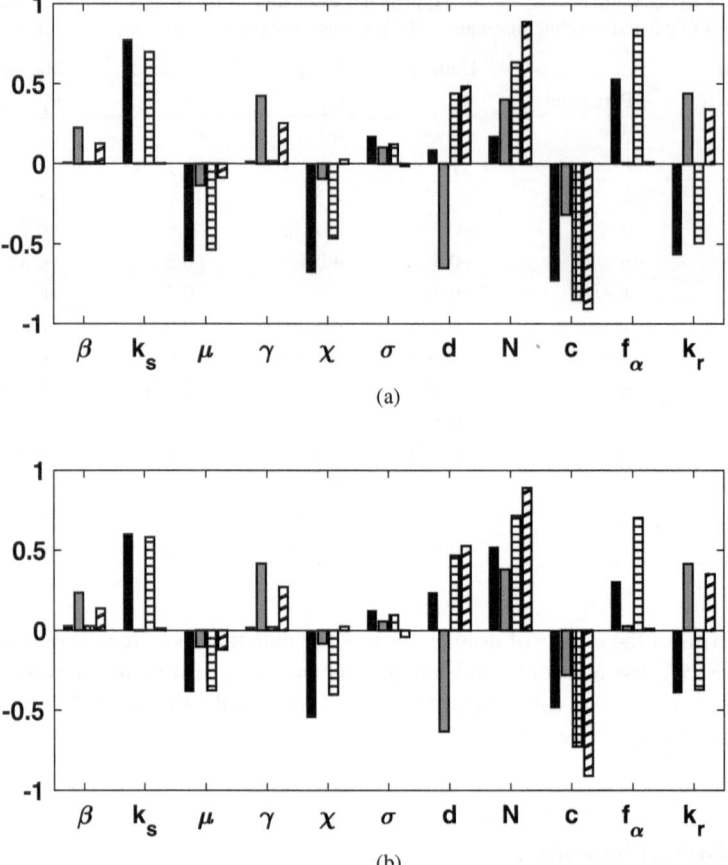

Fig. 7 PRCC results for the Human HIV Model after 200 days of PrEP. The black bars represent the T-cells infected with the DS virus, $L_s + I_s$, the gray bars represent the T-cells infected with the DR virus, $L_r + I_r$, the horizontal striped bars represent the DS virions, V_s, and the diagonal striped bars represent the DR virions, V_r. The label DS indicates the drug-sensitive strain, and DR indicates the drug-resistant strain. (**a**) Human: PrEP before HIV exposure. (**b**) Human: PrEP after HIV infection

In Fig. 7 the PRCC results are shown first for humans given PrEP (CAB-LA) before HIV exposure and second given PrEP 2 weeks after seroconversion. The black bars represent $L_s + I_s$, the gray bars $L_r + I_r$, the horizontal striped bars V_s, and the diagonal striped bars V_r. The parameter sensitivity for the human model does not change dramatically between the PrEP before HIV exposure and the PrEP after HIV exposure. The infected T-cells, both with the DS and DR strains, are sensitive to k_s the infection rate of target cells, μ the natural death rate of T-cells, and χ the transition rate of target cells from latently infected to infected. The DR-infected T-cells and both types of virions are sensitive to the viral-induced death rate of infected cells, although the number of infected T-cells decreases as d increases and

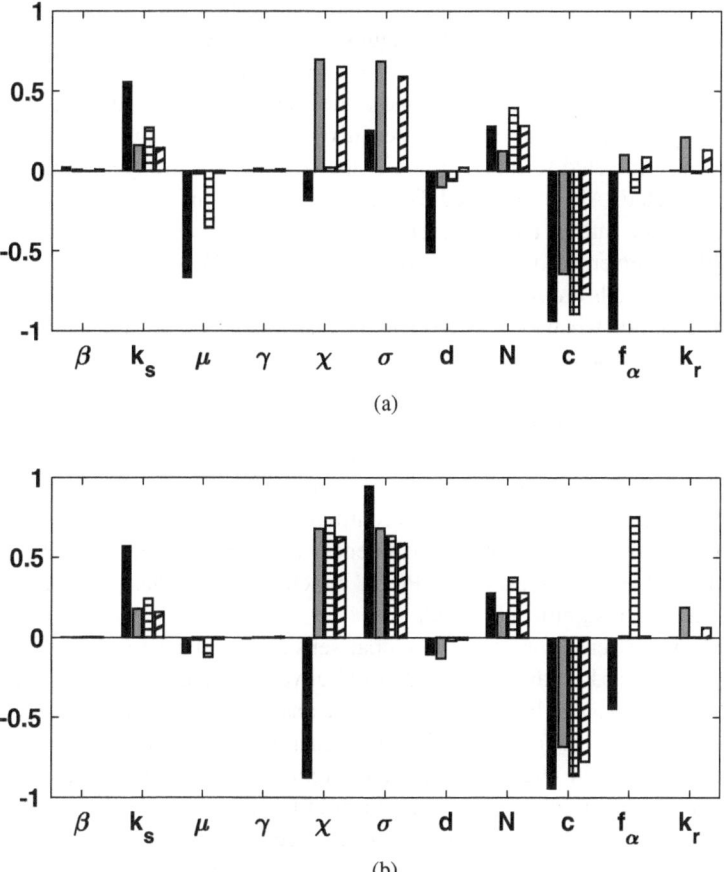

Fig. 8 PRCC results for the Macaque SHIV Model after 200 days of PrEP. The black bars represent the T-cells infected with the DS virus, $L_s + I_s$, the gray bars represent the T-cells infected with the DR virus, $L_r + I_r$, the horizontal striped bars represent the DS virions, V_s, and the diagonal striped bars represent the DR virions, V_r. The label DS indicates the drug-sensitive strain, and DR indicates the drug-resistant strain. (**a**) Macaque: PrEP before SHIV exposure. (**b**) Macaque: PrEP after SHIV infection

the number of virions increases as the infected cell bursts. Both classes of virions are more strongly sensitive to N, the number of virus produced per burst infected cell. When PrEP is given after HIV seroconversion, the infected T-cells become more sensitive to N, but less sensitive to c the clearance rate of virus, f_α the ability of CAB-LA to block HIV, and k_r the infection rate of target cells by the resistant virions. The parameter sensitivity of the virions does not appear to change from the PrEP before HIV exposure to the PrEP after HIV infection scenarios.

In Fig. 8 the PRCC results are shown first for macaques given PrEP before HIV exposure and second given PrEP 2 weeks after seroconversion in accordance with the experiment in [30]. The black bars represent $L_s + I_s$, the gray bars $L_r + I_r$, the horizontal striped bars V_s, and the diagonal striped bars V_r. For PrEP given as a preventative measure to the macaques, both the DS-infected and infectious T-cells and the DS virions are most sensitive to k_s the infection rate of target cells, μ the natural death rate of T-cells, and c the clearance rate of the virus. The cells infected with the DR strain are most sensitive to χ the transition rate of target cells from latently infected to infected and σ the fraction of "stronger" cells moved to latently infected. All the infected T-cell types and virions are equally sensitive to c, the clearance rate of the virus. This sensitivity does not change in the situation when PrEP is given before SHIV exposure or after SHIV seroconversion. Only the DS-infected T-cells are sensitive to CAB-LA, designated by f_α.

In the situation when PrEP is given to the macaques 2 weeks after seroconversion, χ and σ become sensitive for the infected T-cells and their accompanying virions. The infected T-cells become less sensitive to CAB-LA, but the virions become strongly more sensitive to changes in the ability of CAB-LA to block SHIV.

In terms of the effects of drug resistance, we note that the concentration of infected T-cells and virions are sensitive to the fitness of the DR virions to infect a healthy T-cell, k_r. However, the concentration of infected T-cells and virions are not sensitive to the reduction in effectiveness of CAB-LA, represented by ϕ. The parameter ϕ does not appear on the PRCC bar charts since it has a p number greater than 0.5 and an insignificant PRCC.

The forward sensitivity and the global sensitivity appear to give contradictory results with regard to the reduction in effectiveness of CAB-LA toward the DR HIV strain, represented by ϕ. However, these analyses are measuring the sensitivity of two different situations. The forward sensitivity measures how changing the parameter value will change the effective reproduction number, which indicates the sensitivity of the disease-free equilibrium to parameter changes. The forward sensitivity indicates that to achieve a disease-free state increasing the effectiveness of CAB-LA for the drug resistance will be twice as effective as reducing the fitness of the DR virions to infect a healthy T-cell, k_r. But in the global sensitivity after 200 days of PrEP, the reduction of effectiveness of CAB-LA against the DR strain from a baseline of $\phi = 1/2$ is statistically insignificant to the concentration of infected T-cells and the number of virions. On the other hand, both sensitivity analyses indicate that the fitness of the DR virions to infect a healthy T-cell, k_r, is very important to achieving a disease-free equilibrium and after 200 days of PrEP. This indicates that the level of drug resistances and the fitness of the DR virions to infect are both very important to the effectiveness of CAB-LA as PrEP.

7 Treatment Simulations

In our simulations, we are modeling HIV or SHIV infection from the first onset and continuing just through the acute infection stage. We do not include antiretroviral treatment (ART) in our model. In the wild-type systems, we solely have the DS (CAB-LA sensitive) strain of HIV or SHIV, and we assume that there are no mutations occurring, i.e., $m_{sr} = 0$. In the full model, we assume mutations can occur naturally, and this is how drug-resistant HIV or SHIV strains are introduced into the system. In other words, the SHIV or HIV exposures or initial strains upon seroconversion are assumed to be fully drug-sensitive.

In their paper describing their experimental results, Radzio-Basu et al. [30] emphasized that the experimental results for macaques administered PrEP during the acute SHIV infection indicated a strong emergence of drug resistance. In response, the FDA decided to require HIV tests prior to each CAB-LA dose for humans [31]. While caution to avoid the development and propagation of more drug-resistant strains of HIV is commendable, the question still remains: Does this development of drug resistance in macaques truly foretell the similar development of drug resistance in humans? This is a question that is difficult to answer with a dearth of clinical human data, but we are attempting to answer it with our mathematical model and numerical simulations. Hence the goal of this research is to establish a model that can capture the mechanistic behavior of SHIV and HIV virions and their interaction with healthy CD4$^+$ T-cells.

Toward this aim in this section, we validate our model by comparing our simulated results to experimental data for humans and macaques. We use the term validation in the sense of the National Academy of Sciences report [86] where validation is defined to be the process of determining the degree to which a model is an accurate representation of the real world from the perspective of the intended uses of the model. We note that we have not incorporated formal verification or validation methods in this chapter. Instead, our validation is restricted to illustrating graphically that the simulations match experimental results.

In Sect. 7.1 we present our numerical simulations and compare these to experimental data for humans and macaques without any HIV treatment or preventative measures, i.e., without ART or PrEP. In Sect. 7.2 we perform our numerical simulation for macaques administered PrEP before being exposed to SHIV and compare the results to the experiment in [40]. In Sect. 7.3 we perform our numerical simulation for macaques administered PrEP before being exposed to SHIV and compare the results to the experiment in [30].

Once behaviors of SHIV in macaques in pre-clinical settings were captured *in silico* in Sects. 7.1–7.3, human *in silico* clinical trials were run with standard of care in PrEP injection protocols in Sects. 7.4 and 7.5. To see the similarities and differences in the macaque and human simulations, we test several values for the effectiveness of CAB LA on the DR strain and on the fitness of the DR virions to infect healthy T-cells.

7.1   Macaques and Humans: Numerical Validation Without ART or PrEP

We first plot our model simulations against data so that we may validate our model prior to running *in silico* pre-clinical and clinical trials. Our system represents

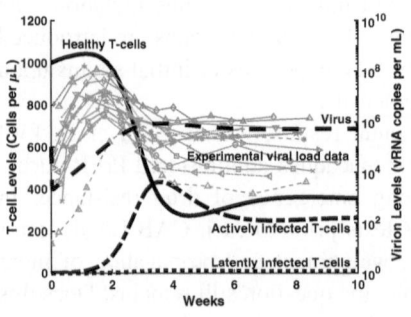
(a) Human: drug-sensitive only without PrEP

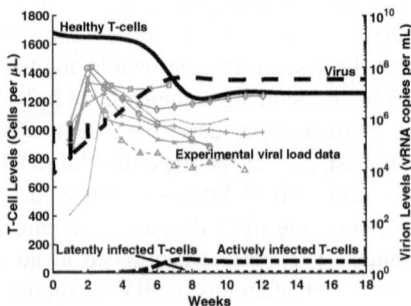
(b) Macaque: drug-sensitive only without PrEP

Fig. 9 Simulation results for drug-sensitive only ($m_{sr} = 0$) infection model given by the system defined in Eq. (2) without ART or PrEP and experimental viral load data. In (**a**) the simulated human (HIV) viral load represented by black dash curve is compared to experimental viral load data from [38], shown in light gray with each individual recorded with a separate symbol. In (**b**) the simulated macaque (SHIV) viral load, represented by dash black curve, is compared to experimental viral load data from [41] given in light gray

wild-type HIV transmission without mutations, treatment for either post-infection ART, or pre-infection PrEP.

Figure 9a displays experimental viral loads from [38] in light gray with marker symbols. The experimental results for each individual human are differentiated by the different markers. The simulated viral loads are plotted as the dashed black curve. In the early stages, viral loads rapidly increase from 10^3 RNA copies per mL to 10^6 or higher, then decline, and reach a plateau around 10^4 to 10^6. The healthy T-cell concentration, represented by the solid black curve, rises slightly in the first 10 days, then drops to around 30% shortly after maximum viral load, then raises to 40%, and stays stable. The actively infected T-cells, represented in black dash-dot curve, increase from 0 to about 400 cells per μL at the maximum viral load, then decline slowly and remain at 250 cells per μL after 40 days. The latently infected T-cells, represented by the black dotted curve remain low throughout the simulation. The combined T-cell concentration (healthy, latently infected, and actively infected) drops by about 35% during this infection process. Overall, wild-type HIV model simulation viral load (DS) results match with the experimental data for humans without ART or PrEP.

Figure 9b displays the comparison between simulation results and experimental data, light gray with marker symbols, in Dobard et al. [41] for macaques. The experimental results for each individual macaque are differentiated by the different markers. The simulated viral loads are plotted as the dashed black curve. In the Dobard et al. [41] experiment the macaques are given a weekly SHIV challenge for the first 3 weeks of the experiment. This exposure, included in our simulation, is captured in Fig. 9b by the three vertical lines on the virus level. In the early stages, viral loads rapidly increase from 10^4 RNA copies per mL to a plateau of 10^8. This is on the higher end of the experimental viral loads in macaques. The healthy

T-cell concentration, represented by the solid black curve, drops over 8 weeks, much slower than in the human simulation shown in Fig. 9a. The actively infected T-cells, represented by the black dash-dot curve, increase from 0 to about 200 cells per mL at the maximum viral load. The latently infected T-cells, represented by the black dotted curve, remain low throughout the simulation. Overall, wild-type SHIV model simulation viral load (DS) results match with the experimental data for macaques without ART or PrEP.

Once our model's sensitive viral load output was calibrated, we varied ϕ, the percentage of drug efficacy on the resistant strain. In addition, to capture a measure of the fitness of the mutated DR strain, we explore the relationship between k_r the infection rate of healthy T-cells by DR virions and k_s the infection rate of healthy T-cells by DS virions. These tests are conducted to better understand the dynamics of the DR viral load. The parameters being varied—ϕ, the effectiveness of CAB-LA for the DR strain, and k_r, the fitness of DR virions to infect a T-cell—are not known in advance. The amount of drug resistance to CAB-LA of the mutated HIV strain is of utmost concern. Parikh et al. [53] note that multiple mutations are required for extensive CAB drug resistance. However, the macaque studies by Radzio-Basu et al. [30] indicate that these mutations are selected when CAB-LA is given to macaques previously infected with SHIV. The fitness of the DR virions to infect is included as parameters to vary. It is often the case that the mutation that makes the HIV strain at least partially drug-resistant may also trade off its ability to invade a T-cell.

In Fig. 10, we show simulation results of the full mutation model without ART and PrEP. For both human and macaques, we set the virus mutation rate from the wild-type DS strain to the mutated DR strain to be $m_{sr} = 10^{-5}$, based on [42]. In Fig. 10a, b, the healthy T-cells are represented by solid black curves. The latently

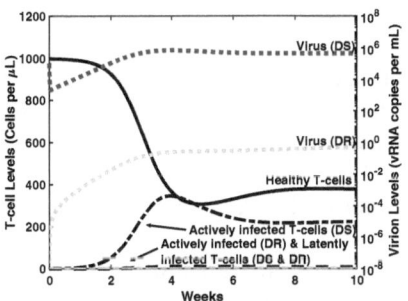
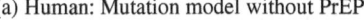
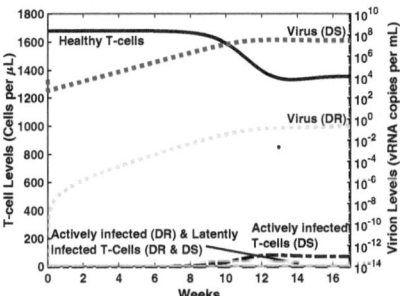

(a) Human: Mutation model without PrEP   (b) Macaque: Mutation model without PrEP

Fig. 10 Simulation results for (**a**) human and (**b**) macaque model with mutations ($m_{sr} \neq 0$) given by the system defined in Eq. (2) without ART or PrEP. Each of the curves is labeled to designate the T-cells in various stages plus the virions. The label DS indicates the drug-sensitive strain, and DR indicates the drug-resistant strain. Solid curves: healthy T-cells. Dash-dot curve: actively infected T-cells. Dash curve: latently infected T-cells. Dark gray: DS, light gray: DR. Dark gray dotted curve: DS virus. Light gray dotted curve: DR virus. The concentration of DR infected T-cells and latently infected T-cells is close to zero in both humans and macaques

infected and actively infected by the DS strain are combined and represented by black dash-dot curves and the DR by the black dashed curves. The DS virions are represented by the gray dotted curves and the DR virions by the black dotted curves.

In humans, the concentrations of healthy, actively, and latently DS-infected T-cells and DS virus have almost identical behaviors in wild-type only ($m_{sr} = 0$) and mutation ($m_{sr} \neq 0$) models during the first 10 weeks. In both Figs. 9a and 10a, the concentration of healthy T-cells changes from 1000 cells per μL to a lowest point, about 300, in 4 weeks and then increases slowly to reach the plateau in 3 weeks. The concentrations of actively DS-infected T-cells increase slowly during the 1st week and then quickly until it reaches its peak at the end of the 4th week. They slowly drop back to a steady state around the 6th week. The DS virus concentration reaches a small peak during the 4th week of infection, then drops slightly, and reaches a steady state. The concentrations of latently DS-infected T-cells and both types of DR-infected T-cells are negligible throughout the simulation.

In macaques, when comparing Figs. 9b and 10b, we notice concentrations of healthy T-cells stay steady around 1680 cells per μL for about 60 days, then drop to 1400 in next 20 days, and remain at that level. The concentration of actively DS-infected T-cells is steady and small and then increases during the same period when the concentration of healthy T-cells drops and then stays at a steady state. The DS virus concentration grows at an exponential rate during the first 60 days after infection and also stays flat afterward. The concentrations of latently DS-infected T-cells and both types of DR-infected T-cell concentrations stay low throughout the simulation.

7.2 PrEP Before SHIV Exposure: Numerical Validation for Macaque Experiment

In Andrews et al. [40], there were multiple experiments with three groups of macaques given three different PrEP regimes before being exposed to SHIV intravenously. The study was intended to evaluate the effectiveness of CAB-LA (as PrEP) against intravenous SHIV challenge. This macaque experimental model was intended to determine whether human studies for CAB-LA as PrEP in people who inject drugs are warranted.

We simulate the experiment where eight macaques are treated with PrEP twice, 4 weeks apart. There were also five control macaques who did not receive PrEP. On day 14, all the macaques are infected with SHIV. In the simulations we vary ϕ between 0.3 and 0.7 and allow $k_r = k_s$ or $k_r = k_s/2$. The parameter sensitivity studies indicated that the concentration of infected cells and virions are sensitive to the fitness of the DR virions to infect. The elasticity studies with the effective reproduction number for the DR strain indicated that there is sensitivity to the effectiveness of CAB-LA on the DR virions. We observe that the DS viral loads

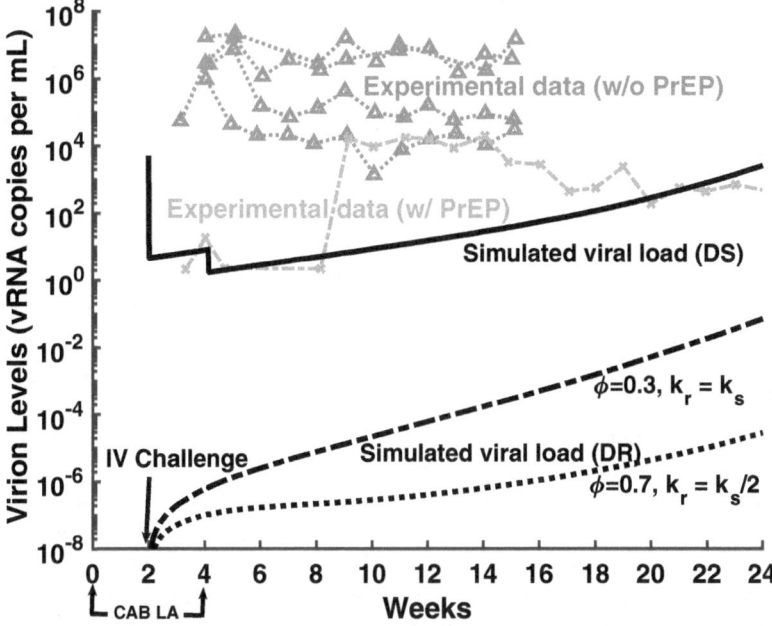

Fig. 11 PrEP injection on days 1 and 28, and SHIV infection on day 14. Macaque SHIV infection viral load data with PrEP treatment [40] compared with numerical simulations. The label DS indicates the drug-sensitive strain, and DR indicates the drug-resistant strain. Solid curve: simulation results for DS virus. Black dot and dash-dot curves: simulation results for DR virus with two most extreme levels of fitness and drug effectiveness. Light gray dash curves with crosses: experimental data for one macaque given PrEP. All other experimental data for macaques given PrEP had virion levels of less than 1 virion/mL and are not shown in this figure. Dark gray dash curves with triangles: experimental data for control macaques (no PrEP)

do not have noticeable changes, but there are obvious changes in the DR viral load. In Fig. 11, we present the experimental data of [40] with the control macaques (no PrEP) and the one macaque with a virion/mL count above 1. The other seven macaques given PrEP had virion concentrations of less than 1, but the exact virion concentration was not given. Also in Fig. 11 we display the numerical simulations for the DS viral load (solid black curve) and the two most extreme cases of the DR viral load (black dotted curve for $\phi = 0.3$, $k_r = k_s$, and black dotted dashed curve for $\phi = 0.7$, $k_r = k_s/2$). Despite varying both the values of ϕ and k_r, the amount of DR virions is negligible when compared to DS virions. The combined viral load is almost identical to the DS viral load. As shown in Fig. 11, the DS viral load solid black curve is comparable to the one macaque received PrEP black dashed curve with triangles plus the seven macaques with even lower virion concentrations. The gray curves with crosses represent the viral load data in the control macaques who did not receive PrEP.

7.3 PrEP After SHIV Infection: Numerical Validation for Macaque Experiment

Radzio-Basu et al. [30] performed experiments infecting eight rhesus macaques intravenously with SHIV. Two of the macaques were used as controls. The other six macaques were given a treatment of CAB-LA 11 days after infection. The experiment was devised to test what would happen when PrEP was initiated during an acute HIV infection. In their experiments, Radzio-Basu et al. [30] found that DR mutations were frequently selected and maintained for several months.

In the Radzio-Basu et al. [30] experiment, the macaques were infected with SHIV on day 1 and then receive PrEP injections on days 11, 39, and 67. We replicated the experiment and then tested several levels of effectiveness of CAB-LA against the DR mutations and fitness for the DR virions to infect T-cells.

We run simulations by varying the effectiveness of CAB-LA, ϕ, between 0.3 and 0.7. To test situations for fitness, we added two scenarios allowing $k_r = k_s$ for no loss of fitness or $k_r = k_s/2$, which corresponds to a 50% loss of fitness. Similar to Fig. 11, we observe that the DS viral loads do not have an obvious change, but there are noticeable changes in the DR viral load. In Fig. 12, we plot our numerical simulations for DS viral load (the solid black curve). We only plot the two most extreme cases of DR viral load with dotted curve for $\phi = 0.3, k_r = k_s$ and the black dash-dot curve for $\phi = 0.7, k_r = k_s/2$. The experimental data of [30] are included in Fig. 12. The dark gray dash curves with triangles represent viral load in macaques without PrEP, and the light gray curves with crosses illustrate the viral load in macaques given PrEP. It is clear to see that the sensitive viral load (solid black curve) agrees with the experimental observations. The amount of DR virions is negligible in comparison to the magnitude of DS virions. There is a noticeable difference between the simulated DR viral load when the DR drug effectiveness and virion fitness ϕ, k_r, and k_s are varied.

7.4 Macaque and Human Simulations: PrEP Before Exposure

Once we observed the DR viral load changes due to the efficacy of the PrEP drug, ϕ, and due to the relationship between the infectivity of DR virions, k_r, and DS virions, k_s, we decided to run our own *in silico* pre-clinical trials in macaques to better understand the role of PrEP injection campaigns on the infectivity of DR and DS SHIV.

First, we designed a simulated trial for an extended version of the experiment in [40] (see Fig. 11) where there macaques were administered an injection of PrEP on day 0, exposed to SHIV on day 14, and then were given a second injection of PrEP on day 28 as shown in Fig. 13. We varied the value of the drug effectiveness against the mutated DR SHIV strain, ϕ, between 30% (top row) and 70% (bottom row). We also set the fitness of the DR virions to infect healthy T-cells to equal the fitness of

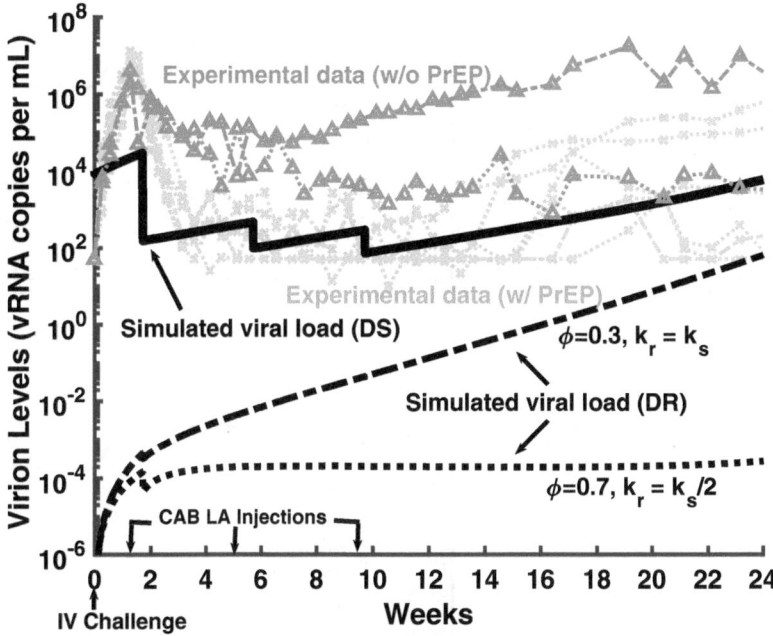

Fig. 12 PrEP injections on days 11, 39, and 67 after SHIV infection. Macaque SHIV infection viral load data with PrEP treatment [30] compared with numerical simulations. The label DS indicates the drug-sensitive strain, and DR indicates the drug-resistant strain. Solid curve: simulation result for DS virus. Dot and dot-dash curves: DR virus with two levels of fitness and drug effectiveness. Light gray dash curves with triangles: experimental data for macaques given PrEP. Dark gray curves with crosses: experimental data for control macaques (no PrEP)

the DS virions, $k_r = k_s$ (first column), i.e., DR virions are equally as infective as DS. Next we set the DR virions to half of the fitness, $k_r = k_s/2$ (second column). We allow the *in silico* trial to last for 24 weeks to allow the system dynamics enough time to respond to the PrEP treatment and SHIV infection.

We note that in all four of these simulations with both the DS and DR strains of SHIV, the healthy T-cell concentrations (black solid curves) dip significantly when the CAB-LA concentration is very low, 27 weeks after the second injection of PrEP. The drop in the healthy T-cell concentration follows the rise in DS virions in each of Fig. 13a–d. The significant difference between the subfigures is the level of DR virions. In Fig. 13a, where CAB-LA is only 30% effective against the DR strain and the DR virions are just as infective as the DS virions, the DR virion level rises above 10^2 vRNA copies per mL. It does not outpace the DS virions, but it is at a level of concern. In Fig. 13c where the effectiveness of CAB-LA against the DR strain has risen to 70% with equal fitness between the DS and DR virions, there is a small increase in the DR virions to about 1 vRNA copy per mL. In Fig. 13b, d where the fitness of the DR virions is a half of the DS virions, the DR strain remains at a

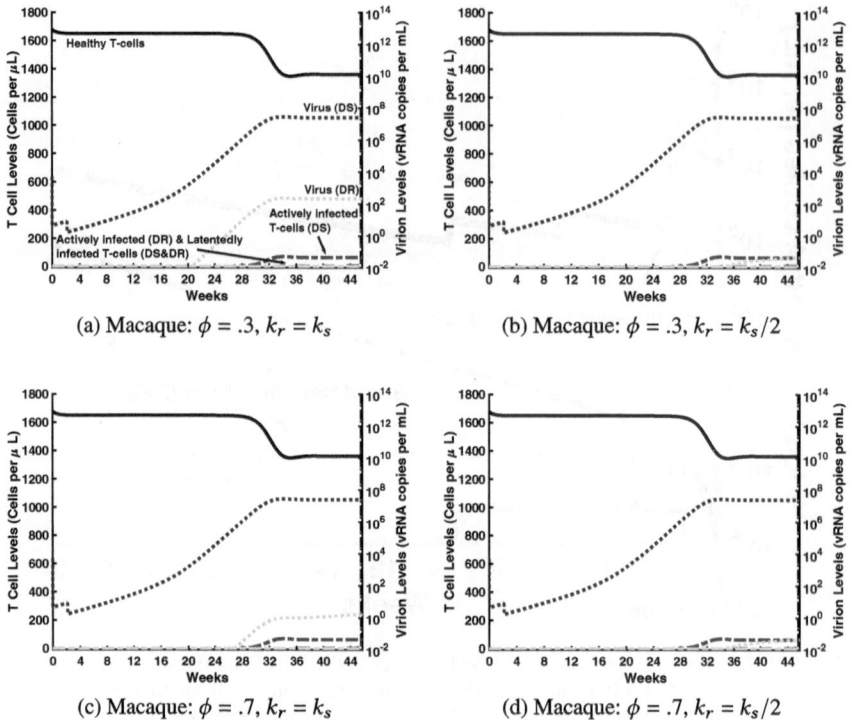

Fig. 13 Macaque *in silico* trial corresponding to the experiment in [40]. PrEP is administered on days 0 and 28. The macaque is exposed to SHIV on day 14. The label DS indicates the drug-sensitive strain, and DR indicates the drug-resistant strain. (**a**) PrEP is 30% effective against SHIV ($\phi = 0.3$). DR virions are equally as infective as DS virions, thus $k_r = k_s$. (**b**) PrEP is 30% effective against SHIV ($\phi = 0.3$). DR virions are half as infective as DS ($k_r = k_s/2$). (**c**) PrEP is 70% effective against SHIV ($\phi = 0.7$). DR virions are equally as infective as DS ($k_r = k_s$). (**d**) PrEP is 70% effective against SHIV ($\phi = 0.7$). DR virions are half as infective as DS ($k_r = k_s/2$). Solid curves: healthy T-cells. Dash-dot curve: actively infected T-cells. Dash curve: latently infected T-cells. Dotted curve: virus. Dark gray: DS, light gray: DR

negligible level. This indicates that the virion fitness may be important as the drug efficacy in determining whether a mutated strain will grow to a significant level.

Once behaviors of SHIV in macaques in pre-clinical settings were captured *in silico*, human *in silico* clinical trials were run with standard of care PrEP injection protocols. Current standard of care for human injectable PrEP is two injections, 1 month apart, followed by an injection continually, every 2 months [28].

In Fig. 14a, where CAB-LA is only 30% effective for the DR strain and the fitness of the DR and DS virions is equal, the mutated DR strain overtakes the DS strain even before the second PrEP injection is administered. In Fig. 14b, where CAB-LA is 70% effective for the DR strain and the fitness of the DR and DS virions is equal, the DR strain takes longer, after the second administration of PrEP on

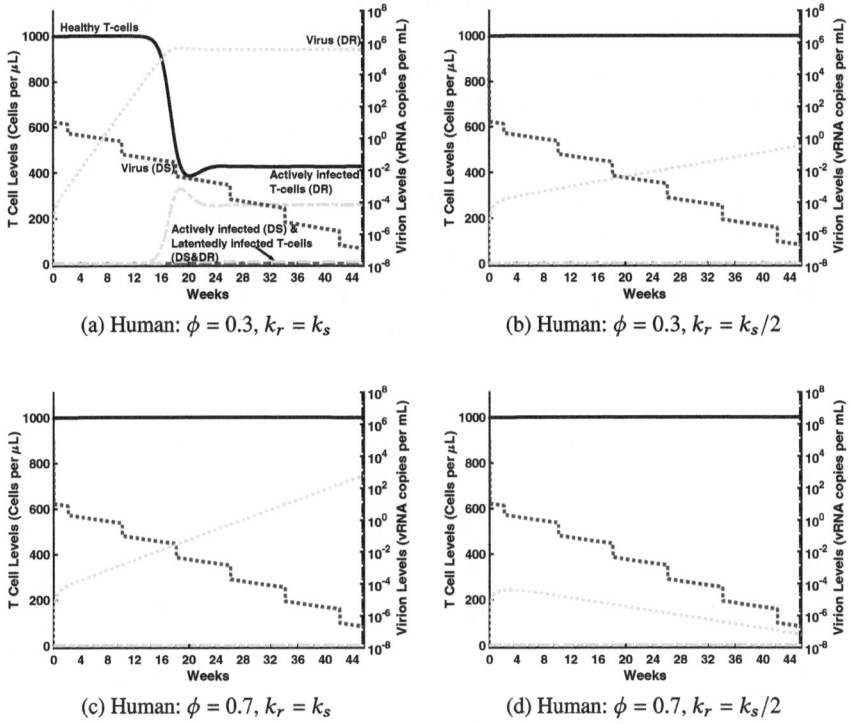

(a) Human: $\phi = 0.3$, $k_r = k_s$

(b) Human: $\phi = 0.3$, $k_r = k_s/2$

(c) Human: $\phi = 0.7$, $k_r = k_s$

(d) Human: $\phi = 0.7$, $k_r = k_s/2$

Fig. 14 Human *in silico* trial. PrEP is administered on days 0 and 28 and then every 2 months. The human is exposed to HIV on day 14. The label DS indicates the drug-sensitive strain, and DR indicates the drug-resistant strain. (**a**) PrEP is 30% effective against HIV ($\phi = 0.3$). DR virions are equally as infective as DS virions ($k_r = k_s$). (**b**) $\phi = 0.3$. DR virions are half as infective as DS ($k_r = k_s/2$). (**c**) $\phi = 0.7$, $k_r = k_s$. (**d**) $\phi = 0.7$, $k_r = k_s/2$. Solid curves: healthy T-cells. Dash-dot curve: actively infected T-cells. Dash curve: latently infected T-cells. Dotted curve: virus. Dark gray: DS, light gray: DR

week 18, to overtake the DS strain. However, in the scenario with $\phi = 0.70$ and $k_r = k_s$ (Fig. 14c), the DR strain is already growing too much to be controlled by a third PrEP injection on week 19 (following the once a month PrEP protocol for macaques). The situation changes if the DR virion fitness is lowered to 50% of the DS virion fitness. In that case, the healthy T-cell count is maintained at a high level for both drug effectiveness levels as shown in Fig. 14b, d for the duration of the *in silico* trial. However, of these two cases, if a third PrEP injection were administered on week 19, the HIV seroconversion would be avoided if the drug were 70% effective against DS virions, but not if the drug were only 30% effective. This indicates that the virion fitness, along with the drug efficacy, is important in determining whether a mutated strain, and at what time, will overpower PrEP. About 50% fitness could have caused this seroconversion.

7.5   Macaque and Human Simulations: PrEP After Infection

Next, we implemented the PrEP injection protocol of [30], where SHIV exposure
occurred on day 0, and then macaques were given PrEP on days 11, 39, and 67, as
shown in Fig. 12. This *in silico* pre-clinical trial then lasted for 30 weeks to follow
the protocol as described in [30]. We similarly varied ϕ between 30% efficacy (top
row) and 70% efficacy (bottom row) and allowed $k_r = k_s$ (first column) and $k_r =$
$0.5 \times k_s$ (second column).

We first observe in Fig. 15 that the healthy T-cell concentration, represented by
the black solid curve, stays high through this whole experiment. The DS virions,
represented by the black dotted curve, react to each of the PrEP injections, but not

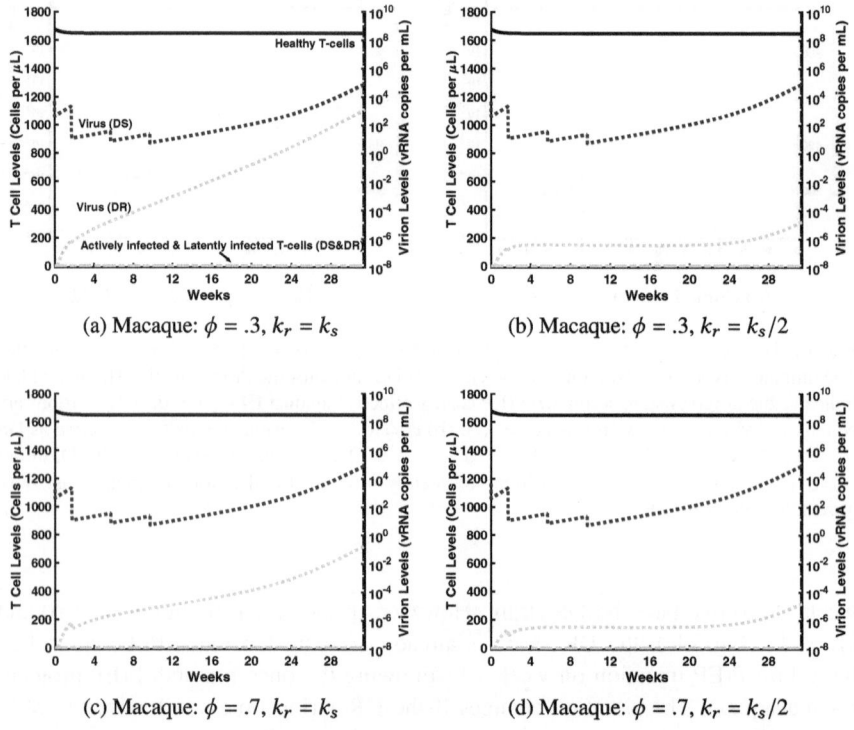

(a) Macaque: $\phi = .3$, $k_r = k_s$ (b) Macaque: $\phi = .3$, $k_r = k_s/2$

(c) Macaque: $\phi = .7$, $k_r = k_s$ (d) Macaque: $\phi = .7$, $k_r = k_s/2$

Fig. 15 Macaque *in silico* trial corresponding to the experiment in [30]. The macaque patient is
exposed to SHIV on day 0, and the PrEP injection is administered on days 11, 39, and 67. The
label DS indicates the drug-sensitive strain, and DR indicates the drug-resistant strain. (**a**) PrEP
is 30% effective against SHIV ($\phi = 0.3$). DR virions are equally as infective as DS virions, thus
$k_r = k_s$. (**b**) PrEP is 30% effective against SHIV ($\phi = 0.3$). DR virions are half as infective as
DS ($k_r = k_s/2$). (**c**) PrEP is 70% effective against SHIV ($\phi = 0.7$). DR virions are equally as
infective as DS ($k_r = k_s$). (**d**) PrEP is 70% effective against SHIV ($\phi = 0.7$). DR virions are half
as infective as DS ($k_r = k_s/2$). Solid curves: healthy T-cells. Dash-dot curve: actively infected
T-cells. Dash curve: latently infected T-cells. Dotted curve: virus. Dark gray: DS, light gray: DR

enough to reduce the overall DS virion level. In each of the *in silico* trials, the initial SHIV dose is 10^4 vRNA copies per mL and grows to 10^5 vRNA copies per mL by the end of 30 weeks. What does change, just as in Fig. 13 based on the Andrews et al. [40] experiment, is the concentration of the DR virions, represented by the gray dotted curve. In Fig. 15a where the CAB-LA is only 30% effective against the DR strain and the DR and DS virions have equal fitness, the DR virions grow to a concentration of 10^3 and are still rising. Note that in macaques, experiments [30] have measured a mean of virion concentration level of 10^4 RNA copies/mL at seroconversion. So this rise in DR virions is significant. In Fig. 15c where the CAB-LA is now 70% effective against the DR strain and the DR and DS virions still have equal fitness, the concentration of DR virions is about 10 vRNA copies per mL. In Fig. 15b, d where the DR virions are 50% as infective as the DS virions, the DR virions have a concentration of about 10^{-4} at the end of 30 weeks despite CAB-LA being 30% or 70% effective in the two subfigures.

Lastly, we observed an *in silico* clinical trial where humans were exposed to HIV on day and administered PrEP on days 14 and 42 and then every 2 months continually, as shown in Fig. 15. We notice initially, in Fig. 16a, that the healthy T-cell population depletes very quickly as the DR virion concentration increases to overtake the system. Since the drug efficacy, $\phi = 0.3$, and the DR virions are just as infective as the DS virions, the DS virion concentration slowly depletes, but the DR virion concentration grows and saturates. We observe subtle differences between trials where the drug is 30% effective and $k_r = k_s/2$ (Fig. 16a) and where the drug is 70% effective and $k_r = k_s$ (Fig. 16b). In both of these cases, the DS virion concentration slowly depletes as a direct result of the PrEP injections, but the DR virion concentration increases linearly. The human patient would have become seropositive in both cases; however, we notice seroconversion happens much quicker in the case where the DR virions are just as infective as the DS virions. In the case where the drug is 70% effective and the DR virions are half as infective as the DS virions, both virion concentrations deplete over time, and the healthy T-cell population remains at a healthy steady state. It is interesting to notice that a combination of an effective, and consistent, drug schedule along with a decreased infectivity of DR virions is necessary to keep HIV seroconversion controlled.

8 Discussion

The FDA approved injectable pre-exposure prophylaxis (PrEP) has a long pharmacologic drug tail that is a plus and a minus. The long half-life is beneficial in reducing the burden of PrEP adherence from a daily pill to bimonthly injections. However, the long pharmacologic drug tail also means that there can be a long period when the CAB level is just high enough to inhibit the growth of the wild-type DS strain, but low enough that the DR strain can grow unchecked. This concern warranted experimentalists to conduct studies on macaque that showed

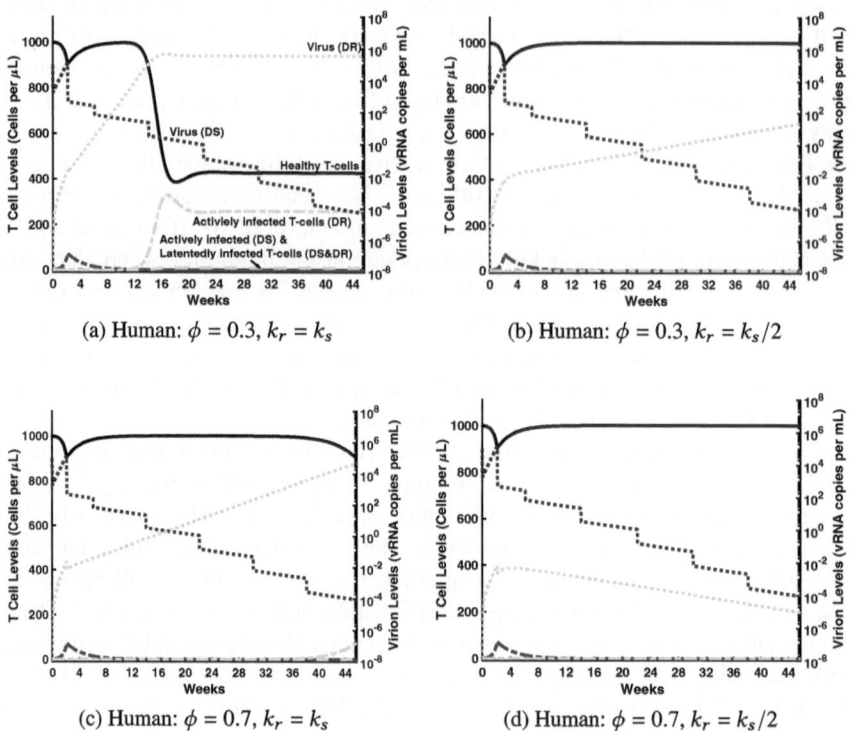

Fig. 16 Human *in silico* trial. Human exposed to HIV on day 0. PrEP injection administered on days 14 and 42 and then every 2 months. The label DS indicates the drug-sensitive strain, and DR indicates the drug-resistant strain. (a) $\phi = 0.3$, $k_r = k_s$. (b) $\phi = 0.3$, $k_r = k_s/2$. (c) $\phi = 0.7$, $k_r = k_s$. (d) $\phi = 0.7$, $k_r = k_s/2$. Solid curves: healthy T-cells. Dash-dot curve: actively infected T-cells. Dash curve: latently infected T-cells. Dotted curve: virus. Dark gray: DS, light gray: DR

that long-acting CAB may encourage the growth of the DR mutated strain [30], despite the fact that multiple INSTI mutations are required for extensive CAB drug resistance [53].

We developed a mathematical model of the within-host HIV infection with naturally occurring mutations that could result in PrEP resistant virions. Our model was validated against data for humans in the early stage of HIV before receiving antiretroviral therapy. We also parameterized our model using studies on macaques. This allowed validation of our model on experimental tests on macaques without antiretroviral therapy and several experiments involving PrEP and SHIV exposures. In this work we were not able to quantify the exact fitness of a mutation that is resistant to CAB-LA to infect a T-cell, k_r. This means we could not pinpoint how close our estimate is for the growth of drug resistance in humans when given CAB-LA and already infected with HIV. However, a study of parameter sensitivity showed that the infected T-cell concentration and the virion concentration were sensitive to the fitness, k_s and k_r, of the DS and DR virions, respectively, to infect a healthy

T-cell. Because we do not have a good estimate of k_r and ϕ the reduction in the effectiveness of CAB-LA, we tested the forward sensitivity and global sensitivity of k_r and ϕ, along with several simulations at various k_r and ϕ levels. The parameter studies showed that the effective reproduction numbers for the DR strain were more sensitive to the reduction in effectiveness of CAB-LA, ϕ, than the infectivity fitness, k_r. This led to our *in silico* trials where we tested combinations of CAB-effectiveness and the fitness of the DR strain. We found that for the macaques, even in the case of 30% effectiveness in PrEP against the DR strain and equal fitness for infecting T-cells, the DR strain did not overtake the DS strain. It did show enough growth in the DR strain to warrant concern. For humans, this drug effectiveness and fitness combination gave more alarming results with the DR strain quickly outcompeting the DS strain.

This work is intended to be a starting point to connect experimental results from SHIV studies on macaques to HIV predictions for humans. Our next step will be to incorporate work estimating fitness costs associated with different types of mutations. We have recently learned of the work by Zanini et al. [87] that estimates the rates of mutation and the spectrum of different kinds of mutations along with fitness costs of HIV. Zanini et al. [87] make these estimates by using whole genome deep-sequencing data of nine untreated patients with HIV-1 [88]. The data contain 6–12 longitudinal samples per patient spanning 5–8 years of infection. In addition, the updated International Antiviral Society–USA (IAS–USA) drug resistance mutations list for HIV from 2019 includes the mutations for cabotegravir [89]. This work provides the mutations on the integrase gene associated with resistance to cabotegravir. While it will require a significant amount of work to combine these sources of information about fitness and specific mutations into a workable model, it should help us give a more accurate picture of the drug-resistant levels to expect when humans take CAB-LA while unknowingly infected with HIV. In addition, as a comparison, we can incorporate the drug mutation model used by Smith et al. [90] in their individual simulation studies.

There are many additional directions to which to continue to expand upon this work. We have lumped all mutations into a single DR HIV strain. One could filter the mutations into beneficial or deleterious mutations with regard to particular mutations taking advantage of tabulated information on key mutations for specific mutations giving rise to HIV drug resistance [89]. The experiments with SHIV positive macaques given 10 days after seroconversion track the specific mutations noted in the daily blood draws [30], so the simulation and experimental results could be aggregated and compared.

We have assumed that each T-cell is infected with only a single strain of the HIV virus. However, it has been shown in experimental studies that many cells in the body can be infected with multiple strains (see spleen studies [91], for one of many experiments). So a natural next step will be to allow co-infection of multiple strains. This can occur through cell-to-cell infection, i.e., viral synapsis, not presently included in our model. Multiple experiments have verified that HIV is spread through synaptic transmissions [92–95]. We could include synaptic transmission in our model using the model of [96] as a starting point.

Retroviruses in general, including HIV-1, are diploid, meaning that each virion contains two genomic RNA molecules. These viral RNA molecules then serve as the templates for proviral DNA synthesis by the means of virus-encoded enzyme reverse transcriptase RT. This creates another means to generate genetic changes and perhaps mutations through recombination. Since two RNA molecules are contained in each virion, the reverse transcription may switch from one template to another [91, 97].

While there are a number of ways to improve the precision of our current model, we believe that the current model results justify the FDA recommendation for extended-release cabotegravir users to take an HIV test before each PrEP injection.

Acknowledgments The work described herein was initiated during the Collaborative Workshop for Women in Mathematical Biology funded and hosted by UnitedHealth Group Optum of Minnetonka, MN and supported by University of Minnesota's Institute for Mathematics and its Applications in June 2022. Additionally, the authors and editors thank the anonymous peer reviewers for their feedback, which strengthened this work.

KG was supported by the National Science Foundation (NSF) under Grant No. DMS-2000044.

References

1. E. Gomez-Lucia, in *Encyclopedia of Virology (Fourth Edition)*, 4th edn., ed. by D.H. Bamford, M. Zuckerman (Academic Press, Oxford, 2021), pp. 56–67. https://doi.org/10.1016/B978-0-12-814515-9.00005-9
2. E.A. Hernandez-Vargas, R.H. Middleton, J. Theo. Biol. **320**, 33 (2013)
3. E. Fiebig, D. Wright, B. Rawal, P. Garrett, R. Schumacher, L. Peddada, C. Heldebrant, R. Smith, A. Conrad, S. Kleinman, et al., AIDS **17**(13), 1871 (2003). https://doi.org/10.1097/00002030-200309050-00005
4. M. Robb, J. Ananworanich, Curr. Opin. HIV AIDS **11**(6), 555 (2016). https://doi.org/10.1097/COH.0000000000000316
5. LibreTexts. https://bio.libretexts.org/Bookshelves/Microbiology/Book%3A_Microbiology_(Boundless)/9%3A_Viruses/9.1%3A_Overview_of_Viruses/9.1B%3A_Nature_of_the_Virion. Accessed 02 Nov 2020
6. E. Lara-Pezzi, N. Rosenthal, in *Heart Development and Regeneration*, ed. by N. Rosenthal, R.P. Harvey (Academic Press, Boston, 2010), pp. 981–997. https://doi.org/10.1016/B978-0-12-381332-9.00046-3
7. S. Margeridon-Thermet, R.W. Shafer, Viruses **2**(12), 2696 (2010). https://doi.org/10.3390/v2122696
8. M. Abram, A. Ferris, W. Shao, W. Alvord, S. Hughes, J. Virol. **84**, 9864 (2010). https://doi.org/10.1128/JVI.00915-10
9. L. Mansky, AIDS Res. Hum. Retrovirus **12**, 307 (1996). https://doi.org/10.1089/aid.1996.12.307
10. E. Svarovskaia, S. Cheslock, W. Zhang, W. Hu, V. Pathak, Front. Biosci. **8**(4), 117 (2003). https://doi.org/10.2741/957
11. M.A. Fischl, D.D. Richman, M.H. Grieco, M.S. Gottlieb, P.A. Volberding, O.L. Laskin, J.M. Leedom, J.E. Groopman, D. Mildvan, R.T. Schooley, et al., N. Engl. J. Med. **317**(4), 185 (1987). https://doi.org/10.1056/NEJM198707233170401

12. M. May, M. Gompels, V. Delpech, K. Porter, F. Poct, M. Johnson, D. Dinn, A. Palfreeman, R. Gilson, B. Gazzard, et al., Br. Med. J. **343**, d6016 (2011). https://doi.org/10.1136/bmj.d6016
13. F. Nakagawa, R. Lodwick, C. Smith, R. Smith, V. Cambiano, J. Lundgren, V. Delpech, A. Phillips, AIDS **26**(3), 335 (2012). https://doi.org/10.1097/QAD.0b013e32834dcec9
14. H. Samji, A. Cescon, R. Hogg, S. Modur, K. Althoff, K. Buchaz, A.N. Burchell, M. Cohen, K.A. Gebo, M.J. Gill, et al., PLoS ONE **8**(12), e81355 (2013). https://doi.org/10.1371/journal.pone.0081355
15. A. Van Sighem, L. Gras, P. Reiss, K. Brinkman, F. de Wolf, ATHENA National Observational Cohort Study, AIDS **24**, 1527 (2010)
16. V. Cambiano, J. O'Connor, A. Phillips, A. Rodger, R. Lodwick, A. Pharris, F. Lampe, F. Nakagawa, C. Smith, M. van de Laar, Euro. Surveill. **18**(48), 20647 (2013). https://doi.org/10.2807/1560-7917.es2013.18.48.20647
17. J. Castilla, J. del Romero, V. Hernando, B. Marincovich, S. García, C. Rodríguez, J. Acquir. Immune. Defic. Syndr. **40**(1), 96 (2005). https://doi.org/10.1097/01.qai.0000157389.78374.45
18. M. Cohen, Y. Chen, M. McCauley, T. Gamble, M. Hosseinipour, N. Kumarasamy, J.G. Hakim, J. Kumwenda, B. Grinsztejn, J.H.S. Pilotto, et al., N. Engl. J. Med. **365**, 493 (2011). https://doi.org/10.1056/NEJMoa1105243
19. R. Granich, C. Gilks, C. Dye, K. De Cock, B. Williams, Lancet **373**, 48 (2009). https://doi.org/10.1016/S0140-6736(08)61697-9
20. T. Porco, J. Martin, K. Page-Shafer, A. Cheng, E. Charlebois, R. Grant, D. Osmond, AIDS **18**(1), 81 (2004). https://doi.org/10.1097/00002030-200401020-00010
21. Preventing perinatal transmission of HIV. https://hivinfo.nih.gov/understanding-hiv/fact-sheets/preventing-perinatal-transmission-hiv. Accessed 21 May 2023
22. Centers for Disease Control and Prevention, MMWR Recomm. Rep. **39**, 1 (1990). https://doi.org/10.1086/672271
23. HIVinfo. https://hivinfo.nih.gov/understanding-hiv/fact-sheets/post-exposure-prophylaxis-pep. Accessed 23 Jan 2023
24. Centers for Disease Control and Prevention. https://www.cdc.gov/nchhstp/newsroom/fact-sheets/hiv/PrEP-for-hiv-prevention-in-the-US-factsheet.html. Accessed 22 Jan 2023
25. Centers for Disease Control and Prevention. https://www.cdc.gov/hiv/basics/prep/about-prep.html. Accessed 22 Jan 2023
26. R. Tetteh, B. Yankey, E. Nartey, M. Lartey, H. Leufkens, A. Dodoo, Drug Saf. **40**, 273 (2017). https://doi.org/10.1007/s40264-017-0505-6
27. U.S. Food and Drug Administration. https://www.fda.gov/news-events/press-announcements/fda-approves-first-injectable-treatment-hiv-pre-exposure-prevention. Accessed 31 Jan 2023
28. World Health Organization. https://www.who.int/news/item/28-07-2022-who-recommends-long-acting-cabotegravir-for-hiv-prevention (2022). Accessed 23 Jan 2023
29. M. Markowitz, I. Frank, R. Grant, K. Mayer, R. Elion, D. Goldstein, C. Fisher, M. Sobieszczyk, J. Gallant, H. Van Tieu, et al., Lancet HIV **4**(8), e331 (2017). https://doi.org/10.1016/S2352-3018(17)30068-1
30. J. Radzio-Basu, O. Council, M.E. Cong, S. Ruone, A. Newton, X. Wei, J. Mitchell, S. Ellis, C.J. Petropoulos, W. Huang, et al., Nat. Commun. **10**(1), 1 (2019). https://doi.org/10.1038/s41467-019-10047-w
31. Centers for Disease Control and Prevention. https://www.cdc.gov/hiv/clinicians/prevention/prep.html. Accessed 24 Jan 2023
32. N. Bai, R. Xu, Math. Biosci. Eng. **18**(2), 1689 (2021). https://doi.org/10.3934/mbe.2021087
33. N.M. Dixit, A.S. Perelson, Proc. Natl. Acad. Sci. USA **102**(23), 8198 (2005). https://doi.org/10.1073/pnas.040749810
34. A. Hill, D. Rosenbloom, M. Nowak, R. Siliciano, Immunol. Rev. **285**(1), 9 (2018). https://doi.org/10.1111/imr.12698
35. A.S. Perelson, P.W. Nelson, SIAM Rev. **41**(1), 3 (1999). https://www.jstor.org/stable/2653164
36. N. Dorratoltaj, R. Nikin-Beers, S. Ciupe, S. Eubank, K. Abbas, PeerJ (2017). https://doi.org/10.7717/peerj.3877

37. H. Moore, W. Gu, Math. Biosci. Eng. **2**(2), 363 (2005). https://doi.org/10.3934/mbe.2005.2.363
38. D.B. Reeves, M. Rolland, B.L. Dearlove, Y. Li, M.L. Robb, J.T. Schiffer, P. Gilbert, E.F. Cardozo-Ojeda, B.T. Mayer, J. R. Soc. Interface **18**(179), 20210314 (2021). https://doi.org/10.1098/rsif.2021.0314
39. N.K. Vaidya, R.M. Ribeiro, C.J. Miller, A.S. Perelson, J. Virol. **84**(9), 4302 (2010). https://doi.org/10.1128/JVI.02284-09
40. C.D. Andrews, L.S. Bernard, A.Y. Poon, H. Mohri, N. Gettie, W.R. Spreen, A. Gettie, K. Russell-Lodrigue, J. Blanchard, Z. Hong, et al., AIDS **31**(4), 461 (2017). https://doi.org/10.1097/QAD.0000000000001343
41. C. Dobard, N. Makarova, K. Nishiura, C. Dinh, A. Holder, M. Sterling, J. Lipscomb, J. Mitchell, F. Deyounks, D. Garber, et al., J. Infect. Dis. **222**(3), 391 (2020). https://doi.org/10.1093/infdis/jiaa095
42. R.A. Saenz, S. Bonhoeffer, Epidemics **5**(1), 34 (2013). https://doi.org/10.1016/j.epidem.2012.11.002
43. F. Kirchhoff, in *Encyclopedia of AIDS*, ed. by T.J. Hope, M. Stevenson, D. Richman (Academic Press, Boston, 2013), pp. 1–9
44. S. Broder, Antiviral Res. **85**(1), 1 (2010). https://doi.org/10.1016/j.antiviral.2009.10.002
45. K. Engelman, A. Engelman, Biochemistry **60**(22), 1731 (2021). https://doi.org/10.1021/acs.biochem.1c00157
46. S.Y. Rhee, P.M. Grant, P.L. Tzou, G. Barrow, P.R. Harrigan, J.P.A. Ioannidis, R.W. Shafer, J. Antimicrob. Chemother. **74**(11), 3135 (2019). https://doi.org/10.1093/jac/dkz256
47. D.A. Cooper, R.T. Steigbigel, J.M. Gatell, J.K. Rockstroh, C. Katlama, P. Yeni, A. Lazzarin, B. Clotet, P.N. Kumar, J.E. Eron, et al., N. Engl. J. Med. **359**(4), 355 (2008). https://doi.org/10.1056/NEJMoa0708978
48. N. Cook, W. Li, D. Berta, M. Badaoui, A. Ballandras-Colas, A. Nans, A. Kotecha, E. Rosta, A. Engelman, P. Cherepanov, Science **367**(6479), 806 (2020). https://doi.org/10.1126/science.aay4919
49. V. Summa, A. Petrocchi, F. Bonelli, B. Crescenzi, M. Donghi, M. Ferrara, F. Fiore, C. Gardelli, O. Gonzalez Paz, D. Hazuda, et al., J. Med. Chem. **51**(18), 5843 (2008). https://doi.org/10.1021/jm800245z
50. K. Shimura, E. Kodama, Y. Sakagami, Y. Matsuzaki, W. Watanabe, K. Yamataka, Y. Watanabe, Y. Ohata, S. Doi, M. Sato, et al., J. Virol. **82**(2), 764 (2008). https://doi.org/10.1128/JVI.01534-07
51. S. Min, I. Song, J. Borland, S. Chen, Y. Lou, T. Fujiwara, S. Piscitelli, Antimicrob. Agents Chemother. **54**(1), 254 (2010). https://doi.org/10.1128/AAC.00842-09
52. M. Tsiang, G. Jones, J. Goldsmith, A. Mulato, D. Hansen, E. Kan, L. Tsai, R. Bam, G. Stepan, K. Stray, et al., Antimicrob. Agents Chemother. **60**(12), 7086 (2016). https://doi.org/10.1128/AAC.01474-16
53. U. Parikh, C. Koss, J. Mellors, Curr. HIV/AIDS Rep. **19**(5), 384 (2022). https://doi.org/10.1007/s11904-022-00616-y
54. World Health Organization. HIV drug resistance strategy, 2021 update. https://www.who.int/publications/i/item/9789240030565 (2021). Accessed 23 May 2023
55. World Health Organization. Surveillance of HIV drug resistance-PrEP users diagnosed with HIV. https://www.who.int/teams/global-hiv-hepatitisand-stis-programmes/hiv/treatment/hiv-drug-resistance/hiv-drug-resistance-surveillance/hiv-drug-resistance-among-prepusers-diagnosed-with-hiv (2022). Accessed 23 May 2023
56. T. McPherson, M. Sobieszczyk, M. Markowitz, Expert Opin. Invest. Drugs **27**, 413 (2018). https://doi.org/10.1080/14656566.2020.1843635
57. W. Spreen, D. Margolis, J. Pottage Jr., Curr. Opin. HIV AIDS **8**, 565 (2013). https://doi.org/10.1097/COH.0000000000000002
58. D. Sidebottom, A. Ekström, S. Strömdahl, BMC Infect. Dis. **18**, 581 (2018). https://doi.org/10.1186/s12879-018-3463-4

59. B. Jilek, M. Zarr, M. Sampah, S. Rabi, C. Bullen, J. Lai, L. Shen, R. Siliciano, Nat. Med. **18**(3), 446 (2012). https://doi.org/10.1038/nm.2649
60. P.R. Somvanshi, K.V. Venkatesh, *Hill Equation* (Springer, New York, 2013), pp. 892–895. https://doi.org/10.1007/978-1-4419-9863-7_946
61. L. Shen, S. Peterson, A. Sedaghat, M. McMahon, M. Callender, H. Zhang, Y. Zhou, E. Pitt, K. Anderson, R. Acosta, E. Siliciano, Nat. Med. **14**, 762 (2008). https://doi.org/10.1038/nm1777
62. J. Shaik, E. Weld, S. Edick, E. Fuchs, S. Riddler, M. Marzinke, R. D'Amico, K. Bakshi, Y. Lou, C. Hendrix, et al., Br. J. Clin. Pharmacol. **88**(4), 1667 (2022). https://doi.org/10.1111/bcp.14980
63. P. Van den Driessche, J. Watmough, Math. Biosci. **180**(1–2), 29 (2002). https://doi.org/10.1016/S0025-5564(02)00108-6
64. S.E. Langford, J. Ananworanich, D.A. Cooper, AIDS Res. Ther. **14**, 11 (2007)
65. J. Lampinen, in *Proceedings of the 2002 Congress on Evolutionary Computation. CEC'02 (Cat. No. 02TH8600)*, vol. 2 (IEEE, Piscataway, 2002), pp. 1468–1473. https://doi.org/10.1109/CEC.2002.1004459
66. M. Aggarwal, M. Hussaini, L. De La Fuente, F. Navarrete, N. Cogan, J. Theor. Biol. **14**(457), 88 (2018). https://doi.org/10.1016/j.jtbi.2018.08.028
67. N. Chitnis, J.M. Hyman, J.M. Cushing, Bull. Math. Biol. **70**, 1272 (2008). https://doi.org/10.1007/s11538-008-9299-0
68. L.E. Jones, A.S. Perelson, J. Acquired Immune Defic. Syndr. **45**(5), 483 (2007). https://doi.org/10.1097/QAI.0b013e3180654836
69. A.R. Kirtane, O. Abouzid, D. Minahan, T. Bensel, A.L. Hill, C. Selinger, A. Bershteyn, M. Craig, S.S. Mo, H. Mazdiyasni, et al., Nat. Commun. **9**(1), 1 (2018). https://doi.org/10.1038/s41467-017-02294-6
70. A. Perelson, A. Neumann, M. Markowitz, J. Leonard, D. Ho, Science **271**, 1582 (1996). https://doi.org/10.1126/science.271.5255.1582
71. M.A. Stafford, L. Corey, Y. Cao, E.S. Daar, D.D. Ho, A.S. Perelson, J. Theo. Biol. **203**(3), 285 (2000). https://doi.org/10.1006/jtbi.2000.1076
72. A.S. Perelson, D.E. Kirschner, R. De Boer, Math. Biosci. **114**(1), 81 (1993). https://doi.org/10.1016/0025-5564(93)90043-A
73. D.B. Reeves, E.R. Duke, T.A. Wagner, S.E. Palmer, A.M. Spivak, J.T. Schiffer, Nat. Commun. **9**, 4811 (2018). https://doi.org/10.1038/s41467-018-06843-5
74. L. Rong, A.S. Perelson, PLoS Comput. Biol. **5**(10), e1000533 (2009). https://doi.org/10.1371/journal.pcbi.1000533
75. A.S. Perelson, P.W. Nelson, Proc. Natl. Acad. Sci. USA **6**, 4 (2015). https://doi.org/10.1073/pnas.1419162112
76. Y. Cao, E.K. Cartwright, G. Silvestri, A.S. Perelson, PLoS Pathog. **14**(10), e1007350 (2018). https://doi.org/10.1371/journal.ppat.1007350
77. I.M. Rouzine, A.D. Weinberger, L.S. Weinberger, Cell **160**(5), 1002 (2015). https://doi.org/10.1016/j.cell.2015.02.017
78. T. Oda, K.S. Kim, Y. Fujita, Y. Ito, T. Miura, S. Iwami, J. Theo. Biol. **509**, 110493 (2021). https://doi.org/10.1016/j.jtbi.2020.110493
79. A. Hara, S. Iwanami, Y. Ito, T. Miura, S. Nakaoka, S. Iwami, J. Theo. Biol. **479**, 29 (2019). https://doi.org/10.1016/j.jtbi.2019.07.005
80. C.T. Barker, N.K. Vaidya, PLoS Comput. Biol. **16**(11), e1008305 (2020). https://doi.org/10.1371/journal.pcbi.1008305
81. H. Mohri, S. Bonhoeffer, S. Monard, A.S. Perelson, D.D. Ho, Science **279**(5354), 1223 (1998). https://doi.org/10.1126/science.279.5354.122
82. J.M. Conway, A.S. Perelson, PLoS Comput. Biol. **12**(1), e1004677 (2016). https://doi.org/10.1371/journal.pcbi.1004677
83. H.Y. Chen, M. Di Mascio, A.S. Perelson, D.D. Ho, L. Zhang, Proc. Natl. Acad. Sci. USA **104**(48), 19079 (2007). https://doi.org/10.1073/pnas.070744910

84. S. Marino, I. Hogue, C.J. Ray, D. Kirschner, J. Theo. Biol. **254**, 178 (2008). https://doi.org/10.1016/j.jtbi.2008.04.011
85. M. McKay, R. Beckman, W. Conover, Technometrics **21**(2), 239 (1979). https://doi.org/10.2307/1268522
86. National Research Council, *Assessing the Reliability of Complex Models: Mathematical and Statistical Foundations of Verification, Validation, and Uncertainty Quantification* (The National Academies Press, Washington, 2012). https://doi.org/10.17226/13395
87. F. Zanini, V. Puller, J. Brodin, J. Albert, R.A. Neher, Virus Evol. **3**(1) (2017). https://doi.org/10.1093/ve/vex003
88. F. Zanini, R. Neher, J. Virol. **87**(21), 11843 (2013). https://doi.org/10.1128/JVI.01529-13
89. A.M. Wensing, V. Calvez, F. Ceccherini-Silberstein, C. Charpentier, H.F. Günthard, R. Paredes, R.W. Shafer, D.D. Richman, Topics Antiviral Med. **27**(3), 111 (2019)
90. J. Smith, L. Bansi-Matharu, V. Cambiano, D. Dimitrov, A. Bershteyn, D. van de Vijver, K. Kripke, P. Revill, M. Boily, G. Meyer-Rath, et al., Lancet HIV **10**, e254 (2023). https://doi.org/10.1016/S2352-3018(22)00365-4
91. A. Jung, R. Maier, G. Vartanian, J.P.and Bocharov, J. Volker, U. Fischer, E. Meese, S. Wain-Hobson, A. Meyerhans, Nature **418**, 144 (2002). https://doi.org/10.1038/418144a
92. L. Agosto, P. Uchil, W. Mothes, Trends Microbiol. **23**(5), 289 (2015). https://doi.org/10.1016/j.tim.2015.02.003
93. P. Chen, W. Hübner, M. Spinelli, B. Chen, J. Virol. **81**(22), 12582 (2007). https://doi.org/10.1128/JVI.00381-07
94. W. Hübner, G.P. McNerney, P. Chen, B.M. Dale, R.E. Gordon, F.Y. Chuang, X.D. Li, D.M. Asmuth, T. Huser, B.K. Chen, Science **323**(5922), 1743 (2009). https://doi.org/10.1126/science.1167525
95. Q. Sattentau, Nat. Rev. Microbiol. **6**(11), 815 (2008). https://doi.org/10.1038/nrmicro1972
96. J. Kreger, N. Komarova, D. Wodarz, PLoS Comput. Biol. **17**(12), e1009713 (2021). https://doi.org/10.1371/journal.pcbi.1009713
97. D.N. Levy, G.M. Aldrovandi, O. Kutsch, G.M. Shaw, Proc. Natl. Acad. Sci. USA **101**(12), 4204 (2004). https://doi.org/10.1073/pnas.0306764101

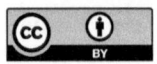

A Survey of Mathematical Modeling of Hormonal Contraception and the Menstrual Cycle

Lihong Zhao, Ruby Kim, Lucy S. Oremland, Mukti Chowkwale, Lisette G. de Pillis, and Heather Z. Brooks

1 Introduction

Developing a mechanistic understanding of the menstrual cycle is important to human health and wellness. Roughly, half of people of reproductive age menstruate. About 5–8% of these individuals experience moderate to severe symptoms of premenstrual syndrome [1]. Many experience issues with their menstrual cycles that affect their health, wellness, and quality of life, including amenorrhea (absence of menstruation), dysmenorrhea (painful menstruation), menorrhagia (excessive

Lihong Zhao and Ruby Kim contributed equally to this work.

L. Zhao (✉)
Department of Applied Mathematics, University of California Merced, Merced, CA, USA

Department of Mathematics, Virginia Tech, Blacksburg, VA, USA
e-mail: lhzhao@vt.edu

R. Kim
Department of Mathematics, University of Michigan, Ann Arbor, MI, USA
e-mail: rshkim@umich.edu

L. S. Oremland
Mathematics and Statistics Department, Skidmore College, Saratoga Springs, NY, USA
e-mail: loremlan@skidmore.edu

M. Chowkwale
Department of Biomedical Engineering, University of Virginia, Charlottesville, VA, USA
e-mail: mc7wt@virginia.edu

L. G. de Pillis · H. Z. Brooks
Department of Mathematics, Harvey Mudd College, Claremont, CA, USA
e-mail: depillis@g.hmc.edu; hzinnbrooks@g.hmc.edu

© The Author(s) 2024
A. N. Ford Versypt et al. (eds.), *Mathematical Modeling for Women's Health*,
The IMA Volumes in Mathematics and its Applications 166,
https://doi.org/10.1007/978-3-031-58516-6_3

menstruation), and diseases directly related to the reproductive system, such as uterine fibroids, endometriosis, and polycystic ovarian syndrome [2]. Some key features of the menstrual cycle are generally robust, with feedback interactions between hormones in the hypothalamus, pituitary gland, and ovaries driving mostly predictable rhythms. At the same time, the system can exhibit a great deal of variation both between different individuals and from cycle to cycle within a single individual.

Mathematical modeling of the menstrual cycle is a relatively recent area of focus, and the development and analysis of mathematical models of this system may still be considered in the early stages. The menstrual cycle is a potentially rich area of study from the perspective of mathematics and dynamical systems. The periodic nature of the hormone fluctuations makes it a natural candidate of study for those who are interested in oscillations and limit cycles. When analyzing a system that evidences naturally cycling behavior, Hopf bifurcations play an important role. A Hopf bifurcation is the process whereby periodic orbits ("self-oscillations") emerge from a fixed point when a parameter crosses a critical value [3]. Identifying the types of bifurcations that are part of a mathematical system of equations yields insight into the structure and mechanisms of a system of equations. Determining how and when a Hopf bifurcation, in particular, may develop is therefore of particular interest in a mathematical model of the menstrual cycle. The existing mathematical work has already made strong contributions in this area. Several early mathematical models in the literature have helped quantify the complex interactions that drive the human menstrual cycle [4–6]. Selgrade and Schlosser [7, 8] worked on a series of models that drove more recent developments in the field, especially as more data became available [9–12].

The dynamics of this system are made both more interesting and more complicated by the introduction of hormonal contraceptives. An individual's autonomy and ability to understand and control their own reproductive health is an important issue in healthcare and in modern society. Contraceptive use is linked to increases in economic empowerment, education, and labor force participation for women [13]. Hormonal contraceptives (including oral contraceptives) remain a leading form of birth control in the United States [14]. However, very few researchers have explicitly incorporated hormonal contraception into their mathematical models.

In this chapter, we provide a survey of the efforts in mechanistic modeling of the menstrual cycle, with a special eye toward the modeling of the effects of hormonal contraception. We begin in Sect. 2 with an overview of the biology of the menstrual cycle, including the different phases and key hormonal drivers of the system. We also describe how these hormones are impacted through hormonal contraception. In Sect. 3, we turn our attention toward existing mathematical models of the menstrual cycle, highlighting the goals and key results from these papers. We discuss strengths and limitations of each of these models. We provide a comparative analysis of these models in Sect. 4. First, we perform a sensitivity analysis that shows how variations in the parameter representing the growth rate of the reserve pool of follicular stimulating hormone have considerable impact on the cycle length across models. This analysis reveals interesting qualitative and quantitative differences on the impact of cycle length in each case. Second, we discuss the different ways in which time delays are incorporated into various differential equation models. We close the section by introducing a simple modification to the existing models to include hormonal contraception and discuss the qualitative responses of each model. To the best of our knowledge, this is the first such comparative analysis of mathematical models of the menstrual cycle. We conclude in Sect. 5 by highlighting existing challenges and important future directions.

2 Biological Background on the Menstrual Cycle

2.1 Stages of the Menstrual Cycle

Here we provide a brief introduction to the stages of the menstrual cycle. A more detailed background of the biology can be found in [15–17]. The expected length of the menstrual cycle is around 29 days [18], although the variability in cycle length is large, with normal cycle lengths ranging from around 15 to 50 days [16, 18]. Day 1 of the menstrual cycle is the first day of menstruation (the "period"), and a menstrual cycle ends at the start of the next period. Both the uterus and ovaries experience changes throughout the menstrual cycle.

The ovarian cycle consists of two primary phases: the *follicular phase* and the *luteal phase*. Based on the traditional assumption of a standard 28-day menstrual cycle, the first phase of the ovarian cycle, called the follicular phase, lasts approximately 14 days. The second phase, called the luteal phase, begins on day 15 and lasts until the end of the ovarian cycle. Studies show that follicular phase lengths vary more than luteal phase lengths [18]. At birth the fetal ovary contains a fixed amount of primordial follicles (small, fluid-filled sacs that contain ova or "eggs"). In each ovarian cycle, a small portion of primordial follicles are activated at the beginning of the follicular phase to begin maturation [19, 20]. During this phase, one of the follicles will develop more quickly (the *primary* or *dominant follicle*), and the others will die out. The primary follicle will continue to develop during the remainder of the follicular phase until it fully matures. The key process during this phase is meiosis, a special type of cell division that produces genetic variation and ensures that all the germ cells needed for sexual reproduction contain the correct number of chromosomes [21]. Two rounds of nuclear division (called meiotic division) are involved in meiosis; the primary follicle develops into a secondary follicle when the first division is completed, and it continues to develop until it fully matures. Around day 14, the fully matured follicle ruptures, and the ovum is released, i.e., *ovulation* occurs. After ovulation, the follicle that previously contained the egg transforms into the *corpus luteum* [22, 23]. This signals the beginning of the luteal phase. If the ovum is fertilized, the second meiotic division completes. If the ovum is not fertilized, the corpus luteum begins to degrade until the start of the next ovarian cycle.

There are three primary phases in the uterine cycle: the menstrual phase, the proliferative phase, and the secretory phase. The menstrual phase occurs in the first 5–7 days of the uterine cycle. The innermost lining of the uterine wall, called the *endometrium*, sheds through the cervix and vagina. This first phase is often commonly referred to as a "period." The phase from the end of the period until ovulation is the proliferative phase. During this phase, the endometrium thickens, while the dominant follicle in the ovary develops. The phase after ovulation is called the secretory phase. If the egg released from the ovary is fertilized, the thickened endometrium is ready for the egg to implant and grow to support the pregnancy. If the egg is not fertilized, the thickened endometrium breaks down and menstruation occurs. At this point, the cycle begins again. The lower section of Fig. 1 provides a visualization of the timeline of these different phases over a standard 28-day menstrual cycle.

2.2 Introduction to the Role of Hormones in the Menstrual Cycle

The menstrual cycle is regulated by several hormones. The key hormonal drivers of the menstrual cycle can be divided into two types: *gonadotrophic hormones* that

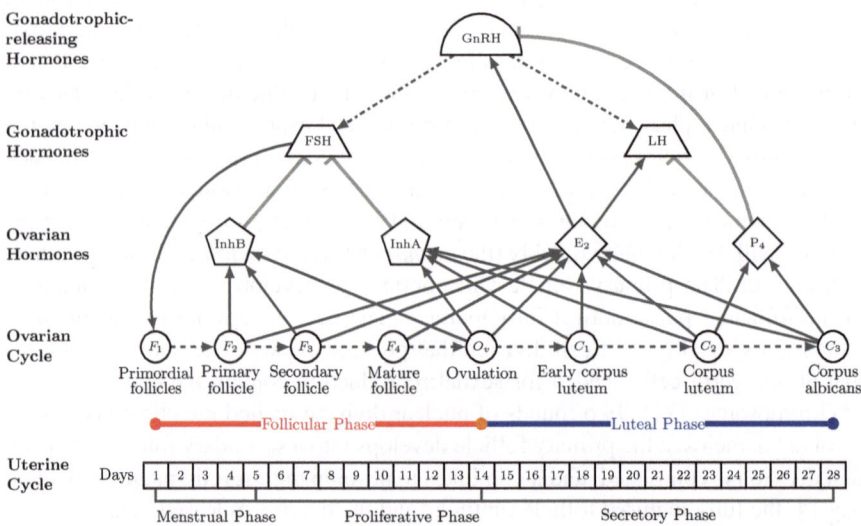

Fig. 1 A schematic of a normal 28-day menstrual cycle. Note that in the hormone/ovarian schematic, black dashed lines indicate transition between different stages in the ovarian cycle, black solid lines indicate hormone production, gray solid lines indicate hormone inhibition, and black dash-dotted lines indicate hormone stimulation. The following abbreviations are used in the schematic: GnRH, gonadotropin-releasing hormone; FSH, follicle stimulating hormone; LH, luteinizing hormone; InhA, inhibin A; InhB, inhibin B; E_2, estradiol; P_4, progesterone

are produced in the pituitary and hypothalamus and travel through the bloodstream and *reproductive or ovarian hormones* that are produced locally in the ovaries. We provide a brief overview of the dynamics and interactions of these hormones here; we refer the reader to [15, 17, 24] for more in-depth discussions of these hormones and their roles.

The primary gonadotrophic hormones regulating the menstrual cycle are *luteinizing hormone* (LH) and *follicular stimulating hormone* (FSH). There are three key reproductive hormones in the menstrual cycle system: estrogen (E_2), progesterone (P_4), and inhibin.

During the menstrual phase of the uterine cycle (roughly the first one-third of the follicular phase of the ovarian cycle), the levels of E_2 and P_4 are very low, which triggers the shedding of the endometrium. FSH enhances the development of the follicles, preparing an egg for ovulation during the follicular phase of the ovarian cycle. The primary follicle secretes E_2 and inhibin as it grows. Both of these hormones suppress further production of FSH via a negative feedback, i.e., increased E_2 and inhibin levels lead to inhibition of FSH [25]. After the period, E_2 increases over the proliferative phase and causes the endometrium to thicken. When E_2 levels are high enough, a signal is sent to the brain, which causes a rapid and significant rise in LH [22]. This surge in LH triggers ovulation [26] and, thus, is often used as a biomarker for ovulation. The corpus luteum produces E_2, P_4, and inhibin during the luteal phase. Right after ovulation, E_2 levels drop [26], and P_4 levels increase, which signals the endometrium to stop thickening and prepare for a fertilized egg. There is a negative feedback of P_4 on the further release of LH from the pituitary. If the ovum is fertilized, the individual becomes pregnant, beginning a new hormonal process. If the ovum is not fertilized, the corpus luteum begins to degrade, which results in a reduction of both E_2 and P_4. The corpus luteum degenerates into corpus albicans toward the end of the menstrual cycle, and the resulting sharp drop in E_2 and P_4 induces menstruation, where the endometrium breaks down. This decline in E_2 leads to an increase in FSH, which marks the beginning of the next menstrual cycle. A diagram showing the relationships of those key hormones and the key elements in the ovarian cycle is shown in Fig. 1. Note that both FSH and LH increase in the early stages of the menstrual cycle, with peaks occurring mid-cycle. The highest levels of E_2 are observed in the first half of the cycle (preceding the LH surge), with a lower secondary peak occurring in the second half. The highest levels of P_4 are observed in the second half of the cycle. Representative curves of hormone levels throughout the menstrual cycle are depicted in Fig. 2.

The reproductive hormone inhibin is involved in negative feedback control of FSH [15]. Inhibin is divided into two types: inhibin A (InhA) and inhibin B (InhB). The contrasting profiles of InhA and InhB suggest that FSH and LH regulate these two inhibins differently [27]. InhA is secreted in the luteal phase by the corpus luteum and peaks in the second half of the menstrual cycle, while InhB is secreted in the follicular phase by developing follicles and peaks in the first half of the menstrual cycle [12]. According to [28], "the appearance of measurable inhibin A can be seen as a marker for a follicle having at least matured to a stage corresponding to the late

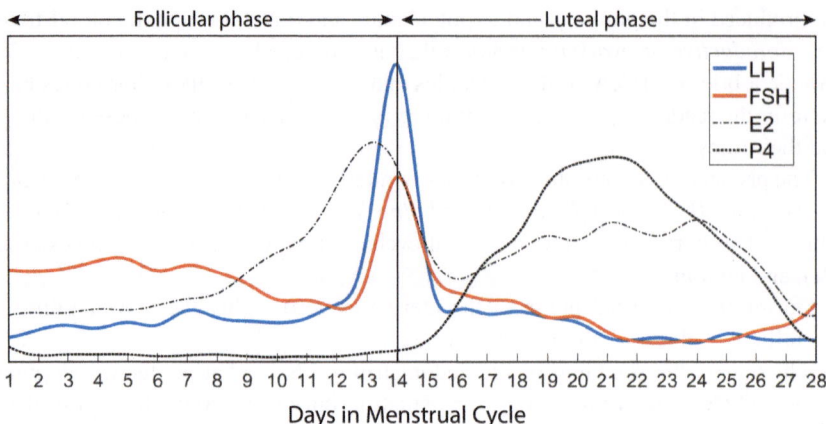

Fig. 2 Schematic time courses of hormone levels throughout the menstrual cycle based on data by Welt et al. [12]

follicular stage in adult women." InhB has been shown to be useful in the evaluation of the ovarian reserve and the assessment of an individual's reproductive capacity [29, 30].

2.3   Hormonal Contraception and Its Effect on the Menstrual Cycle

As a means to prevent pregnancy, hormonal contraception aims to inhibit ovulation or fertilization. The primary biomarkers for ovulation are large increases in P_4 and LH, so contraceptives that work to inhibit ovulation attempt to suppress levels of P_4 and LH [31, 32]. Keeping FSH and E_2 levels low is also important to attain a contraceptive state. If FSH level is low, follicular growth is limited, and follicles cannot mature. Low levels of E_2 mean that LH level cannot increase because of the positive feedback on LH.

Oral contraceptives are the most widely used variant of hormonal contraception in the United States [33]. There are two primary types of oral contraceptives: pills that are synthetic progesterone only (sometimes called "mini pills") and pills that contain both synthetic estrogen and synthetic progesterone ("combination pills") [34]. Synthetic progesterone directly reduces the synthesis of LH, although it also contributes to inducing a contraceptive state more indirectly through its effect on FSH (limiting follicle growth) and through thickening of the cervical mucus [11, 35]. Synthetic estrogen contributes to contraception via suppressing LH (thereby eliminating the LH surge) and by enhancing the effects of synthetic progesterone by increasing receptor effectiveness. Other hormonal contraceptive methods (e.g., vaginal rings, transdermal inserts, and intrauterine devices) also employ the mechanisms of reducing P_4 and LH [33].

3 Mathematical Models of the Menstrual Cycle

There are a number of research groups that have developed mathematical models of the menstrual cycle over the years (cf. [5, 6, 36–40]). Of this collection of modeling efforts, the models first developed by Selgrade, Schlosser, and Harris Clark [7, 8, 41, 42] stand out because they have been used as the foundation for a significant number of subsequent models [43–47]. Another modeling approach was introduced more recently by Röblitz et al. [48] and built upon by George et al. [49]. The model structure created by Röblitz et al. diverges in several ways from the models in the Selgrade lineage. In the following subsections, we introduce the main features and contributions of each of these models. In Fig. 3, we provide a chart depicting the relationship among several of the Selgrade-based models, and we implement and test four of them in this chapter.

3.1 Early Modeling Efforts

Very early efforts in mathematical modeling of the menstrual cycle date back to the 1970s. Shack et al. [4] used first-order differential equations to create a

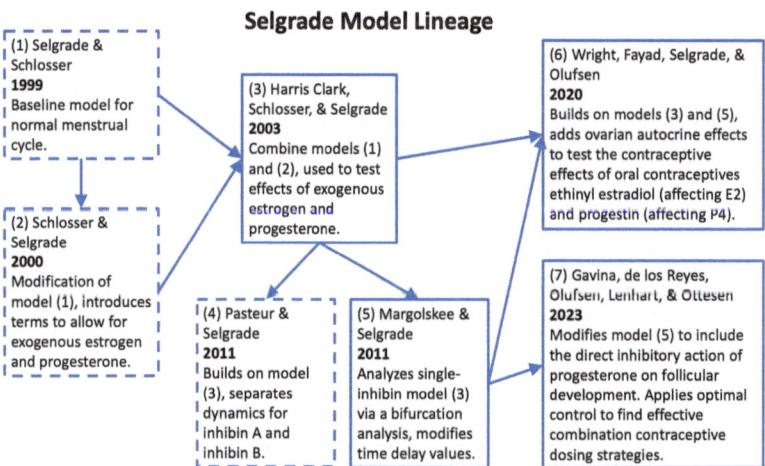

Fig. 3 Selgrade-based mathematical models of the human menstrual cycle. We implement and compare models (3), (5), (6), and (7)

phenomenological computational model of the menstrual cycle. These authors were able to simulate periodic behavior that qualitatively captures some of the hormone fluctuations of the system. In two papers by Bogumil and colleagues [5, 6], the authors observed that short-duration, random events could significantly affect the qualitative behavior of the cycle and that rapid changes in hormone levels were likely responsible for regulating the cycle. In particular, computational simulations of this model were used to study how the introduction of estradiol (E_2) at various points in the cycle might contribute to phase shifts or "resetting" of the cycle, and it was observed in a contemporaneous study that this model did not produce results that agreed with physiologically observed behavior in this context [50]. Biological knowledge of the menstrual cycle has advanced a great deal since these early works; nevertheless, they provide an important foundation for more modern modeling efforts.

Motivated by the problem of exposure to environmental estrogen affecting reproduction, Selgrade and Schlosser developed two mechanistic models that tracked the follicular phase and the luteal phase of the ovarian cycle [7, 8]. The first model [7] used a system of nine linear ordinary differential equations (ODEs) to capture the stages of the cycle that drive the production of the gonadotropin hormones LH and FSH. The production of the ovarian hormones E_2, P_4, and inhibin was modeled as functions of LH and FSH. The ODE system presented in the second model was nonlinear [8] and introduced delays. This model focused on separating the processes of gonadotropin synthesis and gonadotropin release. In this second model, the functions for E_2, P_4, and inhibin were simplified to be functions of time only, with no direct dependence on LH or FSH [7]. While the first model was able to capture the dynamics of E_2, P_4, inhibin, FSH, and LH for normally cycling adult women with no external interference, the second model was extended to allow for the administration of exogenous ovarian hormones E_2 and P_4. The authors stated that their second model could be used as a testbed for exploring a variety of hormone-modulating scenarios. The authors suggested that future uses of this second model could include exploring the role of xenoestrogens in breast and ovarian cancer, better understanding anorexia's role in the cessation of menstruation, and testing the effects of different methods of hormonal birth control [8]. Both the first and second models were fit to hormone-level data from McLachlan et al. [9]. Selgrade and Schlosser's model systems [7, 8] together form the foundation for a number of subsequent modeling efforts [41–43, 51]. We explore several of these models in detail in the subsections below. A detailed comparison of models is provided in Table 1 located in supplementary information at the end of this chapter.

3.2 Available Data Sets on Normal Menstrual Cycles

Two data sets are commonly used in models of the menstrual cycle: McLachlan et al. [9] and Welt et al. [12]. Each of the models we review in Sects. 3.3–3.7 was fit to one of these two data sets. Both studies followed approximately 40 women with a history of regular menstrual cycles (around 25–35 days). Daily blood samples were taken from participants during one complete menstrual cycle in order to measure bloodstream hormone levels (LH, FSH, E_2, P_4, and inhibin). The mean daily hormone levels were reported in each of these studies, with data centered around ovulation (the day of the LH surge). The Welt et al. data were normalized to a 28-day cycle using mean hormone levels in seven phases, including early, mid, and late follicular and luteal phases and the mid-cycle surge [12, 52]. An important distinction between the sets is that Welt et al. reported separate data for inhibin A and inhibin B, while McLachlan et al. reported a measurement of total inhibin since the separate assays for inhibin A and inhibin B were not yet available.

There are other more recent data sets on the normal menstrual cycle (published in or after the year 2000), but these published data sets do not provide the information on all the hormones that may be interesting from a mathematical modeling perspective. For example, Sehested et al. [53] collected daily blood samples from study participants and reported the mean daily levels of inhibin A, inhibin B, FSH, E_2, and LH, but they did not report levels of P_4. In another study, Stricker et al. [54] summarized the mean, median, 5th, and 95th percentiles of the daily levels of LH, FSH, E_2, and P_4 of study participants but did not report inhibin levels.

3.3 Harris Clark et al. Model

The Harris Clark et al. model was presented in both a manuscript and a Ph.D. dissertation [41, 42]. Harris Clark, Schlosser, and Selgrade [41] merged the two models of Schlosser and Selgrade [7, 8] to create a version of the model that included both a system of nonlinear delay differential equations to describe synthesis and release of gonadotropin hormones LH and FSH (as in [8]), as well as the more complicated functional forms for the ovarian hormones E_2, P_4, and inhibin that depend directly on the values of LH and FSH (as in [7]). This model was tuned to published data on normally cycling women (McLachlan et al. data [9]) to predict serum levels of these ovarian and pituitary hormones. The model also allowed for a stable abnormal cycle (in the sense of serum levels of ovarian and pituitary hormones), which can be used to fit for women with polycystic ovarian syndrome. Harris Clark et al. used this model to capture how exogenous administration of E_2 and P_4 impacts the menstrual cycle. They presented numerical experiments that show that hypothetical exogenous P_4 therapy can move a disrupted cycle into a normal cycle and that hypothetical exogenous E_2 can disrupt a normal cycle. We reproduced numerical solutions of this model, which are graphed in Fig. 4. The authors suggested that an exhaustive study of state space should be carried out to determine whether there are additional stable solutions. They pointed out that the following two changes would make the model more biologically realistic: future models should separately model inhibin A and inhibin B, instead of just generic inhibin, and should include a model component that explicitly captures the GnRH

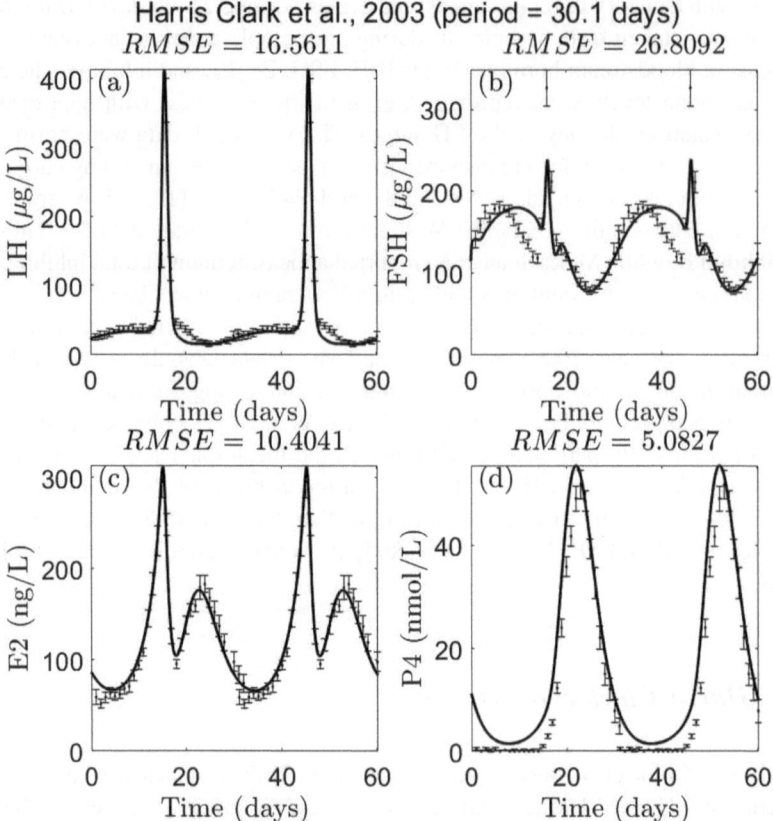

Fig. 4 Numerical solutions to the Harris Clark et al. [41] model plotted against data from McLachlan et al. [9]. The root mean square error (RMSE) for each variable is shown, and the period is computed to be approximately 30.1 days. Note that the units for the hormone levels mirror the units reported in the McLachlan et al. data set

pulse frequency and amplitude, which can be affected by the ovarian hormones. The first of these two challenges was later taken up by Pasteur and Selgrade (described in Sect. 3.4).

3.4 Pasteur and Selgrade Model

Prior to the mid-1990s, separate bioassays for inhibin A and inhibin B were not available [55], so hormone models typically incorporated their effects in a single term. In 2011, Pasteur and Selgrade published the first hormone model to incorporate separate inhibin terms [43]. This model first appeared as part of Pasteur's dissertation [44]. Pasteur and Selgrade estimated parameters for the pituitary and ovarian systems individually and then merged the models. In their multi-inhibin model, one of the key changes was to the functional form of FSH synthesis function. They asserted that because the circulating level of each inhibin has a negative effect on FSH synthesis, they replaced what they called an "unrealistic" quadratic inhibition term (E_2^2) that inhibits FSH with a linear term (E_2). They also modified the governing equation for the reserve pool of FSH. Incorporating multiple inhibins required two separate delay terms: one for inhibin A and one for inhibin B. The authors expanded the ovarian model by adding two new stages at the beginning of the follicular phase and an additional one around the time of ovulation to capture both an early and a mid-cycle peak by inhibin B. They considered the effects of exogenous hormones, noting that increasingly large amounts of exogenous E_2 suppressed the LH surge to an increasingly large degree. After treatment, the hormone concentrations returned to a stable normal cycle after a few months.

Aside from the multi-inhibin model, Pasteur's dissertation [44] also included a five-hormone model that is almost identical to the Harris Clark et al. [41] model, but it was fit to the Welt et al. data instead of the McLachlan et al. data. We do not include Pasteur's five-hormone model in this review. Instead, we point the readers to [56], where Selgrade et al. thoroughly compared these two models and discussed their sensitivity to the data used for parameter fitting.

3.5   Margolskee and Selgrade Model

Margolskee and Selgrade [45] carried out a bifurcation analysis of the Harris Clark et al. [41] model and focused, in particular, on the size of the time delay parameter. In [41], three separate time delay values were used—one for each of the three corresponding ovarian hormones. Margolskee and Selgrade concluded that it is sufficient to include only a delay of $\tau = 1.5$ days for the effect of the inhibin on the pituitary's secretion of FSH. In the Harris Clark et al. [41] model, the delay for the inhibin term was $d_{Ih} = 2.0$. Margolskee and Selgrade found that the delays required for E_2 and P_4 were less than 1 day and could be set to zero. The authors said that this change to the delay values improves the fit to the data for normally cycling women from Welt et al. [12], and they provided extensive bifurcation analysis exploring the effects of changing the length of the inhibin delay. They discussed why the shorter inhibin delay is consistent with biological evidence and how it permits increased ovarian development during the follicular phase of the cycle. We reproduced numerical solutions of this model, which are graphed in Fig. 5.

We note that when the authors fit the model to this new data set [12], a mismatch in units occurred. Unfortunately, all descendants of this model seem to inherit this unit mismatch.

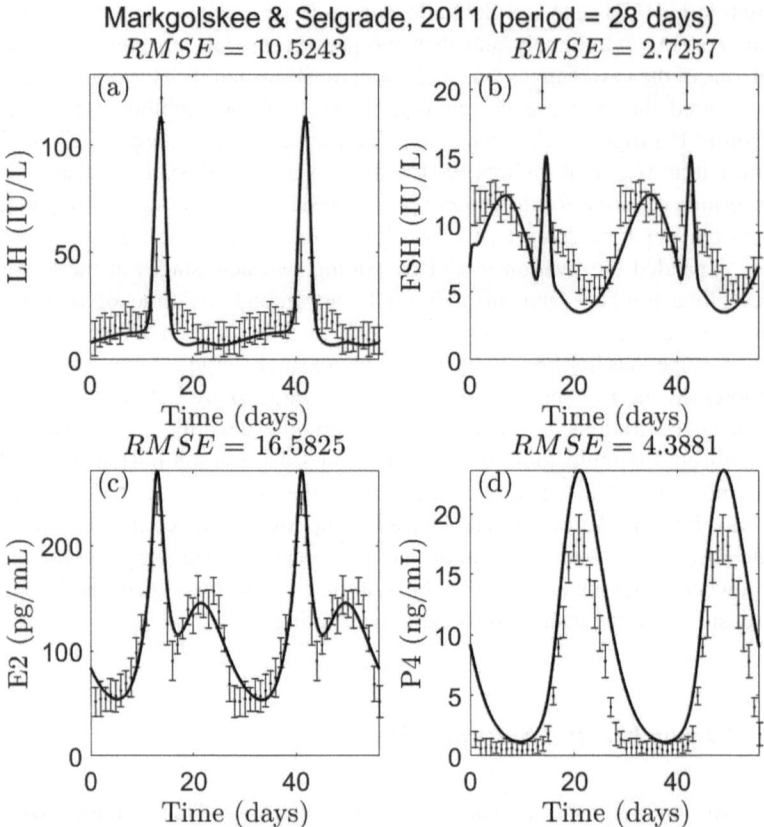

Markgolskee & Selgrade, 2011 (period = 28 days)

Fig. 5 Numerical solutions to the Margolskee and Selgrade [45] model plotted against data from Welt et al. [12]. The root mean square error (RMSE) for each variable is shown, and the period is computed to be approximately 28 days. The units for the hormone levels in these plots reflect the units reported in the Welt et al. data set

3.6 Wright et al. Model

The focus of the Wright et al. [46] model was to explore the effects of oral contraceptive drugs in hormonal control of the menstrual cycle of adult women. This model also included a system of nonlinear delay differential equations with four auxiliary equations representing the ovarian hormones E_2, P_4, and inhibin A. Hormonal contraceptive treatments via oral administration of ethinyl estradiol and progestin were modeled by modifying state variables for blood concentrations of E_2 and P_4. The authors assumed that a contraceptive state is attained if model simulations show a reduction in the LH surge to non-ovulatory levels or a reduction in P_4 levels throughout the cycle. The model extended the Margolskee and Selgrade [45] and Harris Clark et al. [41] models by adding autocrine mechanisms to describe how exogenous estrogen and progesterone could push a normal cycle into a contraceptive state. Model parameters were generally kept at the values used in [45], except for the changes needed for the new model components.

While this model did include more realistic mechanisms to account for exogenous hormones, there were still several simplifying assumptions. The authors assumed that the effect of estrogen on progesterone can be combined into one term P_{app} that does not differentiate between the neuroendocrine and the ovarian systems. They also assumed that P_{app} cannot be larger than P_4. As with the previous works of Harris Clark et al. [41] and Margolskee and Selgrade [45], this model only tracked inhibin A and did not incorporate inhibin B.

Wright et al. showed that the administration of synthetic progesterone, synthetic estrogen, or a combination of these can have a contraceptive effect by preventing ovulation. They concluded that a low dose of both treatments given together is most effective at achieving contraception. They provided numerical experiments that illustrate how the combined contraceptive treatment pushes the system into a non-ovulatory menstrual cycle fairly quickly and that the cycle also returns to normal in short order after the treatment ends. Future work mentioned coupling the model with a model for absorption and metabolism of oral contraceptive drugs. Wright et al. suggested that the addition of absorption and metabolism could help discover minimal effective doses of contraceptive drugs and may lead to patient-specific dosing strategies through pharmacokinetic/pharmacodynamic (PK/PD) modeling based on individual hormone data. We reproduced numerical solutions of this model, which are graphed in Fig. 6.

3.7 Gavina et al. Model

The aim of the Gavina et al. [47] model was to employ optimal control techniques to minimize the total exogenous estrogen or progesterone doses required to enter a contraceptive state. This work was novel in that it considered nonconstant dosage as well as constant dosage of exogenous hormones. This model was also built on the normal menstrual cycle models of Harris Clark et al. [41] and Margolskee and Selgrade [45]. As with those models and the model of Wright et al. [46], this work incorporated only inhibin A (denoted as Inh in their model). The authors justified this assumption by stating that inhibin B is more important when studying reproductive aging. In addition to the system of nonlinear delay differential equations, these authors used auxiliary equations to track E_2, P_4, and Inh. In order to make the optimal control analysis more tractable, the authors made some simplifications from previous models. For example, instead of using the nonlinear term for inhibition and the additional equations employed in Wright et al. [46], these authors opted to use a linear inhibitory term and fewer additions to the Margolskee and Selgrade model; they found that these simplifications reduced computation time when running the optimization code.

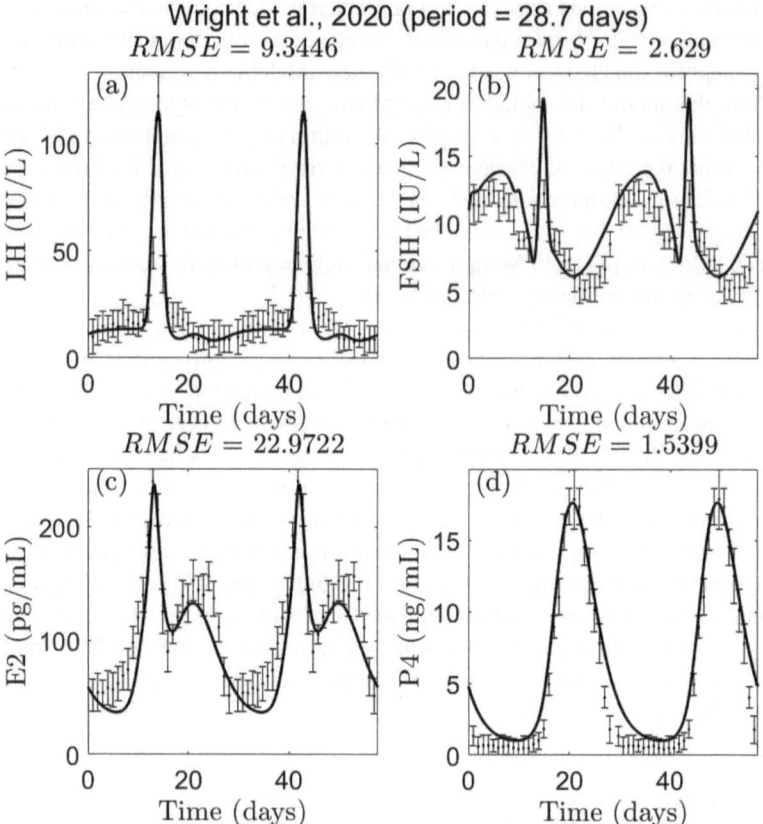

Fig. 6 Numerical solutions to the Wright et al. [46] model plotted against data from Welt et al. [12]. The root mean square error (RMSE) for each variable is shown, and the period is computed to be approximately 28.7 days. The units for the hormone levels in these plots reflect the units reported in the Welt et al. data set

The authors showed that the baseline simplified model without the administration of exogenous hormone reasonably matches the 28-day data from Welt et al. [12] and then proceeded to explore optimal dosing schedules. They observed a reduction in dosage of about 92% in estrogen monotherapy and 43% in progesterone monotherapy. Their simulations showed that delivering the estrogen contraceptive in the mid-follicular phase is the most effective. In addition, they showed that combination therapy significantly lowers doses even further, which is in line with the findings of [46]. While this work made promising steps by studying nonconstant dosing of exogenous hormones, it is important to note that the authors did not discuss whether the optimal dosing schedules are biologically feasible or would be possible to incorporate from a pharmacological perspective. For example, they presented an optimal dosing strategy for delivering synthetic estrogen that requires a large mid-follicular spike in this treatment. While this regimen minimizes the

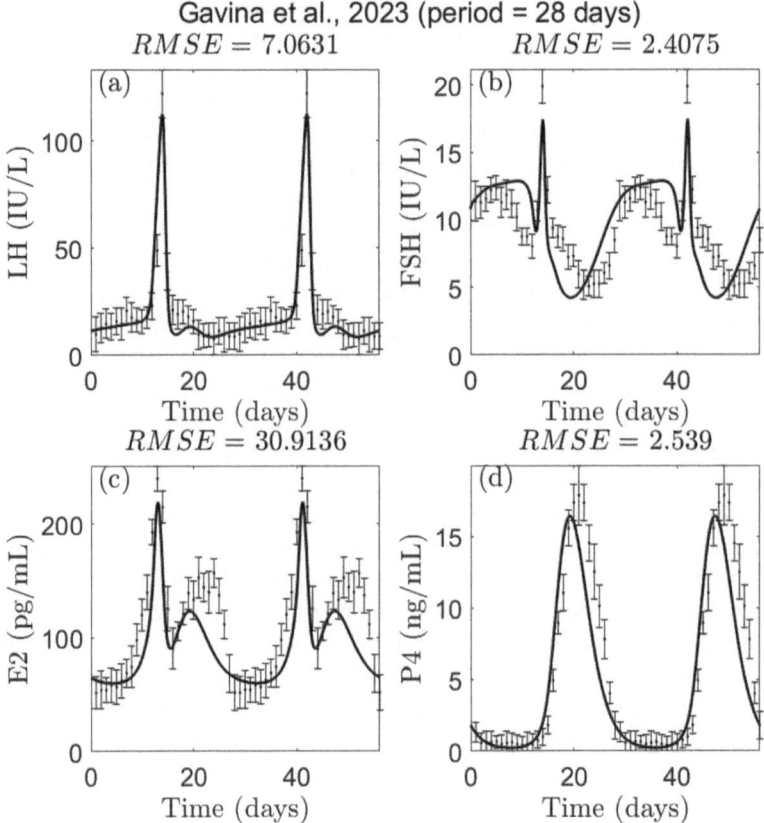

Fig. 7 Numerical solutions to the Gavina et al. [47] model plotted against data from Welt et al. [12]. The root mean square error (RMSE) for each variable is shown, and the period is computed to be approximately 28 days. The units for the hormone levels in these plots reflect the units reported in the Welt et al. data set

total amount of estrogen delivered over the entire cycle, it is unclear whether the magnitude of estrogen applied at this mid-follicular stage would be plausible to administer or if it would create adverse effects for the user. We reproduced numerical solutions of this model, which are graphed in Fig. 7.

3.8 Models Incorporating GnRH

The models discussed in the previous subsections focus on modeling the hormones that are produced in the hypothalamus and pituitary (FSH and LH) and the ovaries (P$_4$, E$_2$, InhA, and InhB). They are potentially well-suited to the incorporation of exogenous hormones. However, these models do not include the gonadotropin-releasing hormone GnRH, which controls the synthesis of FSH and LH. We are aware of two recent models that do incorporate GnRH, which we describe briefly here: Röblitz et al. [48] and George et al. [49].

In 2013, Röblitz et al. [48] published a hormone model with the aim of explicitly exploring GnRH-receptor binding. The focus of this work was to develop a detailed model of GnRH to explore GnRH-receptor binding in response to the introduction of GnRH analogs. Motivated by model equations from Harris Clark [42], Pasteur [44], and Reinecke [51], this model included several novel features. Unlike other models discussed here, this work developed explicit terms for the GnRH system. Moreover, the model eliminated time delays by introducing "effect" compartments, so that the effects of a given hormone could be delayed without incorporating time delays into the differential equations. Finally, the model included separate terms to account for the effects of InhA and InhB. Importantly, their model correctly predicted cycle variations due to doses of a GnRH agonist (which causes the ovaries to stop making estrogen and progesterone) and a GnRH antagonist (which impedes ovulation in *in vitro* fertilization treatment). The authors also communicated an explicit goal to model individual patients, instead of an idealized patient from aggregate data.

In a more recent model by George et al. [49], the authors proposed a simplified model of six nonlinear differential equations to explore the effects of GnRH on the dynamics of the key hormonal drivers in the menstrual cycle. While the model was dramatically simplified from that of Röblitz, the authors were still able to capture some of the qualitative effects of GnRH on the gonadotrophic hormones, including that increasing the width of a GnRH pulse can affect the timing of the release of LH, thereby affecting the timing of when the ovarian hormones are produced.

4 Comparisons Between Existing Menstrual Cycle Models

In this section, we provide a comparative analysis of four menstrual cycle models discussed in Sect. 3: Harris Clark et al. [41], Margolskee and Selgrade [45], Wright et al. [46], and Gavina et al. [47]. For ease of comparison, we selected these four models from the Selgrade lineage, Fig. 3. Our analysis focuses on three areas of interest: a sensitivity analysis of the shared parameter V_{FSH}, a comparison of the ways modelers represent inhibitory delays and the disparate effects of inhibin in the system, and the models' responses to exogenous estrogen and progesterone. The MATLAB codes for the four models we implemented here are available on GitHub: https://github.com/rubyshkim/menstrual-cycle_models.

4.1 Sensitivity to V_{FSH}

By investigating the sensitivity of the models by Harris Clark et al. [41], Margolskee and Selgrade [45], Wright et al. [46], and Gavina et al. [47] to variations in parameter values, we discover that the maximum growth rate of the FSH reserve pool (V_{FSH}) significantly influences the length of the menstrual cycle.

For each model, we vary all parameters one at a time by 25% to determine which parameter values create the most sensitive behaviors when perturbed. Since periodic rhythms are an essential part of the menstrual cycle, we use period length (numerically approximated) as our metric. Of the parameters generating the most sensitivity, we chose to focus on a shared parameter V_{FSH}, which represents the maximum rate of growth of the reserve pool of FSH. Note that although FSH is a measurable hormone, V_{FSH} represents a maximum growth *rate*. V_{FSH}, therefore, is a parameter that is useful in a modeling context, but experimental data on the physiological ranges of V_{FSH} are sparse. Since V_{FSH} is not typically measured in an individual, we focus on the qualitative (rather than the quantitative) model response to changing the values of the rate V_{FSH}. In the models of Selgrade, Schlosser, and Harris Clark [7, 41], the authors cite Odell [58] and state that the nominal values of the follicle growth rates during the follicular phase are chosen by assuming they are proportional to the FSH serum levels. FSH is regulated through negative feedback, and its baseline level can change depending on the availability of regulatory hormones. FSH stimulates the growth of follicles, which modulate inhibins, which in turn then inhibit FSH secretion. Female reproductive aging is accompanied by an increase in circulating FSH due to the loss of follicles, which lowers inhibin levels [12]. In the mathematical models, increasing V_{FSH} from its nominal values shortens the period of the menstrual cycle; see Fig. 8 where the nominal value of V_{FSH} for each model is identified by an orange point. This relationship is consistent with findings that suggest that lower circulating FSH levels lengthen the menstrual cycle [57]. Lowering V_{FSH} from the nominal values generally increases the period in all four models.

However, there is some notable sensitivity in the model behaviors. The model by Harris Clark et al. [41] undergoes a non-smooth change in the approximated period. When V_{FSH} is below just 0.966 of its nominal value of 5700 μg/day, E_2 loses its secondary peak during the luteal phase; example time courses with $V_{FSH} = 5415$ μg/day are provided in Fig. 9. Our simulations of the Wright et al. [46] and Margolskee et al. [45] models also show a sharp transition in approximated period at particular values of V_{FSH} associated with a loss of the E_2 secondary peak. In addition, the length of a regular menstrual cycle is considered to range from 26 to 35 days [16]. A 25% reduction in V_{FSH} results in a loss of periodicity in our simulations of the Gavina et al. [47], Margolskee and Selgrade [45], and Harris Clark et al. [41] models.

4.2 Variation in Incorporating Delays in the Models

In many of the models, discrete time delays are key components that represent the delays between changes in blood hormone levels and their effects on synthesis rate. Among all models we reviewed in this manuscript, only the models of Röblitz et al.

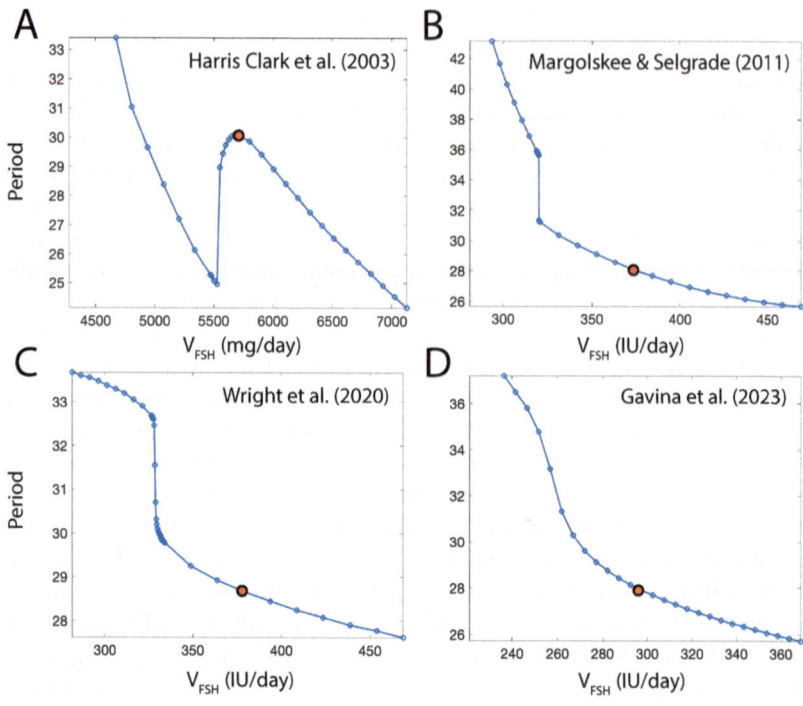

Fig. 8 Sensitivity of the model simulations to V_{FSH}. We approximate the period of numerical solutions over 100 cycles for the models by Harris Clark et al. [41], Margolskee and Selgrade [45], Wright et al. [46], and Gavina et al. [47] for values of the parameter V_{FSH} between 75% and 125% of its nominal value indicated by the orange point in each model. V_{FSH} is the maximum growth rate of reserve pool FSH and generally has an inverse relationship with the period in our simulations. This result is consistent with the findings in [57]. There is notable sensitivity to variations in V_{FSH} in all four models. The axes for each model are chosen to center the nominal value for V_{FSH}

Fig. 9 Example time series from Harris Clark et al. [41] model simulations with $V_{FSH} = 5415\ \mu g/day$ or 95% of its nominal value. Compared with Fig. 4, concentrations of all four hormones are reduced, and FSH and E_2 no longer have a secondary peak during the luteal phase

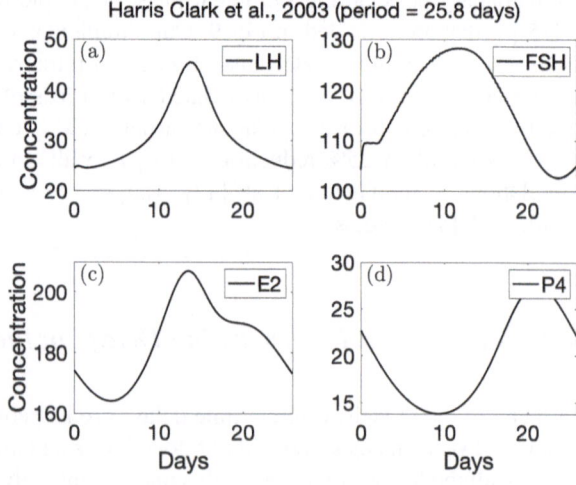

[48] and George et al. [49] do not involve any delay differential equations. Röblitz et al. argued that the delayed inhibitory effect of inhibin B on FSH synthesis in their model was incorporated by the mechanisms in which low GnRH frequencies would stimulate FSH synthesis with adjusted rate constants. To avoid the use of delay differential equations, they introduced a new compartment for effective inhibin A, IhA_e, to account for the delayed inhibitory effect of inhibin A on FSH synthesis.

In the original Selgrade and Schlosser [8] model of the pituitary component of the menstrual cycle that five of the models we reviewed were based on, they included three discrete time delays—one corresponding to each ovarian hormone. First, $d_E = 0.42$ day is the delay in the input function $E_2(t)$, which appears in the LH synthesis term in order to simulate both the rapid rise and fall of LH during the surge, as well as the time difference between the peaks of E_2 and LH. There is a $d_P = 2.9$ day delay in the input function $P_4(t)$, which appears in the LH synthesis term to correctly simulate the timing of the surge in LH synthesis following the changes in serum levels of P_4. Finally, there is a $d_{Ih} = 2$ day delay in the input function $Ih(t)$, which appears in the FSH synthesis term to capture the period of time between the changes in the inhibin blood levels and the FSH synthesis rates.

The Harris Clark et al. [41] model removed the delay in the input function $E_2(t)$ in the LH synthesis term, as their parameter identification indicated that it was insignificant; they kept the delay $d_{Ih} = 2$ day but modified the other delay d_P to be 1 day. The Margolskee and Selgrade [45] model reduced the number of delays to one, incorporating only a $\tau = 1.5$ day delay in the effect of the peptide inhibin on FSH synthesis. This led to an improvement in the fit to the Welt et al. data. They set the other two delays to 0 as the values were less than a day and did not contribute significant additional improvement on data fitting. Similarly, both the Wright et al. [46] and Gavina et al. [47] models kept the discrete time delay, $\tau = 1.5$ day, in the input function $InhA(t)$, which appears in the FSH synthesis term. The Pasteur and Selgrade [43] multi-inhibin model includes four time delays: a $d_E = 0.5086$ day delay in the input function $E_2(t)$ in the LH synthesis term, a $d_P = 0.9156$ day delay in the input function $E_2(t)$ in the LH synthesis term, a $d_{IhA} = 2.5$ day delay in the input function $IhA(t)$ in the FSH synthesis term, and a $d_{IhB} = 1$ day delay in the input function $IhB(t)$ in the FSH synthesis term.

In this collection of models, we see that there are two main approaches taken to capturing the timing of the cascade of events from the onset of changes in blood hormones to the subsequent effect on synthesis: using delays or incorporating more model detail. The models of [8, 41, 45–47] incorporate explicit time delays in strategic parts of the models, and variations in the way the delays are included lie in the fine-tuning of the lengths of the delays. The models of [48, 49] have taken the approach of including more detailed steps in the hormone-synthesis cascade that in turn implicitly produce the necessary time delays. Although software for the numerical solution of delay differential equations is readily available, the delay-free approach may be preferable when implementing numerical solutions because traditional numerical ODE solution methods tend to be stable and easy to implement.

4.3 Subtle Differences in Modeling Inhibins

As mentioned in Sect. 3.2, data from McLachlan et al. [9] and Welt et al. [12] are commonly used in models of the menstrual cycle. It is important to note that these two data sets are different with respect to inhibin, and this impacts modeling choices. Separate bioassays for inhibin A and inhibin B were not available until the mid-1990s. As such, many of the models discussed here represented the total effects of inhibin using one compartment, despite the fact that they peak at different times in the cycle. In this section, we discuss the variations in choices of modeling inhibin and how it connects to these data sets.

Harris Clark et al. [41] fit their model to the McLachlan et al. data set [9], which only reported total inhibin values. Many other authors [43, 45–47] parameterized their models using the Welt et al. data set [12], but the ways they handled the inhibin data are slightly different. Pasteur and Selgrade [43] used the data from the younger age group with 23 women aged 20–34 for parameterization to model both inhibin A and inhibin B. Note that Pasteur et al. parameterized their five-hormone model using both the McLachlan et al. and Welt et al. data in [44, Chapter 3], but they used the inhibin A data from the Welt et al. data only, as they concluded that its profile matched the total inhibin profile from the McLachlan et al. data. Both Gavina et al. [47] and Wright et al. [46] included only inhibin A in their model and used the data from the younger age group (note that in [47] the state variable Inh denotes inhibin A). Margolskee and Selgrade [45] also fit their model to the data from the younger group in the Welt et al. data, but it is unclear how they handled the inhibin data as they did not distinguish inhibin A and inhibin B explicitly in their model.

4.4 Incorporation of Hormonal Contraception in Models

Perhaps the simplest way to incorporate hormonal contraception into a menstrual cycle model is with a constant dosing of exogenous estrogen (e.g., estradiol) and/or progesterone (e.g., progestin). In this section, we compare the qualitative behaviors of each of the models under this simple strategy for modeling hormonal contraception. As discussed in Sect. 2.3, a contraceptive state is marked by the absence of an LH surge. However, contraception may refer to *total contraception*, where LH levels are low and roughly constant, or *biological contraception*, where LH levels increase and decrease in an oscillatory manner but remain low. We capture the dynamics of response to hormonal contraception in each of these models by measuring (a) the amplitude of the LH peak (a biomarker for whether or not ovulation will occur) and (b) the period of P_4 oscillations (a proxy for whether total contraception is achieved).

As expected, with no exogenous progesterone and no exogenous estrogen (i.e., no hormonal contraceptives), all of the models we discuss in this manuscript have a relatively high LH peak and a mean cycle length between 28 and 30 days. This is consistent with the biology of the menstrual cycle in a non-contraceptive state. To visualize the response of each model to the addition of hormonal contraception, we create heatmaps in Fig. 10 for the effect of exogenous estrogen (horizontal axis) and exogenous progesterone (vertical axis) on the maximum value of the LH peak (colors, top panels) and the period of the P_4 oscillations (colors, middle panels) computed using findpeaks in MATLAB. It is advantageous to use the time course of P_4 to estimate the period because it only has one peak (or local maximum) per cycle, while the other variables have multiple peaks. Note that these models have some differences in units and baseline concentrations of variables. For each model, we administer exogenous estrogen up to the mean endogenous estradiol concentration in that model and exogenous progesterone up to 0.3 of the mean endogenous progesterone concentration in that model. These choices are based on the exogenous hormone concentrations used in the models which consider exogenous hormones [41, 46, 47].

First, we explore the behavior of the Harris Clark et al. [41] model to the introduction of exogenous hormones (Fig. 10A and E). We note that this model was not designed primarily with questions of hormonal contraception in mind. In Fig. 10A, we notice that, in the case without any exogenous estrogen, the LH peak increases as the level of exogenous progesterone increases. From a biological perspective, we know that progesterone-only hormonal birth control methods do induce contraceptive states [35], and this is not captured by this model. However, we do see that increasing exogenous estrogen induces a contraceptive state. Biological contraception is achieved for values of e_{dose} larger than 20, while total contraception is achieved only for very large doses of exogenous estrogen. This model has a smaller variance in cycle length (as measured by the period of oscillations in LH) in this parameter space as compared to other models we tested, and we observe a non-monotonic response in cycle length to exogenous estrogen.

As noted in Sect. 3.5, Margolskee and Selgrade [45] modified the model of Harris Clark et al. [41] to perform bifurcation analyses to explore the effects of delays and to fit the data of normally cycling women more closely than previous efforts. The level of the LH peak in this updated model does depend on both exogenous estrogen and exogenous progesterone, but the effect of exogenous estrogen is stronger (Fig. 10B and F). We further observe that contraception is not achieved with progesterone-only treatment, and this model did not achieve total contraception for any values of parameters we tested. As with the model of Harris Clark et al. [41], the authors did not design this model to explore the effects of hormonal contraception, so it is perhaps not surprising to see that these effects are not strongly captured.

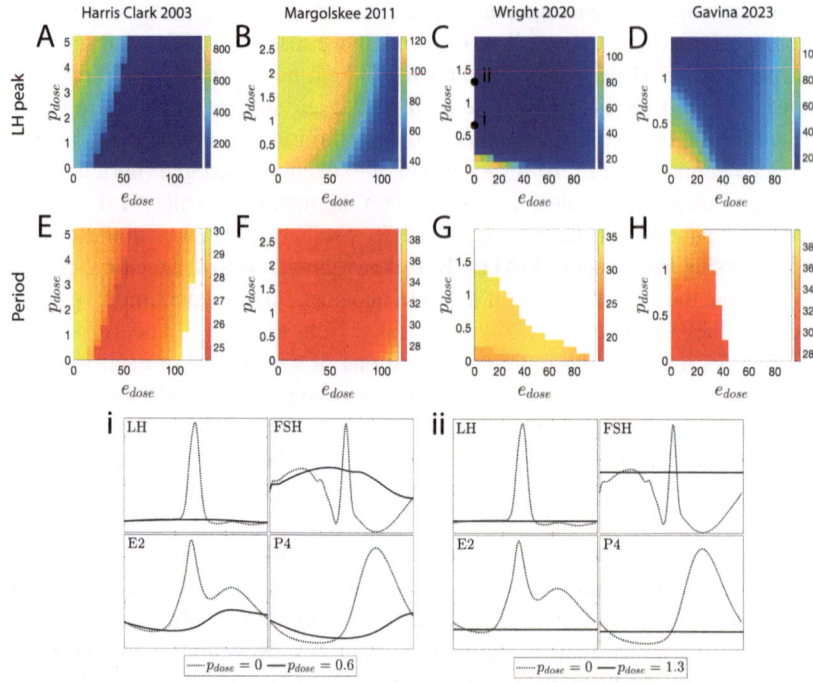

Fig. 10 LH peak concentration and menstrual cycle period for the various models [41, 45–47] with daily administration of exogenous hormones, i.e., hormonal contraceptives. We vary the amount of exogenous estrogen e_{dose} up to its mean concentration (horizontal axis) and exogenous progesterone p_{dose} up to 30% of its mean concentration (vertical axis) in each model. In panels A–D, we visualize the effects of e_{dose} and p_{dose} on LH peak concentrations, with yellow representing high peak concentrations of LH and blue representing low peak concentrations of LH. These blue values represent the parameter ranges where the model is in a contraceptive state. In panels E–H, we visualize the effects of e_{dose} and p_{dose} on the period of P_4 oscillations, with yellow representing a longer cycle length and red representing a shorter cycle length. The white regions in these plots correspond to the absence of oscillations, which indicates that the model is in a total contraceptive state. Schematic plots of example 28-day hormone trajectories corresponding to contraceptive states for the Wright et al. [46] model are provided in panels i ($p_{\text{dose}} = 0.6$) and ii ($p_{\text{dose}} = 1.3$), which correspond to the dots indicated in panel C. Simulations were run for 100 cycles before each computation to help eliminate the effects of transient initialization behavior

In contrast to previous models, Wright et al. [46] developed their model with the intention of including hormonal contraception. In Fig. 10C and G, we see that as both exogenous estrogen (e_{dose}) and exogenous progesterone (p_{dose}) increase, the LH peak decreases, indicating that the system is in a contraceptive state. This occurs for relatively low values for both parameters. We notice that the cycle length, as measured by the period of the oscillations in LH, is relatively stable over a large range of parameter values. Total contraception is achieved for large enough values of e_{dose} and p_{dose}, as indicated by the white region in Fig. 10G. Mathematically, a transition to this parameter regime corresponds to a Hopf bifurcation.

The model of Gavina et al. [47] was also created to explore the effects of hormonal contraception, although from a different perspective than [46]. We notice that the modifications made in this model introduce different qualitative dynamics with constant exogenous hormonal contraception (Fig. 10D and H) than for the Wright et al. [46] model. While we do see the expected behavior of a reduction of LH peak when either exogenous hormone is increased initially, we notice that the LH peak again begins to increase for large values of e_{dose}. We see that LH undergoes a Hopf bifurcation in e_{dose}, with periodic behavior only occurring for small parameter values. Combining these observations, we can conclude that for large values of e_{dose}, LH is being held constant at a high level, which is not conducive to the expected contraceptive state.

5   Outlook and Future Directions

In this survey, we describe the biological mechanisms behind the menstrual cycle and hormonal contraception and review the existing mathematical models of this system. In a comparative analysis of several of these models, we highlight the ways different models are sensitive to parameters, focusing in particular on the maximum growth rate of the reserve pool of follicular stimulating hormone. We discuss the variation in incorporating delays and inhibins into these models. Finally, we explore the qualitative behavior of these models under the incorporation of exogenous hormones to model hormonal contraception. We conclude this chapter by discussing some of the existing challenges in this area and potentially promising future directions.

The complexity of the biological system may seem like a daunting challenge to mathematical modelers. Indeed, even models that try to simplify while including crucial mechanisms can lead to relatively large systems of differential equations with many parameters. The models by Harris Clark et al. [41], Pasteur et al. [43], Margolskee and Selgrade [45], Wright et al. [46], and Gavina et al. [47] all contain on the order of 13 ODEs, 4 auxiliary equations, and 50 parameters. The model of Röblitz et al. [48] considers the system in even more detail, including gonadotrophin-releasing hormone, which facilitates the synthesis and secretion of the gonadotrophic hormones FSH and LH. This model contains 33 ODEs and 114 parameters. These detailed models strive to accurately capture the range of temporal and spatial scales represented in this system. Furthermore, detailed models may provide more flexibility for use in preliminary drug development or testing interventions beyond what the model was originally built for, which is not usually possible in overly simplified phenomenological models. More detailed models do have some limitations, however. Parameter identification and overfitting can be concerns in large systems of differential equations; thus care must be taken when applying or extending these models. Large models are often less amenable to mathematical analysis, which may limit the possible insight into qualitative behaviors or general properties of the models. These different styles of models each have roles to play in gaining a deeper understanding of the menstrual cycle and hormonal contraception. There is still ample opportunity for developing simpler, more analytically tractable models of the menstrual cycle in the style of George et al. [49] that are nevertheless grounded in biological mechanisms.

The parameters in the existing models can present a challenge from the perspective of data fitting and model analysis. While some parameters in the model may be inferred from data, many cannot. When comparing or working with the models of the menstrual cycle discussed in this chapter, it is important to note that even parameters that share the same name or represent the same quantity across different models may not be the same order of magnitude or even the same units. There can also be challenges with a lack of consistency of units and dimensions, even within a single model. For readers who are interested in generalizing, extending, or applying the models we discuss in this chapter, we provide a detailed dimensional analysis on our public GitHub repository (https://github.com/rubyshkim/menstrual-cycle_models/blob/main/SupplementalFiles/SupplementalMaterial.pdf).

One of the interesting features of the menstrual cycle is that it has a high level of variability. Several of the models discussed in this chapter use data from Welt et al. [12], and the models are fit to the mean hormone levels reported across 23 subjects. In the experiments, there is a high level of variability in hormone levels both within and between individuals. For example, in [12, Fig. 5], the authors show hormone levels from two subjects, aged 36 and 47, of two menstrual cycles approximately 10 years apart. These data demonstrate that hormone levels can vary greatly between different subjects: In particular, inhibin A/B and P_4 levels are quite different in individuals of different ages. Even the same subject demonstrates variability of their hormone levels in two cycles. Unfortunately, this variability is not currently incorporated with any of the existing modeling approaches we have reviewed here. While variability can be challenging from the perspective of trying to fit models to data, we believe it also presents an excellent opportunity for further modeling. For example, it may be interesting to study the addition of intrinsic and/or extrinsic sources of noise into these models.

Incorporation of hormonal contraception into mathematical models of the menstrual cycle is still in the early scientific stages. We discuss very few models of the menstrual cycle that incorporate the impact of hormonal contraceptives on the menstrual cycle in Sect. 4.4; of these, only [47] explores time-varying exogenous hormones. This means that at the time of writing, there are no existing mathematical models of the effects of hormonal contraception on the menstrual cycle that incorporate the dynamics of the on/off dosing regimens or the metabolism of the exogenous hormones, even though methods from differential equations and dynamical systems are well-positioned to investigate these questions. One potential area of inquiry would be to develop a model of contraceptive dosing regimens on the menstrual cycle, either by creating a new model or by generalizing an existing model discussed here. This would allow for the exploration of the stability of the contraceptive state achieved by oral hormonal contraceptives using a mechanistic mathematical model of the menstrual cycle. Such a model could provide insight into when a contraceptive state is lost due to inconsistency or changes in hormonal birth control use, which may further inform the advisement of care providers and the choices of birth control users. Another interesting direction of study would be to explore the impact of hormonal stimulation used in cases of infertility. While infertility treatments have increased in their use and success over the last few decades, there are still substantial risks, including high-risk multiple pregnancies and ovarian hyperstimulation [59]. A mathematical model that can successfully incorporate the impact of hormonal contraceptives on the menstrual cycle could be leveraged to explore open questions in this budding field.

Supplementary Information

Table 1 provides a summary of six of the models we reviewed in this chapter. It contains a brief overview of the characteristics of each model (such as the number of differential equations/auxiliary equations/parameters, whether constant time delays are implemented, and how the exogenous hormones are modeled), study focus, some of the contributions, and limitations of each model.

Table 1 Model comparison. Here we name the models in the format "First author + year of publication." Diff.: differential; Aux.: auxiliary; Eq.: equation; Eqs.: equations; Exo. horm.: exogenous hormones. See text for definitions of symbols

Model	Harris Clark 2003	Pasteur 2011	Margolskee 2011	Wright 2020	Gavina 2023	Röblitz 2013
Reference	[41]	[43]	[45]	[46]	[47]	[48]
Other Authors	Schlosser, Selgrade	Selgrade	Selgrade	Fayad, Selgrade, Olufsen	de los Reyes, Olufsen, Lenhart, Ottesen	Stötzel, Deuflhard, Jones, Azulay, van der Graaf, Martin
Year	2003	2011	2011	2020	2023	2013
# of Diff. Eqs.	13	16	13	13	13	33
# of Aux. Eqs.	3	4	3	4	3	0
# of Parameters	$42 + 2$ (exo. horm.)	$58 + 2$ (exo. horm.)	42	$44 + 2$ (exo. horm.)	$42 + 2$ (exo. horm.)	114
Time Delay Constants	Yes. P_4's effect on RP_{LH} synthesis; inhibin's effect on RP_{FSH} synthesis	Yes. E_2's and P_4's effect on RP_{LH} synthesis; inhibin A's effect and inhibin B's effect on RP_{FSH} synthesis	Yes. Inhibin's effect on RP_{FSH} synthesis	Yes. Inhibin A's effect on RP_{FSH} synthesis	Yes. Inhibin A's effect on RP_{FSH} synthesis	No.
# of Hormones	5	6	5	5	5	7
GnRH	No	No	No	No	No	Yes
Total Inhibin	Yes	No	Yes	No	No	No
Inhibin A	No	Yes	No	Yes	Yes	Yes
Inhibin B	No	Yes	No	No	No	Yes
Follicular Phase Stages	3	5	3	3	3	5
Ovulation Stages	2	3	2	2	2	3
Luteal Phase Stages	4	4	4	4	4	4

Exogenous Hormones	Yes. (1) P_4 therapy (P_4 aux. eq.) adds 80, 50, or 30 nmol/L for 5 days at the beginning of the luteal phase of the abnormal cycle; (2) E_2 disruption (E_2 aux. eq.) adds 50 nmol/L either periodically for three cycles (only apply to the first and third cycle) or continuously for two cycles	Yes. Included additional step function term in the auxiliary equation for E_2. Considered three cases: continuous low, medium, and large doses, which correspond to raising the bloodstream E_2 concentration by 25, 75, or 125 pg/mL	No	Yes. Modeled as constant input in the aux. eqs. for E_2 and P_4. An intermediate variable, P_{app}, is used to model P_4's contraceptive effects. Considered daily administration of exogenous hormones over three cycles as well as model behavior before and after treatment period	Yes. Included $E_2^{exo}(t)$ and $P_4^{exo}(t)$ in the aux. eqs. for E_2 and P_4. Added additional term to RP_{FSH} and RcF eqs. to attain contraceptive effect of progesterone. Constant dosage: monotherapy or a combination of P_4^{exo} and E_2^{exo} for 28 days. Nonconstant dosage (optimal control)	Yes. Compared the effect of administering single (at different timepoints in the cycle) and multiple doses of Nafarelin (a GnRH agonist) and Cetrorelix (a GnRH antagonist known to impede ovulation in *in vitro* fertilization treatment)
Data Sets	[9]	[12]	[12]	[10–12]	[11, 12, 60]	[61–64]
Study Focus	Merge the linear systems for the production of pituitary hormones and for the production of ovarian hormones into one nonlinear model, show that it has two locally asymptotically stable periodic solutions, using Hopf bifurcation theory and the software DDEBIFTOOL, and refer them as the normal and abnormal cycle, respectively	Developed a multi-inhibin model, fitted parameters using Welt 1999 data, performed sensitivity and bifurcation analyses. Conducted a dose–response analysis of external estrogen intake, throughout several consecutive menstrual cycles	Carried out a bifurcation analysis using the DDEBIFTOOL for a model that only incorporates time delay in the effect of peptide inhibin on the pituitary's secretion of FSH	Made two major changes to the original models (Harris Clark 2003 and Margolskee 2011) to simulate contraceptive behavior that is consistent with the biology. In this new model, P_{app} limits early follicle development and is enhanced by E_2	Utilized optimal control theory on a modified menstrual cycle model (based on Margolskee 2011 model) to determine the minimum total exogenous estrogen/progesterone dose and timing of administration to induce anovulation (absence of ovulation)	Developed a mathematical model that characterizes the actions of GnRH agonists and antagonists by their different effects on the HPG axis

(continued)

Table 1 (continued)

Model	Harris Clark 2003	Pasteur 2011	Margolskee 2011	Wright 2020	Gavina 2023	Röblitz 2013
Contribution/ Key Results	Removed the delay in E_2's effect on RP_{LH} synthesis from the original system. Used the improved parameter estimation from their dissertation [42] instead of the crude estimates in Schlosser and Selgrade (2000). It was shown that exogenous progesterone treatment can perturb the system from the abnormal cycle to the normal cycle. Demonstrated that exogenous estrogen inputs can cause disruption to the normal cycle and ultimately result in abnormal cycling	Expanded the existing model to include inhibin B. Observed at most one periodic solution, regardless of varying any single parameter from the six most sensitive parameters within biologically realistic ranges. Showed that increasingly high amounts of exogenous E_2 administered continuously suppress the LH surge to an increasingly large degree, upon cessation of the treatment, the modeled hormone concentrations return to the stable normal cycle	Discussed biological and mathematical reasons for a 36 h time delay for the effect of inhibin on the synthesis of FSH. Examined bifurcations with respect to changes in three important parameters that represent the level of E_2 adequate for significant synthesis of LH, mass transfer between the first two stages of ovarian development, and time delay, respectively	Created a physiologically based model of hormonal contraception by adding doses of circulating E_2 and/or P_4 and making adjustments to how they influence other variables. The model can be used to help determine the lowest dose required to achieve contraception	Showed that in monotherapy it is most effective to deliver the estrogen contraceptive in the mid-follicular phase (minimum constant dosage of estrogen of 24.73 pg/mL per day for 28 days vs total dosage of 77.76 pg/mL with optimal nonconstant dosage). Showed that by combining estrogen and progesterone, the dose can be significantly lowered. The results of this study may aid in identifying the minimum dose and treatment schedule that cause anovulation	The model is based on a deterministic modeling of the GnRH pulse pattern and does not involve any delay equations, which shortens computational time. It correctly predicts hormonal changes following administration of single and multiple doses of Nafarelin or Cetrorelix at different stages in the cycle. It gives insight into mechanisms underlying gonadotropin suppression

Limitation	The parameters used were estimated over two periods of data using periodic input functions with period set to 31 days, but their figures suggested that solutions converge to a solution with period 29.5 [42]. Did not carry out complete dose–response analysis on P_4 administration. Did not conduct a strength and duration study of the E_2 disruption. The model does not distinguish inhibin A and inhibin B; it lumps together the pituitary and hypothalamus and does not describe the role of GnRH. Did not provide the initial conditions used	The FSH does not fit well to the data, particularly in the week surrounding ovulation in each cycle. In [44, Chapter 6] they also examined the effect of exogenous progesterone but found no significant reduction in the amplitude of the LH surge. Did not provide the initial conditions and parameter values used in their simulation	They did not show P_4 in their figures demonstrating the effect of inhibin delay on model fit to data; it is not clear whether P_4 from their model fit well to the data. In the figures they showed, FSH and Inh do not fit well to data	The exogenous hormone administration is instantaneous. It is not clear how real drug doses were translated into e_{dose} and p_{dose}. Only tested several different values of e_{dose} and p_{dose}. FSH behavior is not correct, possibly due to the omission of inhibin B. Did not provide the initial conditions used in their simulation presented in the book chapter	Manual adjustments after optimization were carried out to reach the maximum E_2, P_4, Inh, LH, and FSH and to obtain cycle length close to 28 days. They did not provide the value for parameter p_0. They pointed out that Inh B is more significant in studies about reproductive aging and used that to justify why their study only employs Inh A. Only investigated optimization of dosing strategy over one cycle due to the considerable amount of time it takes to obtain results	Followed a semi-mechanistic modeling approach due to the limited data available. Although comprising several organs (hypothalamus, pituitary, blood, and ovaries), the model does not take into account signal transduction on a cellular level

Acknowledgments The work described herein was initiated during the Collaborative Workshop for Women in Mathematical Biology funded and hosted by UnitedHealth Group Optum of Minnetonka, MN and supported by University of Minnesota's Institute for Mathematics and its Applications in June 2022. Additionally, the authors and editors thank the anonymous peer reviewers for their feedback, which strengthened this work.

We would like to thank all the participants of this workshop for their support and feedback, with special acknowledgement to Ashlee Ford Versypt, Rebecca Segal, Blerta Shtylla, and Suzanne Sindi for their leadership in organizing the workshop. LdeP thanks Mette Olfusen and Andrew Wright for providing the MATLAB codes that simulate the model in [46]. We would like to thank Georgia Pope for introducing HZB to some of the papers in this area and for inspiration toward the future directions for this working group.

References

1. K.A. Yonkers, P.S. O'Brien, E. Eriksson, Lancet **371**(9619), 1200 (2008)
2. A. Scambler, G. Scambler, *Menstrual Disorders* (Routledge, Abingdon-on-Thames, England, 2003)
3. J.E. Marsden, M. McCracken, *The Hopf Bifurcation and Its Applications* (Springer, New York, 1976)
4. W.J. Shack, P.Y. Tam, T.J. Lardner, Biophys. J. **11**(10), 835 (1971). https://doi.org/10.1016/S0006-3495(71)86257-4
5. R.J. Bogumil, M. Ferin, J. Rootenberg, L. Speroff, R.L. Vande Wiele, J. Clin. Endocrinol. Metab. **35**(1), 126 (1972)
6. R.J. Bogumil, M. Ferin, R.L. Vande Wiele, J. Clin. Endocrinol. Metab. **35**(1), 144 (1972)
7. J.F. Selgrade, P.M. Schlosser, Fields Inst. Commun. **21**, 429 (1999)
8. P.M. Schlosser, J.F. Selgrade, Environ. Health Perspect. **108**(Suppl 5), 873 (2000)
9. R.I. McLachlan, N.L. Cohen, K.D. Dahl, W.J. Bremner, M.R. Soules, Clin. Endocrinol. **32**(1), 39 (1990)
10. T.M.T. Mulders, T.O.M. Dieben, Fertil. Steril. **75**(5), 865 (2001)
11. A. Obruca, T. Korver, J. Huber, S.R. Killick, B.M. Landgren, M.J. Struijs, Fertil. Steril. **76**(1), 108 (2001)
12. C.K. Welt, D.J. McNicholl, A.E. Taylor, J.E. Hall, J. Clin. Endocrinol. Metab. **84**(1), 105 (1999)
13. J.E. Finlay, M.A. Lee, Milbank. Q. **96**(2), 300 (2018)
14. K. Daniels, J. Daugherty, J. Jones, W. Mosher, National Health Statistics Reports, US National Center for Health Statistics **86** (2015)
15. S.E. Bulun, in *Williams Textbook of Endocrinology*, ed. by K.S. Polonsky, P.R. Larsen, H.M. Kronenberg (Elsevier, Philadelphia, 2011), chap. 17, pp. 581–660
16. M. Mihm, S. Gangooly, S. Muttukrishna, Anim. Reprod. Sci. **124**(3–4), 229 (2011)
17. B.G. Reed, B.R. Carr, *The Normal Menstrual Cycle and the Control of Ovulation* (MDText, South Dartmouth, MA, 2000). http://europepmc.org/books/NBK279054
18. J.R. Bull, S.P. Rowland, E.B. Scherwitzl, R. Scherwitzl, K.G. Danielsson, J. Harper, NPJ Digital Med. **2**(1), 83 (2019)
19. J.B. Kerr, M. Myers, R.A. Anderson, Reproduction **146**(6), R205 (2013)
20. E.A. McGee, A.J.W. Hsueh, Endocr. Rev. **21**(2), 200 (2000)
21. B. Alberts, A. Johnson, J. Lewis, M. Raff, K. Roberts, P. Walter, in *Molecular Biology of the Cell*, 4th edn. (Garland Science, New York, 2002)
22. M.A. Fritz, L. Speroff, *Clinical Gynecologic Endocrinology and Infertility*, 8th edn. (Lippincott Williams & Wilkins, Philadelphia, 2011)
23. F.S. Khan-Dawood, L.T. Goldsmith, G. Weiss, M.Y. Dawood, J. Clin. Endocrinol. Metab. **68**(3), 627 (1989)
24. B.P. Patricio, B.G. Sergio, in *Menstrual Cycle*, ed. by O.I. Lutsenko (IntechOpen, 2019)

25. N.D. Shaw, S.N. Histed, S.S. Srouji, J. Yang, H. Lee, J.E. Hall, J. Clin. Endocrinol. Metab. **95**(4), 1955 (2010)
26. R.E. Jones, K.H. Lopez, *Human Reproductive Biology*, 4th edn. (Academic Press, New York, 2014)
27. C.K. Welt, A.L. Schneyer, J. Clin. Endocrinol. Metab. **86**(1), 330 (2001)
28. A. Sehested, A.A. Juul, A.M. Andersson, J.H. Petersen, T.K. Jensen, J. Müller, N.E. Skakkebaek, J. Clin. Endocrinol. Metab. **85**, 1634 (2000)
29. K.L. Rosewell, T.E. Curry, in *Encyclopedia of Reproduction*, ed. by M.K. Skinner, 2nd edn. (Academic Press, Oxford, 2018), pp. 250–254. https://doi.org/10.1016/B978-0-12-801238-3.64649-4
30. J. Wen, K. Huang, X. Du, H. Zhang, T. Ding, C. Zhang, W. Ma, Y. Zhong, W. Qu, Y. Liu, et al., Front. Endicrinol. **12**, 626534 (2021)
31. J. Kiley, C. Hammond, Clin. Obstet. Gynecol. **50**(4), 868 (2007)
32. R. Rivera, I. Yacobson, D. Grimes, Am. J. Obstet. Gynecol. **181**(5), 1263 (1999)
33. L.P. Shulman, Am. J. Obstet. Gynecol. **205**(4), S9 (2011)
34. J. Jin, J. Am. Med. Assoc. **311**(3), 321 (2014)
35. H.B. Croxatto, Contraception **65**(1), 21 (2002)
36. N.B. Schwartz, in *Mammalian Reproduction*, ed. by H. Gibian, E.J. Plotz (Springer, Berlin, 1970), pp. 97–111
37. J.E.A. McIntosh, R.P. McIntosh, *Mathematical Modelling and Computers in Endocrinology* (Springer, Berlin, 1980)
38. L. Plouffe, S. Luxenberg, Comput. Biomed. Res. **25**(2), 117 (1992). https://doi.org/10.1016/0010-4809(92)90015-3
39. M.C. Kohn, M.L. Hines, J.M. Kootsey, M.D. Feezor, Math. Comput. Model. **19**(6), 75 (1994). https://doi.org/10.1016/0895-7177(94)90190-2
40. M.E. Andersen, H.J. Clewell, J. Gearhart, B.C. Allen, H.A. Barton, J. Toxicol. Environ. Health **52**(3), 189 (1997)
41. L.H. Clark, P.M. Schlosser, J.F. Selgrade, Bull. Math. Biol. **65**(1), 157 (2003)
42. L.A. Harris, Differential Equation Models for the Hormonal Regulation of the Menstrual Cycle. Ph.D. thesis, North Carolina State University (2002)
43. R.D. Pasteur, J.F. Selgrade, in *Understanding the Dynamics of Biological Systems* (Springer, New York, 2011), pp. 39–58
44. R.D. Pasteur II, A Multiple-Inhibin Model of the Human Menstrual Cycle. Ph.D. thesis, North Carolina State University (2008)
45. A. Margolskee, J.F. Selgrade, Math. Biosci. **234**(2), 95 (2011)
46. A.A. Wright, G.N. Fayad, J.F. Selgrade, M.S. Olufsen, PLoS Comput. Biol. **16**(6), e1007848 (2020)
47. B.L.A. Gavina, A.A. de los Reyes V, M.S. Olufsen, S. Lenhart, J.T. Ottesen, PLoS Comput. Biol. **19**(4), e1010073 (2023)
48. S. Röblitz, C. Stötzel, P. Deuflhard, H.M. Jones, D.O. Azulay, P.H. van der Graaf, S.W. Martin, J. Theor. Biol. **321**, 8 (2013)
49. S.S. George, L.O.M. Mercado, C.Y. Oroz, D.X. Tallana-Chimarro, J.R. Melendez-Alvarez, A.L. Murrillo, C.W. Castillo-Garsow, K. Rios-Soto, Technical report, Arizona State University (2018)
50. E.N. Best, Simulation **25**(4), 117 (1975)
51. I. Reinecke, Mathematical Modeling and Simulation of the Female Menstrual Cycle. Ph.D. thesis, Freien Universität Berlin (2009)
52. H.B. Lavoie, K.A. Martin, A.E. Taylor, W.F. Crowley, J.E. Hall, J. Clin. Endocrinol. Metab. **83**(1), 241 (1998)
53. A. Sehested, A. Juul, A.M. Andersson, J.H. Petersen, T.K. Jensen, J. Müller, N.E. Skakkebaek, J. Clin. Endocrinol. Metab. **85**(4), 1634 (2000)
54. R. Stricker, R. Eberhart, M.C. Chevailler, F.A. Quinn, P. Bischof, R. Stricker, Clin. Chem. Lab. Med. **44**(7), 883 (2006)

55. N.P. Groome, P.J. Illingworth, M. O'Brien, R. Pai, F. Rodger, J. Mather, A. McNeilly, J. Clin. Endocrinol. Metab. **81**(4), 1401 (1996)
56. J.F. Selgrade, L.A. Harris, R.D. Pasteur, J. Theor. Biol. **260**(4), 572 (2009)
57. K.S. Ruth, R.N. Beaumont, J. Tyrrell, S.E. Jones, M.A. Tuke, H. Yaghootkar, A.R. Wood, R.M. Freathy, M.N. Weedon, T.M. Frayling, et al., Hum. Reprod. **31**(2), 473 (2016). https://doi.org/10.1093/humrep/dev318
58. W. Odell, *The Reproductive System in Women* (Grune & Stratton, New York, 1979), pp. 1383–1400
59. S.A. Beall, A. DeCherney, Fertil. Steril. **97**(4), 795 (2012)
60. S. Deb, B. Campbell, C. Pincott-Allen, J. Clewes, G. Cumberpatch, N. Raine-Fenning, Ultrasound Obstetr. Gynecol. **39**(5), 574 (2012)
61. S.E. Monroe, M.R. Henzl, M.C. Martin, E. Schriock, V. Lewis, C. Nerenberg, R.B. Jaffe, Fertil. Steril. **43**(8), 361 (1985)
62. S.E. Monroe, Z. Blumenfeld, J.L. Andreyko, E. Schriock, M.R. Henzl, R.B. Jaffe, J. Clin. Endocrinol. Metab. **63**(6), 1334 (1986)
63. I.J.M. Duijkers, C. Klipping, W.N.P. Willemsen, D. Krone, E. Schneider, G. Niebch, R. Hermann, Hum. Reprod. **13**(9), 2392 (1998)
64. I. Leroy, M.F. d'Acremont, S. Brailly-Tabard, R. Frydman, J. deMouzon, P. Bouchard, Fertil. Steril. **62**(3), 461 (1994)

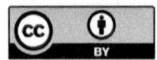

Studying the Effects of Oral Contraceptives on Coagulation Using a Mathematical Modeling Approach

Amy Kent, Karin Leiderman, Anna C. Nelson, Suzanne S. Sindi, Melissa M. Stadt, Lingyun (Ivy) Xiong, and Ying Zhang

1 Introduction

Exogenous hormones are used by hundreds of millions of people worldwide for contraceptives and hormonal replacement therapy. Hormonal contraceptives contain either exclusively progestin—a synthetic progesterone—or a combination of progestin and estrogen in the form of ethinyl estradiol. Combined oral contraceptives (OCs) are classified by the type of progestin and the level of estrogen dose used in the formulation, where the action of progestin prevents ovulation by suppressing luteinizing hormone and estrogen prevents breakthrough bleeding [1]. Progestins used in OCs are grouped by "generations" that correspond to when they

A. Kent
Mathematical Institute, University of Oxford, Oxfordshire, UK

K. Leiderman (✉)
Mathematics Department, Computational Medicine Program, University of North Carolina at Chapel Hill, Chapel Hill, NC, USA
e-mail: karin.leiderman@unc.edu

A. C. Nelson
Department of Mathematics, Duke University, Durham, NC, USA

S. S. Sindi
Department of Applied Mathematics, University of California Merced, Merced, CA, USA

M. M. Stadt
Department of Applied Mathematics, University of Waterloo, Waterloo, ON, Canada

L. (I.) Xiong
Department of Qualitative and Computational Biology, University of Southern California, Los Angeles, CA, USA

Y. Zhang
Department of Mathematics, Brandeis University, Waltham, MA, USA

© The Author(s) 2024
A. N. Ford Versypt et al. (eds.), *Mathematical Modeling for Women's Health*,
The IMA Volumes in Mathematics and its Applications 166,
https://doi.org/10.1007/978-3-031-58516-6_4

first appeared in the formulation. For example, second-generation progestins were used in the 1970s and include levonorgestrel and norgestrel, and third-generation progestins introduced in the 1990s include gestodene, norgestimate, and desogestrel [1, 2].

The use of combined, or combination, oral contraceptives (OCs) and hormone replacement therapies is known to increase the risk of both arterial and venous thrombosis (pathological blood clot formation) [3–8]. While the estrogen component of combination OCs is known to be prothrombotic [4, 5], the progestin formulation has also been shown to affect clotting propensity [9, 10]. Indeed, studies suggest that for a fixed estrogenic dosage, patients on a third-generation OC containing desogestrel and gestoden have a higher risk for venous thrombosis than patients on second-generation OCs that use progestins, such as levonorgestrel and norethisterone [11]. However, the mode of delivery of the OC does not affect the risk for thrombosis, as transdermal and transvaginal forms of contraception also show an increased risk of thrombosis [12, 13]. Exogenous hormones from combined OCs can modulate components of the procoagulant, anticoagulant, and fibrinolytic components of blood coagulation [14–17]. One example is the changes in plasma levels of clotting factors when using OCs [17]. These alterations may elicit a prothrombotic state that is dependent on dose of estrogen and hormonal dose combination [18, 19]. Individuals with deficiencies in endogenous anticoagulant proteins are also more susceptible to thrombosis when taking combination OCs [10, 20, 21]. How various modulations to the clotting system mechanistically contribute to an increased thrombosis risk is not fully understood.

Clotting factors that are modulated while on OCs are components of the blood coagulation network, which is responsible for the generation of the important clotting enzyme thrombin. Blood coagulation involves inhibitors, both positive and negative feedback loops, and must exhibit a robust clotting response given a wide variety of factor levels. Due to these complexities, mathematical modeling can be used to better understand how modulations like exogenous hormones can affect this process. In this study, we are particularly interested in modeling the clotting factors involved in blood coagulation. Our mathematical model uses factor concentration as an input and outputs thrombin concentration over time. For a review of the variety of mathematical models used to describe various components of the blood clotting process, see [22].

One measure that is known to correlate with thrombosis risk is the resistance of the clotting system to the inhibitory effects of activated protein C (APC) [9, 23, 24]. APC is an anticoagulant protein generated during coagulation that serves as a brake on the clotting system to prevent over clotting and spreading of clotting to areas beyond an injury. A clotting system that is more resistant to the effects of APC could therefore be more prothrombotic. The use of OCs is associated with an increased APC resistance [16, 25–27], with patients on third-generation (desogestrel-based) OCs having a more pronounced APC resistance than patients on second-generation (levonorgestrel-based) OCs [16]. Taken together, the use of OCs is thus associated with an increased risk of thrombosis, and the APC resistance (or sensitivity) is one metric to predict this risk.

There are different ways to test for APC resistance, but the most common ways are by comparing an activated partial thromboplastin time (APTT) or an endogenous thrombin potential (ETP) with and without APC [27]. The complete details of these assays are beyond the scope of this chapter, but essentially they test the timing and strength of a clotting response. de Visser et al. [27] performed both kinds of APC resistance tests on hundreds of patients, some of whom were on OCs and some of whom were not. Their study suggested that, in general, clotting factor VIII (FVIII) and clotting factor II (FII, also known as prothrombin), to a lesser extent, are primary determinants of the outcomes in APTT-based tests and that the clotting inhibitor tissue factor pathway inhibitor (TFPI) and protein S are primary determinants in ETP-based tests. Additionally, they suggested that the use of OCs only moderately affected the APTT-based test but strongly affected the ETP-based test and that in the latter case the effects may not be due to clotting factor levels alone. The comparisons were performed on patients on and off OCs, but they were not the same individuals. Furthermore, the correlations were computed using single clotting factors; thus, they would not be able to capture simultaneous contributions from multiple factors.

Midderdorp et al. [17] studied the effects of OCs on clotting factor levels in an elegant cross-over study that reported levels of six clotting factors in 28 patients off OCs and on levogestrel (lev) and desogestrel. This study provided information about how factor levels are *changed* by the OCs, which enables further study regarding the link between OCs, factor levels, and thrombosis risk. However, what was not reported in the paper was the individual patient changes, rather just mean and standard deviation of the study cohorts.

In the current study, we used the data from the Middeldorp et al. [17] study, together with a mathematical model of coagulation, to investigate how factor level changes from OCs affect production and timing of the coagulation enzyme thrombin in addition to an APC sensitivity metric. Because the individual factor level changes were not reported in the Middeldorp et al. [17] study, we used the reported means and standard deviations to generate a large virtual patient population (VPP). We then simulated the effects of lev by adjusting the factor levels by the mean effect of lev reported in [17]. We analyzed the concentration and timing of thrombin generation among the entire VPP after the use of lev and reported the characteristics of patients that had large and small changes in outputs. We computed APC sensitivity ratios and showed that the use of lev, by way of factor level changes alone, increased the systems' sensitivity to APC. Our results suggest that factor changes induced by lev are enough to explain both a change in APC sensitivity and an increased prothrombotic profile.

2   Methods

2.1   Brief Review of Mathematical Model of Flow-Mediated Coagulation

Here we give a brief review of a mathematical model of flow-mediated coagulation [28–32] on which we build for the current study. A more detailed description of the model can be found in the supplementary information at the end of this chapter along with the full model equations in Eqs. (4)–(122), parameters in Tables 2, 3, 4, 5, 6, 7, and 8, and a schematic of the flow-mediated coagulation model in Fig. 12. Further details about this model and its sensitivity to parameters can be found elsewhere [33]. Briefly, the model simulates blood coagulation and platelet deposition under flow. Blood coagulation is a network of biochemical reactions that culminate in the production of the enzyme thrombin. Platelet deposition and aggregation is a biophysical process that initially stops leakage of blood from a vessel. Thrombin is generated on the platelet surfaces and then cleaves the soluble protein fibrinogen into fibrin that turns into a gel and stabilizes the platelets. Here, we are assuming a very small injury completely contained in a blood vessel.

The model simulates the coagulation reactions and platelet deposition at a small injury patch with exposed tissue factor, all occurring in flowing blood (Fig. 12). The reactions occur in two main compartments: the reaction zone (RZ) and the endothelial zone (EZ). Represented schematically in Fig. 1a, the RZ compartment models the region above an injury site, and the EZ compartment models the surrounding region, introduced to account for the effects of flow-mediated transport. Each compartment is assumed to be well-mixed; thus, the time evolution of the concentration of all species is modeled using ordinary differential equations. Different variables are introduced to account for the platelet-bound, membrane-bound, and free concentrations of the relevant enzymes and zymogens (enzyme precursors) within each compartment. The height of the RZ is given by the length scale where diffusive and advective transport are comparable, and the width is taken to be the characteristic size of an intravascular injury, i.e., 10 microns [28]. The EZ height is taken to be the same as the RZ, and the width is dependent on the flow shear rate and protein diffusion coefficients [29]. The coagulation reactions occur in the RZ, where tissue factor (TF) in the subendothelium (SE) is exposed, as depicted in Fig. 12b. The clotting factors, denoted by Roman numerals, and platelets are transported into and out of the RZ. This is represented in the model by a simplified combination of flow and diffusion in the form of a mass transfer coefficient, which characterizes the flow-mediated transport between the two compartments. Clotting factor concentrations in the RZ change due to their involvement in reactions and by transport in and out of the zone. Platelet concentrations are treated similarly. But as platelets build up in the RZ, they are assumed to cover and hinder the enzymatic activity on the subendothelium, and they alter the height and volume of the RZ. The EZ is located adjacent to the RZ, in the direction perpendicular to the flow with height equal to that of the RZ. In the EZ, thrombin that has diffused from the RZ can bind to thrombomodulin (TM) and then protein C in the EZ and activate protein C into APC. This APC either diffuses back into the RZ or is carried away by the flow. APC in the RZ can bind to and inactivate FVa and FVIIIa, which may slow thrombin generation. The inhibitory effect depends on how much FVa and FVIIIa are already in a complex (FVa binds to FXa, and FVIIIa binds FIXa on platelet surfaces) because once bound, they are protected from APC.

2.2 Model Extension: Thrombomodulin and APC Generation in the Reaction Zone

TM is located on endothelial cell (EC) membranes and is generally not embedded in the SE, hence the development of separate compartments in our previous models. Here, we have assumed that portions of ECs protrude into the RZ, thereby providing some mixing of the two zones and their contents. Under this assumption, thrombin in the RZ could readily access and bind to the TM in the RZ, creating active complexes to generate APC *within* the RZ. Compared to our previous models, where APC entered the RZ compartment only by diffusion [29, 31], this assumption should lead to more APC that is available to inactivate FVa and FVIIIa in the RZ. We also assume that platelets cannot cover and inhibit the activity on these protruding EC surfaces, i.e., APC generation occurs even after platelets have covered the subendothelium. Details of the EZ and RZ in models, old and new, are shown schematically in Fig. 1. In the previous model, APC diffusion into the RZ was modeled as $k_{\text{flow}}^c(c_{\text{out}}-c)$, where c is the concentration in the RZ, c_{out} is the upstream concentration, and k_{flow}^c is the rate constant for flow-mediated transport, related to the diffusion coefficient, lumen velocity, and injury size as given in [29].

In this chapter, we allow for protrusion of EC debris into the injury zone. EC protrusion into the RZ compartment allows for direct generation of APC in the RZ *via* TM, in addition to the flow-mediated transport considered previously. To account for the protrusion of EC into the injury zone within the model, we introduce TM and its associated complexes into the RZ compartment. Three new species were added to the model: TM in the RZ TM_{rz}, TM in the RZ that is in complex with thrombin TM_{rz}:E$_2$, and the complex of TM, thrombin, and protein C TM_{rz}:E$_2$:PC. The reactions involved are

$$TM_{rz} + E_2 \underset{k^-}{\overset{k^+}{\rightleftharpoons}} TM_{rz} : E_2, \tag{1}$$

$$TM_{rz} : E_2 + PC \underset{k_{pc}^-}{\overset{k_{pc}^+}{\rightleftharpoons}} TM_{rz} : E_2 : PC \overset{k_{pc}^{cat}}{\longrightarrow} TM_{rz} : E_2 + APC, \tag{2}$$

where Reaction (1) is the binding of thrombin to TM in the RZ and Reaction (2) is the binding of thrombin-bound TM to protein C and the subsequent enzymatic

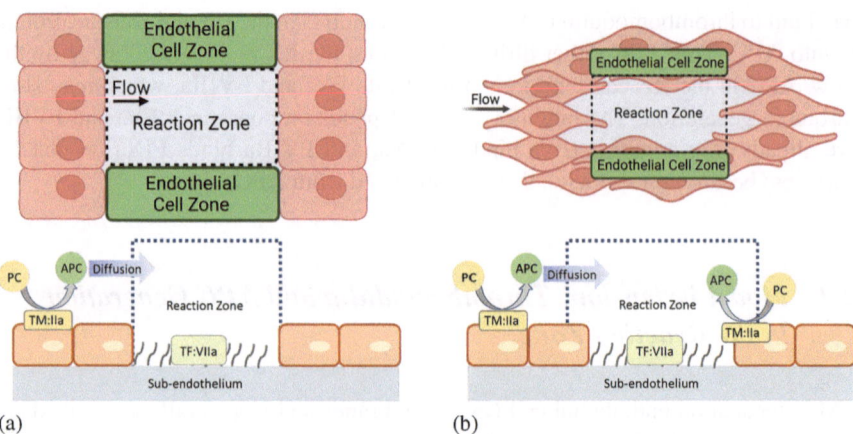

Fig. 1 Model schematics: (**a**) the top and side view of the reaction zone (RZ) and endothelial zone (EZ) in our previous models and (**b**) the updated zones in our current study. Our previous model had distinct RZ and EZ zones that relied on thrombin from the RZ to diffuse to the EZ to make APC, and then the APC had to diffuse back into the RZ to have an inhibitory effect. In the new model, due to protruding ECs into the RZ, thrombin and APC can be generated together in the RZ

cleavage into activated protein C (APC). The kinetic rates in these reactions were taken from a study using the previous mathematical model [29]. The model extension impacts on the model equations are highlighted as bold underlined terms in Eqs. (23), (40), (121), and (122).

2.3 Virtual Patient Population Generation

Based on the data presented in [17], we generated a large virtual patient population. We first generated the smooth kernel density estimates for the population distribution of clotting factors II, V, VII, VIII, and X for patients not on OCs from reported data from [17]. Kernel density functions allowed us to estimate an underlying probability distribution from a sample (see [34] for details). We used the MATLAB function ksdensity to create our kernel density estimates independently for each factor level.

We then created 10,000 virtual patients by randomly and independently sampling factor levels from our kernel density estimates. See Fig. 2 for factor level distributions and kernel density estimates. To determine the factor levels for each virtual patient under lev treatment, we added the mean change in each individual coagulation factor level as reported in [17] (see Table 1). Because virtual patients should represent normal, healthy individuals, we then removed any virtual patients that did not exhibit factor levels within normal physiological range. Specifically, we

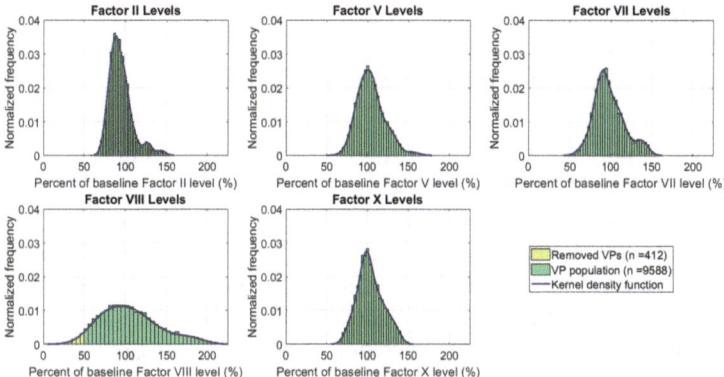

Fig. 2 Factor levels for virtual patients (VPs) were sampled from kernel density estimates computed from data in [17]. Distributions of coagulation factor levels for the virtual patient population before taking OCs. The blue curves indicate the smooth kernel density estimates. Virtual patients with factor VIII level less than 50% were removed ($n = 412$) and are highlighted in yellow. For factor levels while on levonorgestrel, factor levels for each virtual patient were changed by the fixed amount given in Table 1

Table 1 Percent change in factor levels applied to all virtual patients post-exposure to levonorgestrel (lev). These values are the mean percent changes of the patient data reported in [17]

Coagulation factors	Increase after lev (%)
Factor II	12
Factor V	−3
Factor VII	12
Factor VIII	6
Factor X	22

removed the 412 virtual patient samples that had factor VIII levels below 50% (i.e., out of normal physiological range).

2.4 Model Workflow

Each virtual patient had a set of unique factor levels, sampled from the kernel density estimates independently. These factor levels were then used as input (initial conditions) to our mathematical model, which consists of a system of ordinary differential equations that track how each model variable changes in time, under flow. See the supplementary information for more model details, equations, and parameters. The equations of the model were solved numerically to predict concentrations of each species through time. We mainly analyzed thrombin concentrations for this study, but all species concentrations are available for further mechanistic studies.

3   Results

3.1   *Thrombomodulin in the Reaction Zone*

The addition of thrombomodulin (TM) into the RZ is described in Sect. 2.2. Briefly, TM was added to the RZ to enable APC generation by thrombin within the RZ. This feature was added to the model to enhance the sensitivity of the system to APC. Here we study how it alters the clotting dynamics. Figure 3 shows the thrombin concentration after 10 min of activity as a function of the tissue factor density. Tissue factor is the protein embedded in the subendothelium that stimulates the initiation of coagulation and thus thrombin generation. Any single curve in Fig. 3 shows the known threshold dependence of thrombin on tissue factor; the system should have a strong response only when necessary, as clotting is unwarranted without injury. Threshold plots are shown for concentrations of TM in the RZ varying between zero and the concentration assumed in the EZ (500 nM [29]). As the concentration of TM in the RZ increases, an increased tissue factor density is required to attain the same thrombin concentration as we expect, since the APC generation in the RZ has increased. APC inactivates FVa and FVIIIa, which inhibits the formation of two key complexes in the coagulation pathway (Fig. 12): FXa:FVa (prothrombinase), which activates prothrombin to thrombin, and FIXa:FVIIIa (tenase), which activates FX to FXa.

The effects of increased APC generation on clotting dynamics can be explored by considering the time evolution of different factors (Figs. 4 and 5). In Fig. 4, the thrombin lag time (i.e., the time when 1 nM thrombin is generated) increases with increased concentration of TM in the RZ. This is likely due to increased TM in the RZ and the inhibitory effects of the associated increases in APC. Increased APC generation arising from the introduction of TM in the RZ is confirmed in Fig. 4. When no TM is present in the RZ, APC is transported into the RZ from the EZ by diffusion alone, as shown in the model schematic (Fig. 1). Introducing TM into the RZ, thereby allowing APC generation directly within the RZ, increases the amount of total APC generated (Fig. 4).

Fig. 3 Thrombin concentration after 10 min over a range of tissue factor densities that subsequently increase activated protein C (APC) in the reaction zone (RZ). To achieve the same level of thrombin response, more tissue factor is needed as thrombomodulin (TM) is added

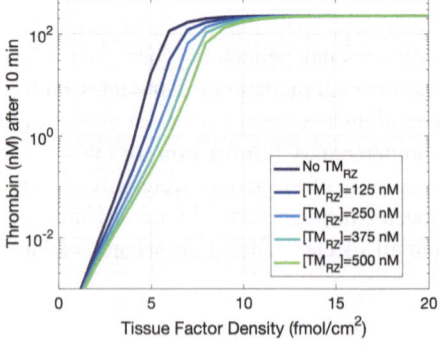

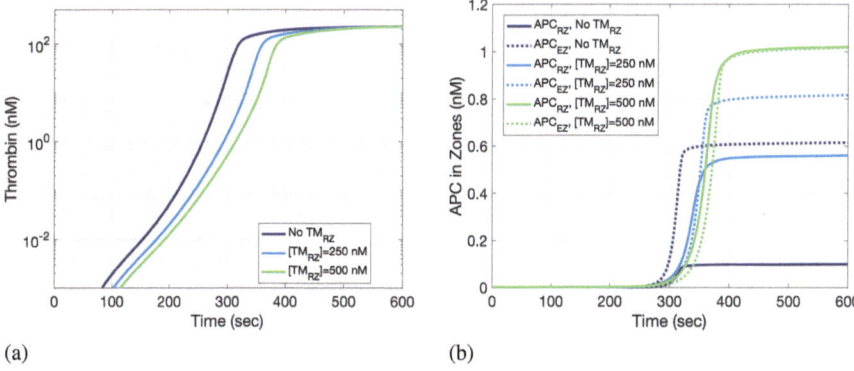

Fig. 4 The effects of thrombomodulin (TM) in the reaction zone (RZ) on (**a**) thrombin and (**b**) activated protein C (APC) concentrations in the RZ and endothelial zone (EZ) over 10 min of clotting activity. TM in the RZ leads to increased thrombin generation and increased APC in the RZ

APC causes a reduction in thrombin generation *via* its inhibitory effects. The specific effect by inactivating factors Va and VIIIa is illustrated in Fig. 5, where the time evolution of total and APC-bound factors Va and VIIIa is shown. Increasing the concentration of TM in the RZ increases the proportion of activated Va and VIIIa.

In summary, the model extension of TM in the RZ led to APC in the RZ that inactivated FVa and FVIIIa, thereby limiting the formation of VIIIa:IXa and Va:Xa complexes, which directly and negatively affected the generation of thrombin. Having established that this extension results in a sensible clotting response, in the next section we use the model to simulate the clotting response of a cohort of virtual patients.

3.2 Predicted Effects of Levonorgestrel on Thrombin Generation

We performed simulations of thrombin generation over 20 min for each of the virtual patients before OC (no OC condition) and after taking lev (lev condition). The mean and 95% confidence intervals of the time series results for select tissue factor (TF) levels are shown in Fig. 6. We can see that at low TF density, i.e., $[TF] = 2$ fmol/cm^2, thrombin generation is minimal, which is in line with the TF threshold behavior of thrombin. This is true for virtual patients on no OCs and on lev. For higher TF levels, thrombin generation increases with increased TF concentrations, again as predicted by the threshold behavior of thrombin on TF.

For TF levels greater than 2 fmol/cm^2, the factor level changes due to lev have a trend that shifts the thrombin curves up and to the left, which means a higher

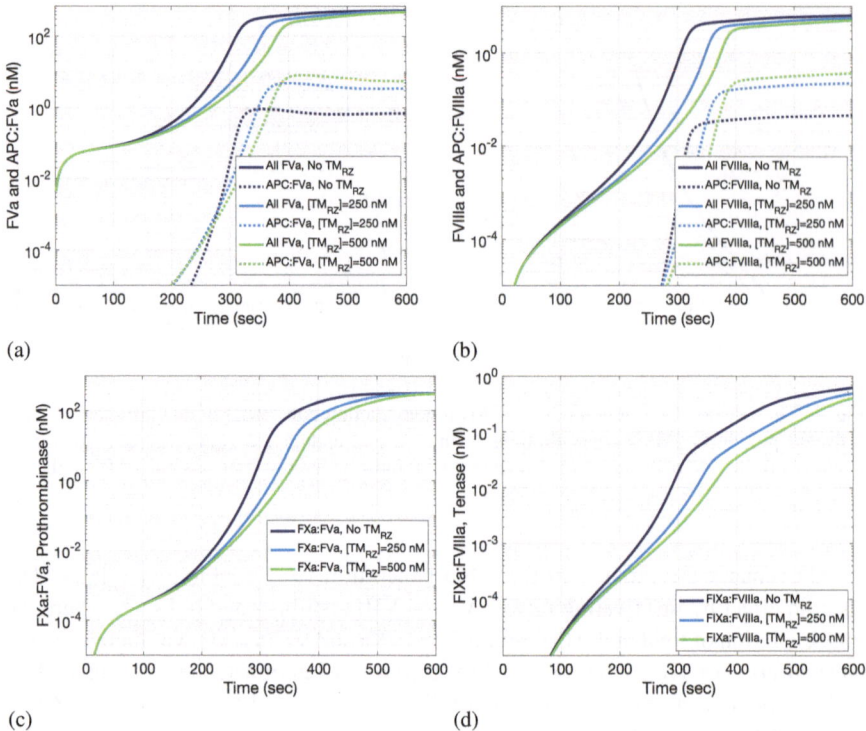

Fig. 5 The effects of activated protein C (APC) generation in the reaction zone (RZ) on (**a**) FVa, (**b**) FVIIIa, (**c**) FXa:FVa, and (**d**) FIXa:FVIIIa over 10 min of clotting activity. APC generation in the RZ leads to inactivation of FVa and FVIIIa and a reduction of prothrombinase and tenase

average thrombin concentration at the end of the simulation as well as a shorter lag time (i.e., the time to reach 1 nM) as compared to the same patients on no OC.

Although average behavior of the population showed a trend of increased thrombin and decreased lag time, we further explored the behavior on an individual patient level. First, we collected all virtual patients whose thrombin levels reached 1 nM within 20 min, for the TF levels of 6, 10, and 14 fmol/cm^2. Next, we examined the changes in lag time and thrombin concentration at 20 min before and after lev usage for each individual virtual patient collected. Figure 7 shows the lag time (top row) and thrombin concentration (bottom row), with these metrics for each individual on lev vs. on no OC. For the lag times, all of the data points lie below the gray dashed line, indicating that all patients had a decreased lag time on lev vs. on no OC. For the thrombin concentration, all of the data points lie above the gray dashed line, indicating that all patients had an increased thrombin concentration on lev vs. on no OC. Furthermore, the largest changes occurred at the lower TF levels. These data suggest that lev induces a heightened thrombotic response.

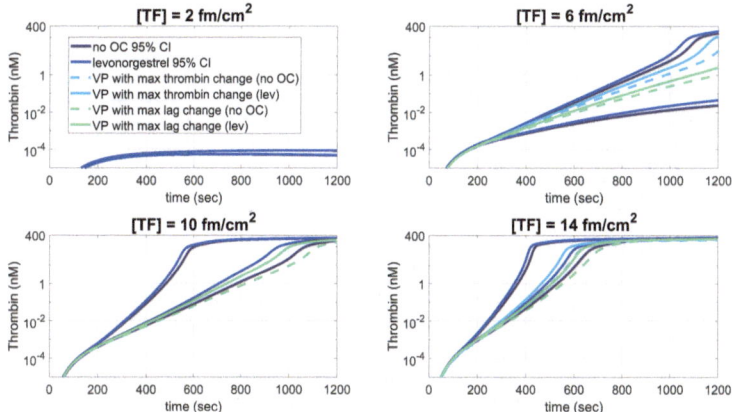

Fig. 6 Simulated thrombin shows an upward and leftward shift on average for patients on lev, with decreased lag times and increased thrombin at 20 min. Simulated thrombin concentration time series results for virtual patients ($n = 9, 588$) with no oral contraceptive (no OC) and levonorgestrel (lev) over 20 min for varied concentrations of tissue factor as given. The 95% confidence intervals (CIs) for the virtual patients on no OC and on lev are shown. The virtual patient individuals with the maximal change in thrombin on no OC and on lev as well as the maximal change in lag time are also plotted. Virtual patient factor level distributions are shown in Fig. 2

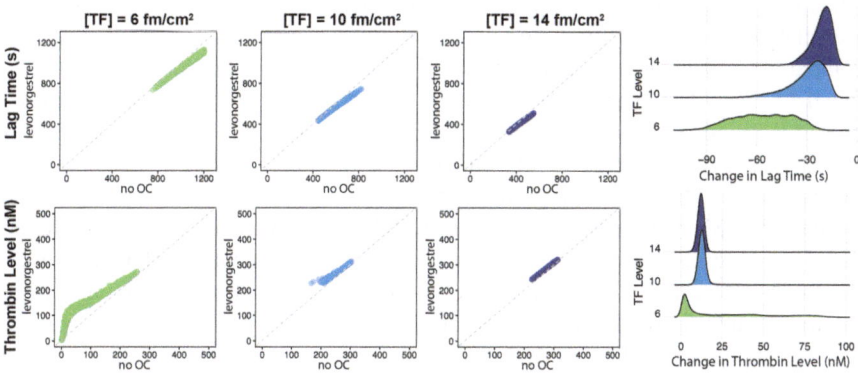

Fig. 7 Use of levonorgestrel (lev) heightens thrombosis response. Scatter plots on the left compare lag time (top) and thrombin concentration after 20 min (bottom) before and after lev usage, for varying concentrations of tissue factor (TF) levels: 6 fmol/cm^2 ($n = 4, 755$), 10 fmol/cm^2 ($n = 9, 588$), and 14 fmol/cm^2 ($n = 9, 588$). Dashed diagonal line indicates matching outcomes before and after lev usage. Density plots on the right show changes in corresponding metrics upon lev usage

3.3 Factor Levels Inducing an Extreme Response

Having established that *all* virtual patients exhibit an increase in thrombin and a decrease in lag time following the use of lev, we now turn to identifying the

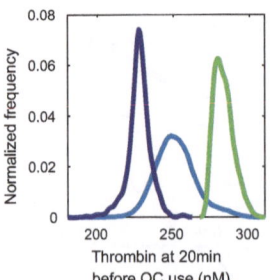 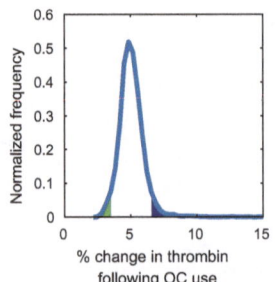 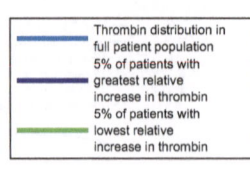

Fig. 8 Patients with lowest thrombin generation before oral contraceptive (OC) use had the largest relative change in thrombin generation after using OCs. Left: distribution in the thrombin concentration at 20 min before OC use for the whole population and subpopulations with the greatest and smallest increases in thrombin generation following OC use. Center: distribution in the percentage change in thrombin following OC use. Tissue factor [TF] $= 10\,\text{fmol/cm}^2$ for these simulations

characteristics of patients with the most extreme changes in their thrombin metrics. For TF $= 10\,\text{fmol/cm}^2$, we considered the distribution of the simulated thrombin concentrations after 20 min on no OCs (the left side of Fig. 8) and the relative increase in thrombin concentration reached after 20 min when on lev compared to when on no OC (the center of Fig. 8). To do this, virtual patients were ordered by the relative increase in thrombin concentration reached after 20 min when on lev compared to when on no OC, discounting 11 patients where the simulated thrombin failed to reach 1 nM thrombin. The light blue curves represent the entire virtual patient population, and the dark blue and green curves represent subpopulations of patients that had the largest 5% and smallest 5% relative increases in thrombin after lev use. We see that the average thrombin concentration for the entire population is near 250 nM, and the mean increase in thrombin generation is 5%. We found that the largest relative changes in thrombin came from patients that had the lowest thrombin concentration prior to lev use. Similarly, the smallest relative changes in thrombin came from patients that had the highest thrombin concentration prior to lev use. This is somewhat intuitive since patients that already have strong thrombin responses prior to OC use are unlikely to have a much stronger increase after OC use. Those patients who had a smaller thrombin response prior to OC use would then likely be able to have larger relative increases in thrombin.

It is interesting to consider the combination of factor levels that characterize the virtual patients that experienced large relative increases in thrombin generation following OC use. We next considered the normalized distributions of factor levels in each subpopulation that exhibited the largest and smallest increases in thrombin concentration (Fig. 9). Relatively low factor VIII levels prior to OC use are observed for patients with the greatest increase in thrombin generation. Factor VIII is associated with increased thrombin generation as when activated, it binds to activated FIX on the platelet surface to form a key complex in the coagulation cascade. Given the greatest increases in thrombin generation were observed for

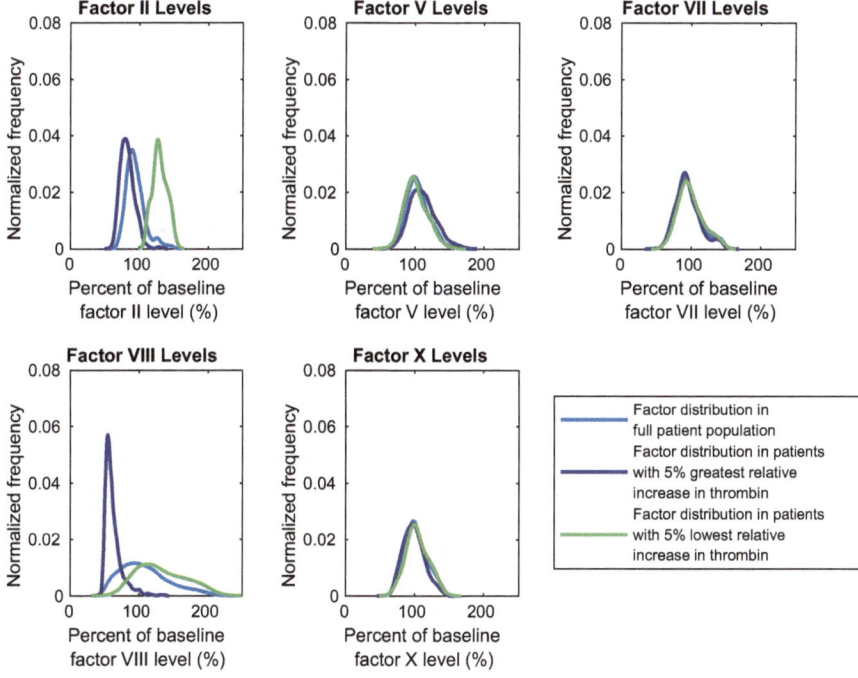

Fig. 9 Patients with lowest levels of FVIII prior to oral contraceptive (OC) use had the largest relative increase in thrombin generation following OC use. Distribution of patient factor levels in the whole population compared with the factor distributions for the 5% of patients exhibiting the largest and smallest percentage increase in thrombin generation following OC use. All tissue factors (TFs) are reported before OC use, and [TF] = 10 fmol/cm^2 for the simulation of the thrombin curves that produced these percentiles

patients with an initially low thrombin response (Fig. 8), reduced FVIII levels may be a key indicator that the virtual patient is at risk of large changes in thrombin generation following OC use.

Conversely, the patients with the smallest increase in thrombin generation following OC use were those with high prothrombin (FII) levels (Fig. 9). Prothrombin is activated by the Va:Xa complex at the platelet surface to form thrombin *in vivo*. Thus, high prothrombin levels before OC use will contribute to a larger initial thrombin concentration, which results in virtual patients with a small relative change in thrombin generation following OC use (Fig. 8).

Distributions in the relative change in lag time following OC use have a similar behavior; virtual patients that undergo the greatest decrease in lag time following OC are those with a lag time at baseline that is longer than average and *vice versa*. The distributions of factor levels for the 5% of virtual patients undergoing the greatest and smallest decrease in lag time are shown in Fig. 10. Low factor VIII levels again signpost the greatest change following OC use, while high factor VIII levels are associated with a small change in lag time. Elevated prothrombin

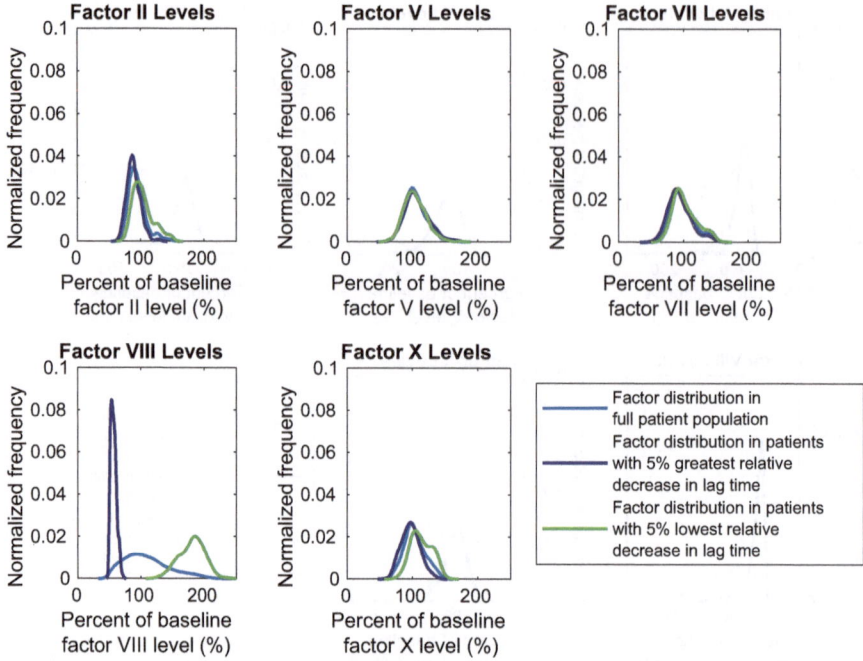

Fig. 10 Patients with lowest levels of FVIII prior to oral contraceptives (OC) use had the largest relative decrease in lag time following OC use. Factor distributions for the 5% of patients exhibiting the greatest and smallest percentage decrease in lag time following OC use. All factors are reported before OC use with tissue factor [TF] = 10 fmol/cm^2

levels are linked to a reduced increase in thrombin generation (Fig. 9); there is less change in thrombin since thrombin is already high with increased prothrombin. Similarly, low prothrombin levels are associated with the greatest relative increase in thrombin; this allows room for greater change in thrombin when thrombin is not as high in the first place. The effect of high prothrombin levels on decreasing changes in lag time is not as pronounced (Fig. 10), and there is no association with low prothrombin and lag time changes. Hence, while a large proportion (80%) of virtual patients are in the subpopulation undergoing the greatest 5% change for both changes in thrombin generation and lag time, factor levels that induce a large change in thrombin generation do not necessitate a large change in lag time. These results provide some insight into the factor levels that bring the greatest increase in thrombin generation and reduction in lag time. Although the TF level is relatively high in this example and there is more variance in these metrics with the lower TF (Fig. 7), the trends in factor levels and relative increases and decreases are similar (not shown). In the future work, a sensitivity analysis could be conducted to systematically identify which combinations of factors are associated with the greatest thrombotic risk.

3.4 APC Sensitivity Metric

To quantify the effect of APC on thrombin generation between OC and non-OC users, we developed a new APC sensitivity ratio. Our ratio is similar to the ETP-based metric [27] described in Sect. 1, because we will use an area under the curve of simulated thrombin. Ours differs from the ETP-based test in that we are not adding exogenous APC. The two cases we compare are a case with TM in the RZ (APC generation in the RZ) and no TM in the RZ (no APC generation in the RZ, so minimal effects of APC). Our metric is defined as

$$\text{APC-sr} = \frac{\int_0^\tau T^{APC+OC}(t)\,dt}{\int_0^\tau T^{OC}(t)\,dt} \bigg/ \frac{\int_0^\tau T^{APC}(t)\,dt}{\int_0^\tau T(t)\,dt}, \tag{3}$$

where τ denotes the termination time for simulating thrombin generation, $T^{APC+OC}(t)$ denotes the thrombin concentration over time for the virtual patient with APC generation in the RZ and taking OC, and $T^{APC}(t)$ denotes the thrombin concentration over time with APC generation in the RZ but without OC. We define $T^{OC}(t)$ to be the thrombin concentration over time without APC generation in the RZ but with OC usage and $T(t)$ to be the thrombin concentration over time without APC generation or OC. The ratio in the numerator of Eq. (3) gives the effect of APC when a virtual patient is on OC, whereas the ratio in the denominator of Eq. (3) represents the effect of APC when a patient is off OC. Taken together, Eq. (3) allows us to explore the inhibitory effect of APC in the presence of OC. Note that in the case where a patient does not use OC and has no APC generation in the RZ, APC-sr $= 1$.

To compute APC-sr for different TF levels, we removed virtual patients that do not reach 1 nM thrombin, so the ratio in Eq. (3) is well-defined. We set τ to be 20 min. Examining the ratios for [TF] $= 6$, 10, and 14 fmol/cm^2 in Fig. 11, we see that the APC sensitivity ratios are increased with OC use since they are always above 1. This means that for all TF levels patients following OC use have a higher APC sensitivity (Fig. 11) than non-OC users and therefore may have an increased risk of thrombosis. In comparing the ratio as the TF level is increased, we see that APC-sr decreases on average (see red dots in Fig. 11). This shows that patients are less sensitive to APC when TF level is high.

4 Discussion

In this study we used a mathematical model of flow-mediated coagulation to study the effects of the OC lev on thrombin generation. To simulate the effects of the OC, we used clotting factor levels and their changes due to OC use, measured in 28 patients as part of a cross-over study [17]. Based on the clotting factor levels for the patients prior to OC use, we generated a large virtual patient population with

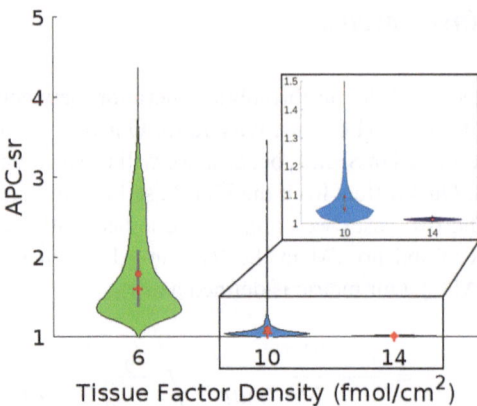

Fig. 11 The activated protein C (APC) sensitivity ratio (APC-sr) is greater than 1 for all TF levels for patients on oral contraceptives. For TF = 6 fmol/cm^2 (n = 4,755), 10 fmol/cm^2 (N = 9,588), and 14 fmol/cm^2 (n = 9,588), the APC-sr is calculated using Eq. (3) with the sample mean represented as a solid red dot and the sample median represented as a red cross. The thick gray bar in the center represents the interquartile range. Patients with simulated thrombin curves that did not reach 1 nM thrombin by 20 min were removed from these calculations

the same mean and standard deviation as the reported data. Next, we represented the effects of the OCs on that virtual patient population by changing the clotting factor levels according to mean changes reported in the real patient data. The clotting factor levels were used as initial conditions for our mathematical model that simulates thrombin generation under flow. After analyzing the outputs of the virtual population before and after OC use, we found that the changes in clotting factor levels due to OC use always increased thrombin generation and decreased the lag time (sped up the process), with these changes being more pronounced at a low to moderate TF level. We concluded from this that the changes in factor levels alone can heighten the prothrombotic state of the clotting system in our model. Additionally, to test the system's sensitivity to APC, we extended our previous mathematical model to include thrombomodulin, and thus APC generation within the reaction zone so that APC was not only confined to generation in the endothelial zone assumed to be distinct and adjacent to the reaction zone. In previous studies, where it was confined to the adjacent endothelial zone, APC has little to no inhibitory effect [29]. As seen in the TF threshold and thrombin plots, our model shows susceptibility to inhibition by APC. With this new model, we were then able to study the APC sensitivity that may occur with OC use. Indeed, we showed that the changes in clotting factor levels alone were enough to increase the APC sensitivity (as shown by an increased APC sensitivity ratio). Previous studies have shown only minor to moderate changes in APC sensitivity ratios due to changes in factor levels [27]. However, the assays in that study used high tissue factor levels, which possibly masked differences in relatively small factor level changes. Our model in this study focused on varied tissue factor levels to allow for more sensitivity in the system.

We have shown here that the effects of lev on our virtual patient population, in the form of clotting factor level changes alone, contribute to a prothrombotic state. However, there are some limitations of this study. We did not allow for variation in the changes with lev beyond the means reported in the data. In the future we plan to develop statistical methods to refine that assumption. Additionally, the cross-over study included data from the same patients on another OC (third generation) that we did not study here. In fact, it has been shown that the use of third-generation versus second-generation OCs is associated with an increased resistance. In the current study, we have created virtual patient populations based on the patients' distributions of factor levels in the data, and then we changed the levels of individual patients by the same value (the mean change reported). The third-generation OC leads to a further increase in FII and FVII and a further decrease in FV. Thus, based on our results with lev, we speculate that changing the levels of the virtual patients, simply by the mean of the desogestrel data [17], should give us a similar, albeit slightly enhanced, result in terms of APC sensitivity. Investigating the effects of third-generation OCs using more sophisticated techniques to sample the data is a primary focus of our immediate future work.

Supplementary Information

The model includes the coagulation reactions shown in Fig. 12a. The reactions involve many coagulation proteins: inactive enzyme precursors (zymogens), active enzymes, and inactive and active cofactors. Active cofactors are not enzymes themselves but act to make the enzymes to which they are bound more effective than if they would be alone. In Fig. 12a, the zymogens are FVII, FIX, FX, FXI, and FII (prothrombin), which have respective active enzymes FVIIa, FIXa, FXa, FXIa, and FIIa (thrombin). The inactive/active cofactor pairs are FV/FVh/FVa and FVIII/FVIIIa. It is also shown that many of the coagulation reactions occur only on a cellular surface, some on the subendothelium (SE), some on the endothelium (EC), and others on an activated platelet's surface (APS). There are three critical surface-bound enzyme-cofactor complexes: TF:FVIIa on the SE ("extrinsic tenase"), plt-FVIIIa:FIXa ("intrinsic tenase," which we refer to simply as tenase), and plt-FVa:FXa ("prothrombinase") on an APS. Their substrates (i.e., the proteins that the enzyme complexes activate) must also be bound to the cellular surface to become activated [35].

The mathematical model simulates the clotting response due to a small injury to a vessel wall. The response is monitored in a reaction zone (RZ) above a region where tissue factor (TF) in the SE is exposed to flowing blood (Fig. 12b). Within the RZ, coagulation protein concentrations are assumed to change due to transport into and out of the RZ and due to their involvement in the coagulation reactions depicted in Fig. 12a. Similarly, platelet concentrations change as platelets adhere to the injured wall, become activated, and are transported into and out of the RZ. The height of the RZ and the rate of platelet and protein transport into and out of the RZ depend on the

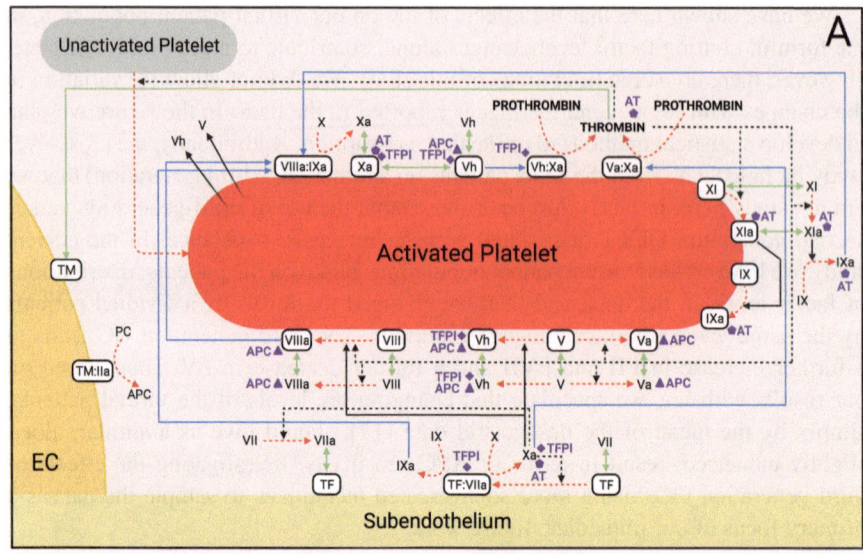

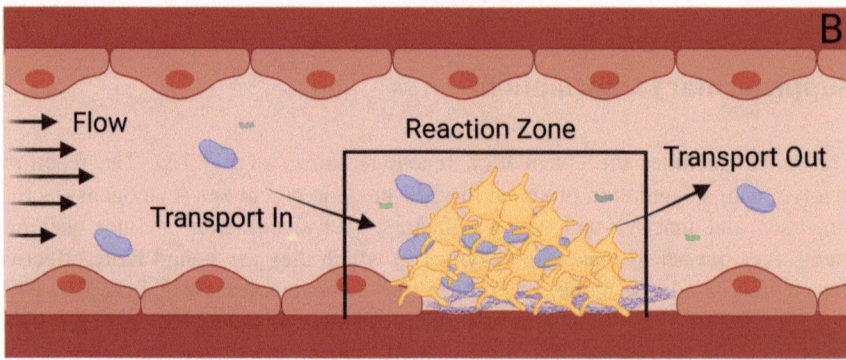

Fig. 12 Schematic of flow-mediated coagulation model. (**a**) Schematic of coagulation reactions included in the model. Dashed red arrows show cellular or chemical activation processes. Blue arrows show chemical transport in the fluid or on a surface. Green arrows depict binding and unbinding from cell surfaces. White boxes denote surface-bound species. Solid black lines show enzyme action in a forward direction, while dashed black lines show feedback action of enzymes. Black lines with a fade indicate release from the platelet. Purple shapes show inhibitors. (**b**) Schematic of the reaction zone. Notation: The lowercase letter "a" on any species means that it is in an "activated" form, e.g., FX and FXa are clotting factor X and activated clotting factor FX. EC: endothelial cell. See the text for other species definitions

flow's shear rate and on the species' diffusivities. Each species in the RZ is assumed to be uniformly distributed (well-mixed) and is described by its concentration, whose dynamics are tracked through an ordinary differential equation. Adjacent to the RZ, in the direction perpendicular to the flow, is an endothelial zone with height equal to that of the RZ and width dependent on the flow shear rate and protein diffusion coefficients [36]. Each species in the endothelial zone is also assumed to be well-mixed. Endothelial cells also protrude into the RZ, and any reaction in the endothelial zone can also occur in the RZ.

Platelets are either (i) unactivated, unattached, and so free to move with the fluid or (ii) activated, bound to the SE or to other activated platelets (APs), and therefore stationary. Platelet activation occurs by contact with the SE, by exposure to thrombin, or by contact with other APs. The last of these is used as a surrogate for activation by platelet-released ADP, which we do not explicitly track in this model. Activation results in the release of platelet-derived FV with no FVa functionality or resistance to APC. Additionally, activation upregulates binding sites for coagulation proteins involved in surface-bound reactions. We characterize each coagulation protein not only by its chemical identity but also by whether it is in the fluid, bound to the SE or bound to an APS. Proteins bound to a surface are stationary, whereas proteins in the plasma move with the fluid. During a transition from SE to APS, or *vice versa*, a protein is subjected to flow and thus might be carried downstream.

An in-house FORTRAN program is used to set up the system of differential equations, set parameter values, and run the simulation. It uses the software package DLSODE [37] to solve the differential equations. Simulation sampling was carried out via a Python wrapper of the FORTRAN program. Graphical processing of simulation results was performed with MATLAB.

For each simulation, we specify the initial plasma concentrations of platelet and protein species, the rate constants for all reactions, the numbers of specific binding sites for coagulation factors on each APS, the dimensions of the injury, the flow velocity near the injured wall, the diffusion coefficients for all fluid-phase species, and the density of exposed TF. The outputs of the simulation are the concentration of every protein species in the RZ at each instant of time from initiation of the injury until the completion of the simulation and the concentrations of platelets attached either directly to the SE or to other platelets.

We have listed the full model equations for all species in Eqs. (4)–(122). Critical parameters are listed in Tables 2, 3, 4, 5, 6, 7, and 8. The model detailed includes extensions of our previous work [28–32]. New terms are in bold and underlined in Eqs. (23), (40), (121), and (122). The model consists of 119 species (and their corresponding ordinary differential equations) and 239 parameters including kinetic rates and initial/upstream concentrations. The solution of the model equations was carried out with our in-house FORTRAN code that uses DLSODE for the numerical solution of the differential equations; each run of the model that simulates 40 min of clotting activity takes less than 10 s on a Linux-based laptop. Simulations of this model (in the absence of heparin) can be performed with our online coagulation simulator ClotSims available at https://clotsims.app.

$$\frac{d}{dt} z_7 = -k_7^{on} z_7 [TF]^{avail} + k_7^{off} z_7^m - k_{z7:e_2}^+ z_7 e_2 \tag{4}$$

$$+ k_{z7:e_2}^- [Z_7 : E_2] - k_{z7:e_{10}}^+ z_7 e_{10} + k_{z7:e_{10}}^- [Z_7 : E_{10}]$$

$$+ k_{flow} (z_7^{up} - z_7) - k_{z7:e_9}^+ z_7 e_9 + k_{z7:e_9}^- [Z_7 : E_9]$$

Table 2 Normal concentrations and surface binding site numbers

Species	Values and units	Notes
Prothrombin	$1.4\,\mu$M	a
Factor V	$0.01\,\mu$M	b
Factor VII	$0.01\,\mu$M	a
Factor VIIa	$0.1\,$nM	c
Factor VIII	$1.0\,$nM	a
Factor IX	$0.09\,\mu$M	a
Factor X	$0.17\,\mu$M	a
Factor XI	$30.0\,$nM	a
TFPI	$0.5\,$nM	d
Protein C	$65\,$nM	e
Platelet count	$2.5 \times 10^5\,\mu$l	e
N_2	1000/plt	f
N_2^*	1000/plt	f
N_5	3000/plt	g
N_8	450/plt	h
N_9	250/plt	i
N_9^*	250/plt	i
N_{10}	2700/plt	j
N_{11}	1500/plt	k
N_{11}^*	250/plt	k
n_5	3000/plt	l
p_{PLAS}	$0.167\,$nM	m
AT	$2.4\,$nM	n

(a) From [38]. (b) From [39]. (c) [40] suggests that normal plasma concentration of fVIIa is about 1% of the normal fVII concentration. (d) From [41]. (e) From [42]. (f) Estimated as described in the text of the supplementary information. (g) From [43]. (h) From [44]. (i) From [45]. (j) From [46]. (k) From [47, 48]. (l) Number of fV molecules released per activated platelet [49]. (m) Maximum concentration of platelets in a $2\,\mu$m high reaction zone assuming that 20 platelets can cover a $10\,\mu$m $\times$ $10\,\mu$m injured surface [50]. (n) From [51]

$$\frac{d}{dt}e_7 = -k_7^{on} e_7 [TF]^{avail} + k_7^{off} e_7^m + k_{z7:e2}^{cat}[Z_7 : E_2] \tag{5}$$

$$+ k_{z7:e10}^{cat}[Z_7 : E_{10}] + k_{flow}(e_7^{up} - e_7) + k_{z7:e9}^{cat}[Z_7 : E_9]$$

$$\frac{d}{dt}z_{10} = -k_{10}^{on} z_{10} p_{10}^{avail} + k_{10}^{off} z_{10}^m - k_{z10:e_7^m}^{+} e_7^m \tag{6}$$

$$+ k_{z10:e_7^m}^{-}[Z_{10} : E_7^m] + k_{flow}(z_{10}^{up} - z_{10})$$

Table 3 Binding to platelet surfaces

Reaction	Reactants	Products	$(M^{-1}sec^{-1})$	(sec^{-1})	Note
Factor IX	Z_9, P_9	Z_9^m	$k_9^{on} = 1.0 \times 10^7$	$k_9^{off} = 2.5 \times 10^{-2}$	a
Factor IXa	E_9, P_9	E_9^m	$k_9^{on} = 1.0 \times 10^7$	$k_9^{off} = 2.5 \times 10^{-2}$	a
Factor IXa	E_9, P_9^*	$E_9^{m,*}$	$k_9^{on} = 1.0 \times 10^7$	$k_9^{off} = 2.5 \times 10^{-2}$	b
Factor X	Z_{10}, P_{10}	Z_{10}^m	$k_{10}^{on} = 1.0 \times 10^7$	$k_{10}^{off} = 2.5 \times 10^{-2}$	a
Factor Xa	E_{10}, P_{10}	E_{10}^m	$k_{10}^{on} = 1.0 \times 10^7$	$k_{10}^{off} = 2.5 \times 10^{-2}$	a
Factor V	Z_5, P_5	Z_5^m	$k_5^{on} = 5.7 \times 10^7$	$k_5^{off} = 0.17$	c
Factor Vh	E_5^h, P_5	E_5^{hm}	$k_5^{on} = 5.7 \times 10^7$	$k_5^{off} = 0.17$	c
Factor Va	E_5, P_5	E_5^m	$k_5^{on} = 5.7 \times 10^7$	$k_5^{off} = 0.17$	c
Factor VIII	Z_8, P_8	Z_8^m	$k_8^{on} = 5.0 \times 10^7$	$k_8^{off} = 0.17$	d
Factor VIIIa	E_8, P_8	E_8^m	$k_8^{on} = 5.0 \times 10^7$	$k_8^{off} = 0.17$	d
Factor II	Z_2, P_2	Z_2^m	$k_2^{on} = 1.0 \times 10^7$	$k_2^{off} = 5.9$	e
Factor IIa	E_2, P_2	E_2^m	$k_2^{*,on} = 1.0 \times 10^7$	$k_2^{*,off} = 0.2$	f
Factor XI	Z_{11}, P_{11}	Z_{11}^m	$k_{z_{11}}^{on} = 1.0 \times 10^7$	$k_{z_{11}}^{off} = 0.1$	g
Factor XIa	E_{11}, P_{11}^*	E_{11}^m	$k_{e_{11}}^{on} = 1.0 \times 10^7$	$k_{e_{11}}^{off} = 0.017$	h

(a) For fIX binding to platelets, $K_d = 2.5 \times 10^{-9}$ M [45], and for fX binding to platelets, K_d has approximately the same value [43]. For fX binding to PCPS vesicles, the on-rate is about 10^7 $M^{-1}sec^{-1}$, and the off-rate is about 1.0 sec^{-1} [52] giving a dissociation constant of about 10^{-7} M. To estimate on- and off-rates for the higher affinity binding of fX to platelets, we keep the on-rate the same as for vesicles and adjust the off-rate to give the correct dissociation constant. The rates for fIX binding with platelets are taken to be the same as for fX binding. (b) We assume binding constants for fIXa binding to the specific fIXa binding sites are the same as for shared sites. (c) fV binds with high affinity to phospholipids (PCPS) [52], and we use the same rate constants reported there to describe fV binding to platelets. (d) The K_d for fVIII binding with platelets is taken from [44]. We set the off-rate k_8^{off} for fVIII binding to platelets equal to that for fV binding to platelets and calculate the on-rate k_8^{on}. (e) For prothrombin interactions with platelets, K_d is reported to be 5.9×10^{-7} M [53]. We choose k_2^{off} and set $k_2^{on} = k_2^{off}/K_d$. (f) Estimated as described in the text of the supplementary information. (g) $K_d = 10\,nM$ [54]. (h) $K_d = 1.7\,nM$ [48]

$$\frac{d}{dt}e_{10} = -k_{10}^{on}e_{10}p_{10}^{avail} + k_{10}^{off}e_{10}^m + k_{z_{10}:e_7^m}^{cat}[Z_{10} : E_7^m] \tag{7}$$

$$+ (k_{z_7:e_{10}}^{cat} + k_{z_7:e_{10}}^-)[Z_7 : E_{10}] - k_{z_7:e_{10}}^+ e_{10}z_7$$

$$+ (k_{z_7^m:e_{10}}^{cat} + k_{z_7^m:e_{10}}^-)[Z_7^m : E_{10}] - k_{z_7^m:e_{10}}^+ e_{10}z_7^m$$

$$- k_{TFPI:e_{10}}^+ e_{10}[TFPI] + k_{TFPI:e_{10}^m}^-[TFPI : E_{10}]$$

Table 4 Reactions on subendothelium. Notation: activation (of -, by -); binding (of -, with -)

Reaction	Reactants	Complex	Product	(M^{-1}sec^{-1})	(sec^{-1})	(sec^{-1})	Notes
Activation							
(TF:VII,fXa)	E_{10}, Z_7^m	$Z_7^m : E_{10}$	E_7^m	$k_{z_7^m:e10}^+ = 5.0 \times 10^6$	$k_{z_7^m:e10}^- = 1.0$	$k_{z_7^m:e10}^{cat} = 5.0$	a
(TF:VII, fIIa)	E_2, Z_7^m	$Z_7^m : E_2$	E_7^m	$k_{z_7^m:e_2}^+ = 3.92 \times 10^5$	$k_{z_7^m:e_2}^- = 1.0$	$k_{z_7^m:e_2}^{cat} = 6.1 \times 10^{-2}$	b
(fX, TF:VIIa)	E_7^m, Z_{10}	$Z_{10} : E_7^m$	E_{10}	$k_{z10:e_7^m}^+ = 5.0 \times 10^6$	$k_{z10:e_7^m}^- = 1.0$	$k_{z10:e_7^m}^{cat} = 1.15$	c
(fIX, TF:VIIa)	E_7^m, Z_9	$Z_9 : E_7^m$	E_9	$k_{z9:e_7^m}^+ = 9.4 \times 10^6$	$k_{z9:e_7^m}^- = 1.0$	$k_{z9:e_7^m}^{cat} = 1.15$	d
Binding							
(fVII, TF)	Z_7, TF		Z_7^m	$k_7^{on} = 5.0 \times 10^7$	$k_7^{off} = 5.0 \times 10^{-3}$		e
(fVIIa, TF)	E_7, TF		E_7^m	$k_7^{on} = 5.0 \times 10^7$	$k_7^{off} = 5.0 \times 10^{-3}$		e

(a) $K_M = 1.2 \times 10^{-6}$ M [55]. (b) $K_M = 2.7 \times 10^{-6}$ M [55]. (c) $K_M = 4.5 \times 10^{-7}$ M [38]. (d) $K_M = 2.4 \times 10^{-7}$ M [56]. (e) $K_d = 1.0 \times 10^{-10}$ M [57]

Table 5 Reactions in the plasma. All of the reactions are activation reactions. Notation: activation (of -, by -)

Reaction	Reactants	Complex	Product	$(\mathrm{M}^{-1}\mathrm{sec}^{-1})$	(sec^{-1})	(sec^{-1})	Note
(fVII, fXa)	Z_7, E_{10}	$Z_7 : E_{10}$	E_7	$k^+_{z7:e10} = 5 \times 10^6$	$k^-_{z7:e10} = 1.0$	$k^{cat}_{z7:e10} = 5.0$	a
(fVII, fIIa)	Z_7, E_2	$Z_7 : E_2$	E_7	$k^+_{z7:e2} = 3.92 \times 10^5$	$k^-_{z7:e2} = 1.0$	$k^{cat}_{z7:e2} = 6.1 \times 10^{-2}$	b
(fV, fIIa)	Z_5, E_2	$Z_5 : E_2$	E_5	$k^+_{z5:e2} = 1.73 \times 10^7$	$k^-_{z5:e2} = 1.0$	$k^{cat}_{z5:e2} = 0.23$	c
(fVIII, fIIa)	Z_8, E_2	$Z_8 : E_2$	E_8	$k^+_{z8:e2} = 2.64 \times 10^7$	$k^-_{z8:e2} = 1.0$	$k^{cat}_{z8:e2} = 0.9$	d
(fXI-fXI, fIIa)	Z_{11}, E_2	$Z_{11} : E_2$	E^h_{11}	$k^+_{z11:e2} = 2.0 \times 10^7$	$k^-_{z11:e2} = 1.0$	$k^{cat}_{z11:e2} = 1.3 \times 10^{-4}$	e
(fIX, fXIa)	Z_9, E^h_{11}	$Z_9 : E^h_{11}$	E_9	$k^+_{z9:e^h_{11}} = 0.6 \times (10)^7$	$k^-_{z9:e^h_{11}} = 1.0$	$k^{cat}_{z9:e^h_{11}} = 0.21$	f

(a) $K_M = 1.2 \times 10^{-6}$ M [55]. (b) $K_M = 2.7 \times 10^{-6}$ M [55] (c) $K_M = 7.17 \times 10^{-8}$ M [58]. (d) [59], $K_M = 2 \times 10^{-7}$ M [60]. (e) $K_M = 50\,\mathrm{nM}$ [61]. Rate constants apply also for thrombin activation of XIa-XI. (f) $K_M = 0.2\,\mu\mathrm{M}$ [62, 63] and for activation of IX by XIa-XIa

Table 6 Platelet transitions

Reactants	Reactants	Products	$(\text{M}^{-1}\text{sec}^{-1})$	(sec^{-1})	Note
Unactivated platelet adhering to SE	PL, SE	PL_a^s	$k_{\text{adh}}^+ = 2 \times 10^{10}$	$k_{\text{adh}}^- = 0$	a
Activated platelet adhering to SE	PL_a^v, SE	PL_a^v	$k_{\text{adh}}^+ = 2 \times 10^{10}$	$k_{\text{adh}}^- = 0$	a
Platelet activation by platelet in solution	PL, PL_a^v	$2PL_a^v$	$k_{plt}^{\text{act}} = 3 \times 10^8$		b
Platelet activation on SE	PL, PL_a^s	PL_a^v, PL_a^s	$k_{plt}^{\text{act}} = 3 \times 10^8$		b
Platelet activation by thrombin	PL, E_2	PL_a^v		$k_{e_2}^{\text{act}} = 0.50$	b

(a) Estimated from data in [64, 65] as described in [28]. (b) Estimated from data in [66] as described in [28]. SE = subendothelium

Table 7 Reactions on platelet surfaces. Notation: activation (of -, by -); binding (of -, with -)

Reaction	Reactants	Complex	Product	$(M^{-1}sec^{-1})$	(sec^{-1})	(sec^{-1})	Note
Activation							
(V, Xa)	Z_5^m, E_{10}^m	$Z_5^m : E_{10}^m$	E_5^{hm}	$k_{z_5^m:e_{10}^m}^+ = 1.0 \times 10^8$	$k_{z_5^m:e_{10}^m}^- = 1.0$	$k_{z_5^m:e_{10}^m}^{cat} = 4.6 \times 10^{-2}$	a
(V, IIa)	Z_5^m, E_2^m	$Z_5^m : E_2^m$	E_5^m	$k_{z_5^m:e_2^m}^+ = 1.73 \times 10^7$	$k_{z_5^m:e_2^m}^- = 1.0$	$k_{z_5^m:e_2^m}^{cat} = 0.23$	b
(Vh, IIa)	E_5^{hm}, E_2^m	$E_5^{hm} : E_2^m$	E_5^m	$k_{z_5^m:e_2^m}^+ = 1.73 \times 10^7$	$k_{z_5^m:e_2^m}^- = 1.0$	$k_{z_5^m:e_2^m}^{cat} = 0.23$	b
(VIII, Xa)	Z_8^m, E_{10}^m	$Z_8^m : E_{10}^m$	E_8^m	$k_{z_8^m:e_{10}^m}^+ = 5.1 \times 10^7$	$k_{z_8^m:e_{10}^m}^- = 1.0$	$k_{z_8^m:e_{10}^m}^{cat} = 2.3 \times 10^{-2}$	c
(VIII, IIa)	Z_8^m, E_2^m	$Z_8^m : E_2^m$	E_8^m	$k_{z_8^m:e_2^m}^+ = 2.64 \times 10^7$	$k_{z_8^m:e_2^m}^- = 1.0$	$k_{z_8^m:e_2^m}^{cat} = 0.9$	d
(X, VIIIa:IXa)	Z_{10}^m, TEN	$Z_{10}^m : TEN$	E_{10}^m	$k_{z_{10}^m:ten}^+ = 1.31 \times 10^8$	$k_{z_{10}^m:ten}^- = 1.0$	$k_{z_{10}^m:ten}^{cat} = 20.0$	f
(X, VIIIa:IXa*)	Z_{10}^m, TEN^*	$Z_{10}^m : TEN^*$	E_{10}^m	$k_{z_{10}^m:ten}^+ = 1.31 \times 10^8$	$k_{z_{10}^m:ten}^- = 1.0$	$k_{z_{10}^m:ten}^{cat} = 20.0$	f
(II, Vh:Xa)	Z_2^m, $PROh$	$Z_2^m : PROh$	E_2^m	$k_{z_2^m:pro}^+ = 1.03 \times 10^8$	$k_{z_2^m:pro}^- = 1.0$	$k_{z_2^m:pro}^{cat} = 30.0$	g
(II, Va:Xa)	Z_2^m, PRC	$Z_2^m : PRO$	E_2^m	$k_{z_2^m:pro}^+ = 1.03 \times 10^8$	$k_{z_2^m:pro}^- = 1.0$	$k_{z_2^m:pro}^{cat} = 30.0$	g
(XI-XI, IIa)	Z_{11}^m, E_2^m	$Z_{11}^m : E_2^m$	E_{11}^{hm}	$k_{z_{11}^m:e_2^m}^+ = 2.0 \times 10^7$	$k_{z_{11}^m:e_2^m}^- = 1.0$	$k_{z_{11}^m:e_2^m}^{cat} = 1.3 \times 10^{-4}$	h
(IX, XIa)	Z_9^m, E_{11}^{hm}	$Z_9^m : E_{11}^m$	E_9	$k_{z_9^m:e_{11}}^+ = 0.6 \times 10^7$	$k_{z_9^m:e_{11}}^- = 1.0$	$k_{z_9^m:e_{11}}^{cat} = 0.21$	i

(continued)

Table 7 (continued)

Reaction	Reactants	Complex	Product	$(M^{-1} sec^{-1})$	(sec^{-1})	(sec^{-1})	Note
Binding							
(VIIIa, IXa)	E_8^m, E_9^m		TEN	$k_{ten}^+ = 1.0 \times 10^8$	$k_{ten}^- = 0.01$		e
(VIIIa, IXa*)	E_8^m, $E_9^{m,*}$		TEN^*	$k_{ten}^+ = 1.0 \times 10^8$	$k_{ten}^- = 0.01$		e
(Vh, Xa)	E_5^{hm}, E_{10}^m		$PROh$	$k_{pro}^+ = 1.0 \times 10^8$	$k_{pro}^- = 0.01$		e
(Va, Xa)	E_5^m, E_{10}^m		PRO	$k_{pro}^+ = 1.0 \times 10^8$	$k_{pro}^- = 0.01$		e

(a) $K_M = 10.4 \times 10^{-9}$ M [67]. (b) The rate constants for thrombin activation of fV on platelets are assumed to be the same as in plasma. (c) $K_M = 2.0 \times 10^{-8}$ M [60]. (d) The rate constants for thrombin activation of fVIII on platelets are assumed to be the same as in plasma. (e) The formation of the tenase and prothrombinase complexes is assumed to be very fast with $K_d = 1.0 \times 10^{-10}$ M [68]. (f) $K_M = 1.6 \times 10^{-7}$ M [69]. (g) $K_M = 3.0 \times 10^{-7}$ M [70]. (h) $K_M = 50$ nM [61]. Rate constants apply also for thrombin activation of Plt-XIa-XI. (i) $K_M = 0.2 \mu$M [62, 63]. Rate constants apply also for activation of platelet-bound IX by Plt-XIa-XIa

Table 8 Inhibition reactions. Notation: inactivation (of -, by -); activation (of -, by -); binding (of -, with -)

Reaction	Reactants	Product	$(M^{-1}sec^{-1})$	(sec^{-1})	(sec^{-1})	Note
Inactivation						
(IXa, AT-III)	E_9, AT	$E_9 : AT$		$k^{AT}_{e_9} = 4.8 \times 10^2$		a
(Xa, AT-III)	E_{10}, AT	$E_{10} : AT$		$k^{AT}_{e_{10}} = 3.5 \times 10^3$		a
(IIa, AT-III)	E_2, AT	$E_2 : AT$		$k^{AT}_{e_2} = 1.4 \times 10^4$		a
(XIa, AT-III)	E_{11}, AT	$E_{11} : AT$		$k^{AT}_{e_{11}} = 2.4 \times 10^2$		a
(XIa:AT, AT-III)	$E_{11} : AT$, AT	$AT : E_{11} : AT$		$k^{AT}_{e_{11}} = 2.4 \times 10^2$		a
(APC, Va)	APC, E^m_5	None	$k^+_{e^m_5 : APC} = 1.2 \times 10^8$	$k^-_{e^m_5 : APC} = 1.0$	$k^{cat}_{e^m_5 : APC} = 0.5$	c
(APC, VIIIa)	APC, E^m_8	None	$k^+_{e^m_8 : APC} = 1.2 \times 10^8$	$k^-_{e^m_8 : APC} = 1.0$	$k^{cat}_{e^m_8 : APC} = 0.5$	c
Binding						
(TFPI, Xa)	$TFPI$, E_{10}	$TFPIa$	$k^+_{tfpia:e_{10}} = 1.6 \times 10^7$	$k^-_{tfpia:e_{10}} = 3.3 \times 10^{-4}$		d
(TFPI, Vh)	TFPI, E^h_5	$TFPI : E^h_5$	$k^+_{tfpi:e5h} = 0.05 \times 10^9$	$k^-_{tfpi:e5h} = 0.0045$		e
(TFPI:Xa, Vh)	TFPIa, E^h_5	$E^h_5 : TFPI : E_{10}$	$k^+_{tfpi:e5h} = 0.05 \times 10^9$	$k^-_{tfpi:e5h} = 0.0045$		e
(TFPI:Vh, Xa)	$TFPI : E^h_5$, E_{10}	$E^h_5 : TFPI : E_{10}$	$k^+_{tfpia:e_{10}} = 1.6 \times 10^7$	$k^-_{tfpia:e_{10}} = 3.3 \times 10^{-4}$		d

(continued)

Table 8 (continued)

Reaction	Reactants	Product	(M^{-1} sec^{-1})	(sec^{-1})	(sec^{-1})	Note
(Xa:Vh, TFPI)	$E_{10}:E_5^h, TFPIa$	$TFPI:E_{10}:E_5^h$	$k_{tfpibprohv10}^+ = 1.6 \times 10^7$	$k_{tfpibprohv10}^- = 3.3 \times 10^{-4}$		d
(TFPIa, TF:VIIa)	$TFPIa, E_7^m$	$TFPIa:E_7^m$	$k_{tfpia:e10}^+ = 1.6 \times 10^7$	$k_{tfpia:e10}^- = 3.3 \times 10^{-4}$		f
(TM, Thrombin)	TM, E_2^{ec}	$TM:E_2^{ec}$	$k_{TM}^{on} = 1.0 \times 10^8$	$k_{TM}^{off} = 5.0 \times 10^{-2}$		g
(TM_{RZ}, Thrombin)	TM_{RZ}, E_2	$TM_{RZ}:E_2$	$k_{TM_{RZ}}^{on} = 1.0 \times 10^8$	$k_{TM_{RZ}}^{off} = 5.0 \times 10^{-2}$		g
Activation						
(PC, $TM_{RZ}:E_2$)	$PC, TM_{RZ}:E_2$	APC	$k_{pc}^+ = 1.7 \times 10^6$	$k_{pc}^- = 1.0$	$k_{pc}^{cat} = 0.16$	i
(PC, TM:E_2^{ec})	$PC, TM:E_2^{ec}$	APC	$k_{PC:TM:e_2^{ec}}^+ = 1.7 \times 10^6$	$k_{PC:TM:e_2^{ec}}^- = 1.0$	$k_{PC:TM:e_2^{ec}}^{cat} = 0.16$	i

(a) From [51]. (b) From [71]. (c) For inhibition of fVa by APC, $K_M = 12.5 \times 10^{-9}$ [72]. We assume the same reaction rates for the inhibition of fVIIIa by APC. (d) From [73]. (e) From [74]. (f) From [75]. (g) $K_d = 0.5$ nM and $[PC] = 65$ nM [76]. (h) From [77]. (i) $k_{PC:TM:e_2^{ec}} = 0.167$ sec^{-1}, $K_M = 0.7 \times 10^{-6}$ M [78]

$$+ k_{flow}(e_{10}^{up} - e_{10}) - k_{diff}(e_{10} - e_{10}^{ec})$$

$$- k_{TFPI:e_5^h:e_{10}}^{+}[TFPI : E_5^h]e_{10}$$

$$- k_{TFPI:e_5^h:e_{10}}^{-}[E_{10} : TFPI : E_5^h]$$

$$- k_{TFPI:e_5^{hm}:e_{10}}^{+}[TFPI : E_5^{hm}]e_{10}$$

$$+ k_{TFPI:e_5^{hm}:e_{10}}^{-}[E_{10} : TFPI : E_5^{hm}]$$

$$\frac{d}{dt}z_{10}^m = k_{10}^{on}z_{10}p_{10}^{avail} - k_{10}^{off}z_{10}^m + k_{z_{10}^m:TEN}^{+}z_{10}^m[TEN] \tag{8}$$

$$+ k_{z_{10}^m:TEN}^{-}[Z_{10}^m : TEN] - k_{z_{10}^m:TEN}^{+}z_{10}^m[TEN^*]$$

$$+ k_{z_{10}^m:TEN}^{-}[Z_{10}^m : TEN^*]$$

$$\frac{d}{dt}e_{10}^m = k_{10}^{on}e_{10}p_{10}^{avail} - k_{10}^{off}e_{10}^m + k_{z_{10}^m:TEN}^{cat}[Z_{10}^m : TEN] \tag{9}$$

$$+ (k_{z_5^m:e_{10}^m}^{cat} + k_{z_5^m:e_{10}^m}^{-})[Z_5^m : E_{10}^m] - k_{z_5^m:e_{10}^m}^{+}e_{10}^m z_5^m$$

$$+ (k_{z_8^m e_{10}^m}^{cat} + k_{z_8^m e_{10}^m}^{-})[Z_8^m : E_{10^m}] - k_{z_8^m:e_{10}^m}^{+}e_{10}^m z_8^m$$

$$+ k_{e_5^m:e_{10}^m}^{-}[PRO] - k_{e_5^m:e_{10}^m}^{p}e_{10}^m e_5^m$$

$$+ k_{z_{10}^m:TEN}^{cat}[Z_{10}^m : TEN^*] - k_{e_5^{hm}:e_{10}^m}^{+}e_{10}^m e_5^{hm}$$

$$+ k_{e_5^{hm}:e_{10}^m}^{-}PRO^h - k_{TFPI:e_{10}^m}^{+}e_{10}^m TFPI$$

$$+ k_{TFPI:e_{10}^m}^{-}[TFPI : E_{10}^m]$$

$$- k_{TFPI:e_5^{hm}:e_{10}^m}^{+}[TFPI : E_5^{hm}]e_{10}^m$$

$$+ k_{TFPI:e_5^{hm}:e_{10}^m}^{-}[E_{10}^m : TFPI : E_5^{hm}]$$

$$- k_{TFPI:e_{10}:e_5^{hm}}^{+}[TFPI : E_5^{hm}]e_{10}^m$$

$$+ k_{TFPI:e_{10}:e_5^{hm}}^{-}[TFPI : PRO_{v5}^h] - k_{e_{10}^m}^{AT}e_{10}^m[AT]$$

$$\frac{d}{dt}z_5 = - k_5^{on}z_5p_5^{avail} + k_5^{off}z_5^m - k_{z_5:e_2}^{+}z_5e_2 \tag{10}$$

$$+ k_{z_5:e_2}^{-}[Z_5 : E_2] + k_{flow}(z_5^{up} - z_5)$$

$$+ n_5(k_{adh}^{+}p_{PLAS}^{avail} + k_{plt}^{act}([PL_a^v] + [PL_a^s])$$

$$+ k_{e2}^{act}\frac{e_2}{e_2 + 0.001})[PL]$$

$$\frac{d}{dt}e_5 = -k_5^{on}e_5p_5^{avail} + k_5^{off}e_5^m + k_{z5:e2}^{cat}[Z_5 : E_2]$$ (11)

$$+ k_{flow}(e_5^{up} - e_5) + k_{e5:APC}^{-}[APC : E_5]$$

$$- k_{e5:APC}^{+}e_5[APC] + k_{e_5^h:e_2}^{cat}[E_5^h : E_2]$$

$$\frac{d}{dt}z_5^m = k_5^{on}z_5p_5^{avail} - k_5^{off}z_5^m - k_{z_5^m e_{10}^m}^{+}z_5^m e_{10}^m$$ (12)

$$+ k_{z_5^m:e_{10}^m}^{-}[Z_5^m : E_{10}^m] - k_{z_5^m:e_2^m}^{+}z_5^m e_2^m + k_{z_5^m e_2^m}^{-}[Z_5^m : E_2^m]$$

$$\frac{d}{dt}e_5^m = k_5^{on}e_5p_5^{avail} - k_5^{off}e_5^m + k_{z_5^m:e_2^m}^{cat}[Z_5^m : E_2^m]$$ (13)

$$+ k_{e_5^m:APC}^{-}[APC : E_5^m] - k_{e_5^m:APC}^{+}e_5^m[APC]$$

$$- k_{e_5^m:e_{10}^m}^{+}e_5^m e_{10}^m + k_{e_5^m:e_{10}^m}^{-}[PRO] + k_{e_5^{hm}:e_2^m}^{cat}[E_5^{hm} : E_2^m]$$

$$\frac{d}{dt}z_8 = -k_8^{on}z_8p_8^{avail} + k_8^{off}z_8^m + k_{flow}(z_8^{up} - z_8)$$ (14)

$$- k_{z8:e_2}^{+}z_8e_2 + k_{z8:e_2}^{-}[Z_8 : E_2]$$

$$\frac{d}{dt}e_8 = -k_8^{on}e_8p_8^{avail} + k_8^{off}e_8^m + k_{flow}(e_8^{up} - e_8)$$ (15)

$$+ k_{z8:e_2}^{cat} - 0.005e_8 + k_{e8:APC}^{-}[APC : E_8]$$

$$- k_{e8:APC}^{+}e_8[APC]$$

$$\frac{d}{dt}z_8^m = k_8^{on}z_8p_8^{avail} - k_8^{off}z_8^m - k_{z_8^m:e_{10}^m}^{+}z_8^m e_{10}^m$$ (16)

$$+ k_{z_8^m:e_{10}^m}^{-}[Z_8^m : E_{10}^m] - k_{z_8^m:e_2^m}^{+}z_8^m e_2^m + k_{z_8^m:e_2^m}^{-}[Z_8^m : E_2^m]$$

$$\frac{d}{dt}e_8^m = k_8^{on}e_8p_8^{avail} - k_8^{off}e_8^m + k_{z_8^m:e_{10}^m}^{cat}[Z_8^m : E_{10}^m]$$ (17)

$$+ k_{z_8^m:e_2^m}^{cat}[Z_8^m : E_2^m] - k_{e_8^m:APC}^{+}e_8^m[APC]$$

$$+ k_{e_8^m:APC}^{-}[APC : E_8^m] - k_{e_8^m:e_9^m}^{+}e_9^m e_8^m - 0.005e_8^m$$

$$+ k_{e_8^m:e_9^m}^{-}[TEN] - k_{e_8^m:e_9^m}^{+}e_8^m e_9^m + k_{e_8^m:e_9^m}^{-}[TEN]$$

$$\frac{d}{dt}z_9 = k_{flow}(z_9^{up} - z_9) - k_9^{on}p_9^{avail}z_9 + k_9^{off}z_9^m$$ (18)

$$- k_{z9:e_7^m}^{+}z_9e_7^m + k_{z9:e_7^m}^{-}[Z_9 : E_7^m]$$

$$- k_{z9:e_{11}^h}^{+}e_{11}^h + k_{z9:e_{11}^h}^{-}[Z_9 : E_{11}^h]$$

$$- k_{z9:e_{11}}^{+}z_9e_{11} + k_{z9:e_{11}}^{-}[Z_9 : E_{11}]$$

$$\frac{d}{dt}e_9 = k_{flow}(e_9^{up} - e_9) - k_9^{on} p_9^{avail} e_9 + k_9^{off} e_9^m \tag{19}$$

$$+ k_{z9:e_7^m}^{cat}[Z_9 : E_7^m] - k_{z7:e9}^+ z_7 e_9$$

$$+ (k_{z7:e9}^{cat} + k_{z7:e9}^-)[Z_7 : E_9]$$

$$+ (k_{z_7^m:e9}^{cat} + k_{z_7^m:e9}^-)[Z_7^m : E_9] - k_{z_7^m:e9}^+ z_7^m e_9$$

$$- k_9^{on} p_9^{*,avail} e_9 + k_9^{off} e_9^{m*} - k_{diff}(e_9 - e_9^{ec})$$

$$+ k_{z9:e_{11}^h}^{cat}[Z_9 : E_{11}^h] + k_{z9:e11}^{cat}[Z_9 : E_{11}] - k_{e9}^{AT} e_9[AT]$$

$$\frac{d}{dt}z_9^m = k_9^{on} p_9^{avail} z_9 - k_9^{off} z_9^m - k_{z_9^m:e_{11}^{h,m}}^+ z_9^m e_{11}^{h,m} \tag{20}$$

$$+ k_{z_9^m e_{11}^{h,m}}^-[Z_9^m : E_{11}^{h,m}] - k_{z_9^m:e_{11}^{m*}}^+ z_9^m e_{11}^{m*}$$

$$+ k_{z_9^m e_{11}^{m*}}^-[Z_9^m : E_{11}^{m*}]$$

$$\frac{d}{dt}e_9^m = k_9^{on} p_9^{avail} e_9 - k_9^{off} e_9^m - k_{e_8^m:e_9^m}^+ e_8^m e_9^m \tag{21}$$

$$+ k_{e_8^m:e_9^m}^-[TEN] + k_{z_9^m:e_{11}^{h,m}}^{cat}[Z_9^m : E_{11}^{h,m}]$$

$$+ k_{z_9^m:e_{11}^{m*}}^{cat}[Z_9^m : E_{11}^{m*}] - k_{e_9^m}^{AT} e_9^m[AT]$$

$$\frac{d}{dt}z_2 = - k_2^{on} p_2^{avail} z_2 + k_2^{off} z_2^m + k_{flow}(z_2^{up} - z_2) \tag{22}$$

$$\frac{d}{dt}e_2 = k_{flow}(e_2^{up} - e_2) - k_{2*}^{on} p_2^{*,avail} e_2 + k_{2*}^{off} e_2^m \tag{23}$$

$$+ k_{z_2^m:PRO}^{cat}[Z_2^m : PRO] - k_{z5:e_2^p} z_5 e_2$$

$$+ (k_{z5:e_2}^{cat} + k_{z5:e_2}^-)[Z_5 : E_2] - k_{z8:e_2}^+ z_8 e_2$$

$$+ (k_{z8:e_2}^{cat} + k_{z8:e_2}^-)[Z_8 : E_2] - k_{z7:e_2}^+ z_7 e_2$$

$$+ (k_{z_7^m:e_2}^{cat} + k_{z7:e_2}^-)[Z_7 : E_2] - k_{z_7^m e_2}^+ z_7^m e_2$$

$$+ (k_{z_7^m:e_2}^{cat} + k_{z_7^m:e_2}^-)[Z_7^m : E_2] - k_{diff}(e_2 - e_2^{ec})$$

$$- k_{z11:e_2}^+ z_{11} + (k_{z11:e_2}^- + k_{z11:e_2}^{cat})[Z_{11} : E_2]$$

$$- k_{e_{11}^h:e_2}^+ e_{11}^h e_2 + (k_{e_{11}^h:e_2}^- + k_{e_{11}^h} : e_2)^{cat}[E_{11}^h : E_2]$$

$$+ k_{z_2^m:PRO^h}^{cat}[Z_2^m : PRO^h] - k_{e_5^h:e_2}^+ e_2 e_5^h$$

$$+ k^-_{e^h_5:e_2}[E^h_5 : E_2] + k^{cat}_{e^h_5:e_2}[E^h_5 : E_2] - k^{AT}_{e_2}e_2[AT]$$

$$\underline{- k^{on}_{TM_{RZ}}e_2[TM^{avail}_{RZ}] + k^{off}_{TM_{RZ}}[TM_{RZ} : E_2]}$$

$$\frac{d}{dt}z^m_2 = k^{on}_2 p^{avail}_2 z_2 - k^{off}_2 z^m_2 - k^+_{z^m_2:PRO}z^m_2[PRO] \tag{24}$$

$$+ k^-_{z^m_2:PRO}[Z^m_2 : PRO] - k^+_{z^m_2:PRO^h}z^m_2 PRO^h$$

$$+ k^-_{z^m_2:PRO^h}[Z^m_2 : PRO^h]$$

$$\frac{d}{dt}e^m_2 = k^{on}_{2*} p^{avail}_2 e_2 - k^{off}_{2*} e^m_2 \tag{25}$$

$$+ (k^{cat}_{z^m_5:e^m_2} + k^-_{z^m_5:e^m_2})[Z^m_5 : E^m_2] - k^+_{z^m_5:e^m_2}z^m_5 e^m_2$$

$$+ (k^{cat}_{z^m_8:e^m_2} + k^-_{z^m_8:e^m_2})[Z^m_8 : E^m_2] - k^+_{z^m_8:E^m_2}z^m_8 e^m_2$$

$$- k^+_{z^m_{11}:e^m_2}z^m_{11} e^m_2 - k^+_{e^{h,m*}_{11}:e^m_2}e^{h,m*}_{11} e^m_2$$

$$+ (k^-_{z^m_{11}:e^m_2} + k^{cat}_{z^m_{11}:e^m_2})[Z^m_{11} : E^m_2]$$

$$+ (k^-_{e^{h,m*}_{11}:e^m_2} + k^{cat}_{e^{h,m*}_{11}:e^m_2})[E^{hms}_{11} : E^m_2]$$

$$- k^+_{e^{hm}_5:e^m_2}e^m_2 e^{hm}_5 + k^-_{e^{hm}_5:e^m_2}[E^{hm}_5 : E^m_2]$$

$$+ k^{cat}_{e^{hm}_5:e^m_2}[E^{hm}_5 : E^m_2] - k^+_{PRO^h:e^m_2}PRO^h e^m_2$$

$$+ k^-_{PRO^h:e^m_2}[PRO^h : E^m_2]$$

$$+ k^{cat}_{PRO^h:e^m_2}[PRO^h : E^m_2] - k^{AT}_{e^m_2}e^m_2[AT]$$

$$\frac{d}{dt}[TEN] = - k^-_{e^m_8:e^m_9}[TEN] + k_{e^m_8:e^m_9}e^m_8 e^m_9 \tag{26}$$

$$+ (k^{cat}_{z^m_{10}:TEN} + k^-_{z^m_{10}:TEN})[Z^m_{10} : TEN]$$

$$- k^+_{z^m_{10}:TEN}z^m_{10}[TEN]$$

$$\frac{d}{dt}[PRO] = - k^-_{e^m_5:e^m_{10}}[PRO] + k^+_{e^m_5:e^m_{10}}e^m_{10} e^m_5 \tag{27}$$

$$+ (k^{cat}_{z^m_2:PRO} + k^-_{z^m_2:PRO})[Z^m_2 : PRO]$$

$$- k^+_{z^m_2:PRO}z^m_2[PRO] + k^{cat}_{PRO^h:e^m_2}[PRO^h : E^m_2]$$

$$\frac{d}{dt}[PL^s_a] = k^+_{adh}p^{avail}_{PLAS}[PL] - k^-_{adh}[PL^s_a] + k^+_{adh}[PL^v_a]p^{avail}_{PLAS} \tag{28}$$

$$\frac{d}{dt}[PL] = k^p_{flow}([PL]^{up} - [PL]) - k^+_{adh}p^{avail}_{PLAS} \tag{29}$$

$$+ (k^{act}_{plt}([PL^v_a] + [PL^s_a]) + k^{act}_{e2}\frac{e_2}{e_2 + 0.001})[PL]$$

$$\frac{d}{dt}[PL^v_a] = k^-_{adh}[PL^s_a] - k^+_{adh}[PL^v_a]p^{avail}_{PLAS} \tag{30}$$

$$+ (k^{act}_{plt}([PL^v_a] + [PL^s_a]) + k^{act}_{e2}\frac{e_2}{e_2 + 0.001})[PL]$$

$$\frac{d}{dt}z^m_7 = k^{on}_7 z_7[TF]^{avail} - k^{off}_7 z^m_7 - k^+_{z^m_7:e10}z^m_7 e_{10} \tag{31}$$

$$- k^+_{z^m_7:e2}z^m_7 e_2 + k^-_{z^m_7:e10}[Z^m_7 : E_{10}]$$

$$+ k^-_{z^m_7:e2}[Z^m_7 : E_2] - k^+_{z^m_7:e9}z^m_7 e_9$$

$$+ k^-_{z^m_7:e9}[Z^m_7 : E_9] - z^m_7 \frac{d}{dt}[PL^s_a]\frac{1}{p^{avail}_{PLAS}}$$

$$\frac{d}{dt}e^m_7 = k^{on}_7 e_7[TF]^{avail} + k^-_{TFPI:e10:e^m_7}[TFPI : E^m_7] \tag{32}$$

$$- k^{off}_7 e^m_7 k^+_{TFPI:e10:e^m_7}e^m_7[TFPI : E_{10}]$$

$$+ k^{cat}_{z^m_7:e10}[Z^m_7 : E_{10}] + k^{cat}_{z^m_7:e2}[Z^m_7 : E_2]$$

$$+ (k^{cat}_{z10:e^m_7} + k^-_{z10:e^m_7})[Z_{10} : E^m_7]$$

$$- k^+_{z10:e^m_7}e^m_7 z_{10} - k^+_{z9:e^m_7}e^m_7 z_9$$

$$+ (k^{cat}_{z9:e^m_7} + k^-_{z9:e^m_7})[Z_9 : E^m_7]$$

$$+ k^{cat}_{z^m_7:e9}[Z^m_7 : E_9] - e^m_7 \frac{d}{dt}[PL^s_a]\frac{1}{p^{avail}_{PLAS}}$$

$$\frac{d}{dt}[TFPI] = - k^+_{TFPI:e10}e_{10}[TFPI] + k^-_{TFPI:e10}[TFPI : E_{10}] \tag{33}$$

$$+ k_{flow}([TFPI]^{up} - [TFPI])$$

$$- k^+_{TFPI:e^{hm}_5}e^{hm}_5[TFPI]$$

$$+ k^-_{TFPI:e^{hm}_5}[TFPI : E^{hm}_5] - k^+_{TFPI:e^h_5}e^h_5[TFPI]$$

$$+ k^-_{TFPI:e^h_5}[TFPI : E^h_5] - k^+_{TFPI:e^m_{10}}e^m_{10}[TFPI]$$

$$+ k^-_{TFPI:e^m_{10}}[TFPI : E^m_{10}]$$

$$- k^+_{TFPI:PRO^h_{v10}} PRO^h[TFPI]$$

$$+ k^-_{TFPI:PRO^h_{v10}} [TFPI : PRO^h_{v10}]$$

$$- k^+_{TFPI:PRO^h_{v5}} PRO^h[TFPI]$$

$$+ k^-_{TFPI:PRO^h_{v5}} [TFPI : PRO^h_{v5}]$$

$$\frac{d}{dt}[TFPI : E_{10}] = k^+_{TFPI:e_{10}} e_{10}[TFPI] - k^-_{TFPI:e_{10}}[TFPI : E_{10}] \tag{34}$$

$$+ k^-_{TFPI:e_{10}:e^m_7}[TFPI : E_{10} : E^m_7]$$

$$- k^+_{TFPI:e_{10}:e^m_7} e^m_7[TFPI : E_{10}]$$

$$+ k_{flow}([TFPI : E_{10}]^{up} - [TFPI : E_{10}])$$

$$- k^+_{TFPI:e_{10}:e^h_5}[TFPI : E_{10}]e^h_5$$

$$+ k^-_{TFPI:e_{10}:e^h_5}[E_{10} : TFPI : E^h_5]$$

$$- k^{on}_{10}[TFPI : E_{10}]p^{avail}_{10} + k^{off}_{10}[TFPI : E^m_{10}]$$

$$\frac{d}{dt}[TFPI : E_{10} : E^m_7] = - k^-_{TFPI:e_{10}:e^m_7}[TFPI : E_{10} : E^m_7] \tag{35}$$

$$+ k^+_{TFPI:e_{10}:e^m_7} e^m_7[TFPI : E_{10}]$$

$$- [TFPI : E_{10} : E^m_7]\frac{d}{dt}[PL^s_a]\frac{1}{p^{avail}_{PLAS}}$$

$$\frac{d}{dt}[Z_7 : E_2] = k_{flow}([Z_7 : E_2]^{up} - [Z_7 : E_2]) + k^+_{z7:e_2} e_2 z_7 \tag{36}$$

$$- (k^{cat}_{z7:e_2} + k^-_{z7:e_2})[Z_7 : E_2]$$

$$\frac{d}{dt}[Z_7 : E_{10}] = k^+_{z7:e_{10}} e_{10} z_7 - (k^{cat}_{z7:e_{10}} + k^-_{z7:e_{10}})[Z_7 : E_{10}] \tag{37}$$

$$+ k_{flow}([Z_7 : E_{10}]^{up} - [Z_7 : E_{10}])$$

$$\frac{d}{dt}[Z^m_7 : E_{10}] = k^+_{z^m_7:e_{10}} e_{10} z^m_7 - (k^{cat}_{z^m_7:e_{10}} + k^-_{z^m_7:e_{10}})[Z^m_7 : E_{10}] \tag{38}$$

$$- [Z^m_7 : E_{10}]\frac{d}{dt}[PL^s_a]\frac{1}{p^{avail}_{PLAS}}$$

$$\frac{d}{dt}[Z_7^m : E_2] = k_{z_7^m:e_2}^+ e_2 z_7^m - (k_{z_7^m:e_2}^{cat} + k_{z_7^m:e_2}^-)[Z_7^m : E_2] \tag{39}$$

$$- [Z_7^m : E_2]\frac{d}{dt}[PL_a^s]\frac{1}{p_{PLAS}^{avail}}$$

$$\frac{d}{dt}[APC] = (k_{e_5^m:APC}^{cat} + k_{e_5^m:APC}^-)[APC : E_5^m] - k_{e_5^m:APC}^{cat}e_5^m \tag{40}$$

$$+ (k_{e_8^m:APC}^{cat} + k_{e_8^m:APC}^-)[APC : E_8^m]$$

$$- k_{e_8^m:APC}^+ e_8^m[APC]$$

$$+ k_{flow}([APC]^{up} - [APC])$$

$$- k_{diff}([APC] - [APC^{ec}]) - k_{e_5:APC}^+ e_5[APC]$$

$$+ (k_{e_5:APC}^{cat} + k_{e_5:APC}^-)[APC:E_5] - k_{e_5^{hm}:APC}^+ e_5^{hm} APC$$

$$+ (k_{e_8:APC}^{cat} + k_{e_8:APC}^-)[APC : E_8] - k_{e_8:APC}^+ e_8[APC]$$

$$+ k_{e_5^{hm}:APC}^-[APC : E_5^{hm}] + k_{e_5^{hm}:APC}^{cat}[APC : E_5^{hm}]$$

$$- k_{e_5^h:APC}^+ e_5^h APC + k_{e_5^h:APC}^-[APC : E_5^h]$$

$$+ k_{e_5^h:APC}^{cat}[APC : E_5^h]+ \underline{k_{pc}^{cat}[TM_{RZ} : E_2 : PC]}$$

$$\frac{d}{dt}[Z_{10} : E_7^m] = k_{z_{10}:e_7^m}^+ e_7^m z_{10} - (k_{z_{10}:e_7^m}^{cat} + k_{z_{10}:e_7^m}^-)[Z_{10} : E_7^m] \tag{41}$$

$$- [Z_{10} : E_7^m]\frac{d}{dt}[PL_a^s]\frac{1}{p_{PLAS}^{avail}}$$

$$\frac{d}{dt}[Z_{10}^m : TEN] = k_{z_{10}^m:TEN}^+ z_{10}^m[TEN] \tag{42}$$

$$- (k_{z_{10}^m:TEN}^{cat} + k_{z_{10}^m:TEN}^-)[Z_{10^m} : TEN]$$

$$\frac{d}{dt}[Z_5 : E_2] = k_{z_5:e_2}^+ e_2 z_5 - (k_{z_5:e_2}^{cat} + k_{z_5:e_2}^-)[Z_5 : E_2] \tag{43}$$

$$+ k_{flow}([Z_5 : E_2]^{up} - [Z_5 : E_2])$$

$$\frac{d}{dt}[Z_5^m : e_{10}^m] = k_{z_5^m:e_{10}}^+ e_{10}^m z_5^m - (k_{z_5^m:e_{10}}^{cat} + k_{z_5^m:e_{10}}^-)[Z_5^m : E_{10}^m] \tag{44}$$

$$\frac{d}{dt}[Z_5^m : E_2^m] = k_{z_5^m:e_2^m}^+ e_2^m z_5^m - (k_{z_5^m:e_2^m}^{cat} + k_{z_5^m:e_2^m}^-)[Z_5^m : E_2^m] \tag{45}$$

$$\frac{d}{dt}[Z_8^m : E_{10}^m] = k_{z_8^m:e_{10}}^+ e_{10}^m z_8^m - (k_{z_8^m:e_{10}}^{cat} + k_{z_8^m:e_{10}}^-)[Z_8^m : E_{10}^m] \tag{46}$$

$$\frac{d}{dt}[Z_8^m : E_2^m] = k_{z_8^m:e_2^m}^+ e_2^m z_8^m - (k_{z_8^m:e_2^m}^{cat} + k_{z_8^m:e_2^m}^-)[Z_8^m : E_2^m] \tag{47}$$

$$\frac{d}{dt}[Z_8 : E_2] = k_{z_8:e_2}^+ e_2 z_8 - (k_{z_8:e_2}^{cat} + k_{z_8:e_2}^-)[Z_8 : E_2] \tag{48}$$
$$+ k_{flow}([Z_8 : E_2]^{up} - [Z_8 : E_2])$$

$$\frac{d}{dt}[APC : E_8^m] = k_{e_8^m:APC}^+ e_8^m [APC] \tag{49}$$
$$- (k_{e_8^m:APC}^{cat} + k_{e_8^m:APC}^-)[APC : E_8^m]$$

$$\frac{d}{dt}[Z_9 : E_7^m] = k_{z_9:e_7^m}^+ e_7^m z_9 - (k_{z_9:e_7^m}^{cat} + k_{z_9:e_7^m}^-)[Z_9 : E_7^m] \tag{50}$$
$$- [Z_9 : E_7^m]\frac{d}{dt}[PL_a^s]\frac{1}{p_{PLAS}^{avail}}$$

$$\frac{d}{dt}[Z_2^m : PRO] = k_{z_2^m:PRO}^+ z_2^m [PRO] \tag{51}$$
$$- (k_{z_2^m:PRO}^{cat} + k_{z_2^m:PRO}^-)[Z_2^m : PRO]$$

$$\frac{d}{dt}[APC : E_5^m] = k_{e_5^m:APC}^+ e_5^m [APC] \tag{52}$$
$$- (k_{e_5^m:APC}^{cat} + k_{e_5^m:APC}^-)[APC : E_5^m]$$

$$\frac{d}{dt}[Z_7 : E_9] = k_{z_7:e_9}^+ e_9 z_7 - (k_{z_7:e_9}^{cat} + k_{z_7:e_9}^-)[Z_7 : E_9] \tag{53}$$

$$\frac{d}{dt}[Z_7^m : E_9] = k_{z_7^m:e_9}^+ e_9 z_7^m - (k_{z_7^m:e_9}^{cat} + k_{z_7^m:e_9}^-)[Z_7^m : E_9] \tag{54}$$
$$- [Z_7^m : E_9]\frac{d}{dt}[PL_a^s]\frac{1}{p_{PLAS}^{avail}}$$

$$\frac{d}{dt}[TF] = -[TF]\frac{d}{dt}[PL_a^s]\frac{1}{p_{PLAS}^{avail}} \tag{55}$$

$$\frac{d}{dt}e_9^{m*} = k_9^{on} p_9^{*,avail} e_9 - k_9^{off} e_9^{m*} + k_{e_8^m:e_9^m}^-[TEN^*] \tag{56}$$
$$- k_{e_8^m:e_9^m}^+ e_8^m e_9^{m*} - k_{e_9^m}^{AT} e_9^{m*}[AT]$$

$$\frac{d}{dt}[TEN^*] = -k^-_{e^m_8:e^m_9}[TEN^*] + k^+_{e^m_8:e^m_9}e^m_8 e^{m*}_9 \tag{57}$$

$$+ (k^{cat}_{z^m_{10}:TEN} + k^-_{z^m_{10}:TEN})[Z^m_{10} : TEN^*]$$

$$+ k^+_{z^m_{10}:TEN}[TEN^*]z^m_{10}$$

$$\frac{d}{dt}[Z^m_{10} : TEN^*] = k^+_{z^m_{10}:TEN}[TEN^*]z^m_{10} \tag{58}$$

$$- (k^{cat}_{z^m_{10}:TEN} + k_{z^m_{10}} : TEN)^-[Z^m_{10} : TEN^*]$$

$$\frac{d}{dt}e^{ec}_2 = k_{diff}(e_2 - e^{ec}_2) + k_{flow}(e^{ec,up}_2 - e^{ec}_2) \tag{59}$$

$$- k^{on}_{TM}e^{ec}_2[TM]^{avail} + k^{off}_{TM}[TM : E^{ec}_2] - k^{AT}_{e^{ec}_2}e^{ec}_2[AT]$$

$$\frac{d}{dt}[APC^{ec}] = k_{flow}([APC]^{up} - [APC^{ec}]) \tag{60}$$

$$+ k_{diff}([APC] - [APC^{ec}])$$

$$+ k^{cat}_{PC:TM:e_2}[TM : E^{ec}_2 : APC]$$

$$\frac{d}{dt}e^{ec}_9 = k_{diff}(e_9 - e^{ec}_9) + k_{flow}(e^{up}_9 - e^{ec}_9) - k^{AT}_{e^{ec}_9}e^{ec}_9[AT] \tag{61}$$

$$\frac{d}{dt}e^{ec}_{10} = k_{diff}(e_{10} - e^{ec}_{10}) + k_{flow}(e^{up}_{10} - e^{ec}_{10}) - k^{AT}_{e^{ec}_{10}}e^{ec}_{10}[AT] \tag{62}$$

$$\frac{d}{dt}[TM : E^{ec}_2] = k^+_{TM}[E^{ec}_2](1 - [TM : E^{ec}_2] - k^+_{PC:TM:e_2}[TM : E^{ec}_2] \tag{63}$$

$$- [TM : E^{ec}_2 : APC]) - k^-_{TM}[TM : E^{ec}_2]$$

$$+ (k^-_{PC:TM:e_2} + k^{cat}_{PC:TM:e_2})[TM : E^{ec}_2 : APC]$$

$$\frac{d}{dt}[TM : E^{ec}_2 : APC] = k^+_{PC:TM:e_2}[TM : E^{ec}_2] \tag{64}$$

$$- (k^-_{PC:TM:e_2} + k^{cat}_{PC:TM:e_2})[TM : E^{ec}_2 : APC]$$

$$\frac{d}{dt}[APC : E_5] = - (k^{cat}_{e_5:APC} + k^-_{e_5:APC})[APC : E_5] + k^+_{e_5:APC}e_5[APC] \tag{65}$$

$$\frac{d}{dt}[APC : E_8] = - (k^{cat}_{e_8:APC} + k^-_{e_8:APC})[APC : E_8] + k^+_{e_8:APC}e_8[APC] \tag{66}$$

$$\frac{d}{dt}z_{11} = k_{flow}(z_{11}^{up} - z_{11}) - k_{z_{11}}^{on} z_{11} p_{11}^{avail} + k_{z_{11}}^{off} z_{11}^m \tag{67}$$

$$- k_{z_{11}:e_{11}^h}^+ z_{11} e_{11}^h + k_{z_{11}:e_{11}^h}^- [Z_{11} : E_{11}^h]$$

$$- k_{z_{11}:e_{11}}^+ z_{11} e_{11} + k_{z_{11}:e_{11}}^- [Z_{11} : E_{11}]$$

$$- k_{z_{11}:e_2}^+ z_{11} e_2 + k_{z_{11}:e_2}^- [Z_{11} : E_2]$$

$$\frac{d}{dt}e_{11} = k_{flow}(e_{11}^{up} - e_{11}) - k_{e_{11}}^{on,s} e_{11} p_{111}^{avail} + k_{e_{11}}^{off,s} e_{11}^{m*} \tag{68}$$

$$- k_{z_9:e_{11}}^+ z_9 e_{11} + (k_{z_9:e_{11}}^- + k_{z_9:e_{11}}^{cat})[Z_9 : E_{11}]$$

$$- k_{z_{11}:e_{11}}^+ z_{11} e_{11} + (k_{z_{11}:e_{11}}^- + k_{z_{11}:e_{11}}^{cat})[Z_{11} : E_{11}]$$

$$+ k_{e_{11}^h:e_{11}^h}^{cat} [E_{11}^h : E_{11}^h] - k_{e_{11}^h:e_{11}}^+ e_{11}^h e_{11}$$

$$+ (k_{e_{11}^h:e_{11}}^- + 2k_{e_{11}^h:e_{11}}^{cat})[E_{11}^h : E_{11}]$$

$$+ k_{e_{11}^h:e_2}^{cat} [E_{11}^h : E_2] - k_{e_{11}}^{AT} e_{11}[AT]$$

$$\frac{d}{dt}z_{11}^m = k_{z_{11}}^{on} z_{11} p_{11}^{avail} - k_{z_{11}}^{off} z_{11}^m - k_{z_{11}^m:e_{11}^{h,m}}^+ z_{11}^m e_{11}^{h,m} \tag{69}$$

$$+ k_{z_{11}^m:e_{11}^{h,m}}^- [Z_{11}^m : E_{11hm}] - k_{z_{11}^m:e_{11}^{m*}}^+ z_{11}^m e_{11}^{m*}$$

$$+ k_{z_{11}^m:e_{11}^{m*}}^- [Z_{11}^m : E_{11}^{m*}] - k_{z_{11}^m:E_2^m}^+ z_{11}^m e_2^m$$

$$+ k_{z_{11}^m:e_2^m}^- [Z_{11}^m : E_2^m]$$

$$\frac{d}{dt}e_{11}^{m*} = k_{e_{11}}^{on*} e_{11} p_{111}^{avail} - k_{e_{11}}^{off*} e_{11}^{m*} - k_{z_9^m:e_{11}^{m*}}^+ e_{11}^{m*} z_9^m \tag{70}$$

$$+ (k_{z_9^m:e_{11}^{m*}}^- + k_{z_9^m:e_{11}^{m*}}^{cat})[Z_9^m : E_{11}^{m*}]$$

$$- k_{z_{11}^m:e_{11}^{m*}}^+ z_{11}^m e_{11}^{m*} + (k_{z_{11}^m:e_{11}^{m*}}^-$$

$$+ k_{z_{11}^m:e_{11}^{m*}}^{cat})[Z_{11}^m : E_{11}^{m*}] + k_{e_{11}^{h,m*}:e_{11}^{m*}}^+ e_{11}^{h,m*} e_{11}^{m*}$$

$$+ (k_{e_{11}^{h,m*}:e_{11}^{m*}}^- + 2k_{e_{11}^{h,m*}:e_{11}^{m*}}^{cat})[E_{11}^{hms} : E_{11}^{h,m}]$$

$$+ k_{e_{11}^{h,m*}:e_2^m}^{cat} [E_{11}^{hms} : E_2^m] + k_{e_{11}}^{AT} e_{11}^{m*}[AT]$$

$$\frac{d}{dt}[Z_{11}^m : E_2^m] = k_{z_{11}^m:e_2^m}^+ z_{11}^m e_2^m - (k_{z_{11}^m:e_2^m}^- + k_{z_{11}^m:e_2^m}^{cat})[Z_{11}^m : E_2^m] \tag{71}$$

$$\frac{d}{dt}[Z_9^m : E_{11}^{m*}] = k_{z_9^m:e_{11}^{m*}}^+ z_9^m e_{11}^m s - (k_{z_9^m:e_{11}^{m*}}^- + k_{z_9^m:e_{11}^{m*}}^{cat})[Z_9^m : E_{11}^{m*}] \tag{72}$$

$$\frac{d}{dt}[Z_{11} : E_2] = k_{flow}([Z_{11} : E_2]^{up} - [Z_{11} : E_2]) + k^+_{z_{11}:e_2} z_{11} e_2 \tag{73}$$

$$- (k^-_{z_{11}:e_2} + k^{cat}_{z_{11}:e_2})[Z_{11} : E_2]$$

$$\frac{d}{dt}[Z_9 : E_{11}] = k_{flow}([Z_9 : E_{11}]^{up} - [Z_9 : E_{11}]) + k^+_{z_9:e_{11}} z_9 e_{11} \tag{74}$$

$$- (k^-_{z_9:E_{11}} + k^{cat}_{z_9:e_{11}})[Z_9 : E_{11}]$$

$$\frac{d}{dt}[Z_{11} : E_{11}] = k_{flow}([Z_{11} : E_{11}]^{up} - [Z_{11} : E_{11}]) \tag{75}$$

$$+ k^+_{z_{11}:e_{11}} z_{11} e_{11} - (k^-_{z_{11}:e_{11}} + k^{cat}_{z_{11}:e_{11}})[Z_{11} : E_{11}]$$

$$\frac{d}{dt}[Z_9 : E_{11}^h] = k_{flow}([Z_9 : E_{11}^h]^{up} - [Z_9 : E_{11}^h]) + k^+_{z_9:e_{11}^h} z_9 e_{11}^h \tag{76}$$

$$- (k^-_{z_9:e_{11}^h} + k^{cat}_{z_9:e_{11}^h})[Z_9 : E_{11}^h]$$

$$\frac{d}{dt}[Z_9^m : E_{11}^{h,m}] = k^+_{z_9^m:e_{11}^{h,m}} z_9^m e_{11}^{h,m} - (k^-_{z_9^m:e_{11}^h m} + k^{cat}_{z_9^m:e_{11}^h m})[Z_9^m : E_{11}^{h,m}] \tag{77}$$

$$\frac{d}{dt}[Z_{11} : E_{11}^h] = k_{flow}([Z_{11} : E_{11}^h]^{up} - [Z_{11} : E_{11}^h]) \tag{78}$$

$$+ k_{z_{11}:e_{11}^{hp}} z_{11} e_{11}^h - (k^-_{z_{11}:e_{11}^h} + k^{cat}_{z_{11}:e_{11}^h})[Z_{11} : E_{11}^h]$$

$$\frac{d}{dt}[E_{11}^h : E_{11}^h] = k_{flow}([E_{11}^h : E_{11}^h]^{up} - [E_{11}^h : E_{11}^h]) \tag{79}$$

$$+ k^+_{e_{11}^h:e_{11}^h} e_{11}^h e_{11}^h - (k^-_{e_{11}^h:e_{11}^h} + k^{cat}_{e_{11}^h:e_{11}^h})[E_{11}^h : E_{11}^h]$$

$$\frac{d}{dt}[E_{11}^h : E_{11}] = k_{flow}([E_{11}^h : E_{11}]^{up} - [E_{11}^h : E_{11}]) + k^+_{e_{11}^h:e_{11}} e_{11}^h \tag{80}$$

$$- (k^-_{e_{11}^h:e_{11}} + k^{cat}_{e_{11}^h:e_{11}})[E_{11}^h : E_{11}]$$

$$\frac{d}{dt}[E_{11}^h : E_2] = k_{flow}([E_{11}^h : E_2]^{up} - [E_{11}^h : E_2]) + k^+_{e_{11}^h:e_2} e_{11}^h e_2 \tag{81}$$

$$- (k^-_{e_{11}^h:e_2} + k^{cat}_{e_{11}^h:e_2})[E_{11}^h : E_2]$$

$$\frac{d}{dt}[Z_{11}^m : E_{11}^{h,m}] = k^+_{z_{11}^m:e_{11}^{h,m}} z_{11}^m e_{11}^{h,m} - (k^-_{z_{11}^m:e_{11}^{h,m}} + k^{cat}_{z_{11}^m:e_{11}^{h,m}})[Z_{11}^m : E_{11}^{h,m}] \tag{82}$$

$$\frac{d}{dt}[Z_{11}^m : F_{11}^{m*}] = k^+_{z_{11}^m:e_{11}^{m*}} z_{11}^m e_{11}^{m*} - (k^-_{z_{11}^m:e_{11}^{m*}} + k^{cat}_{z_{11}^m:e_{11}^{m}s})[Z_{11}^m : E_{11}^{m*}] \tag{83}$$

$$\frac{d}{dt}[E_{11}^{hms} : E_{11}^{h,m}] = k^+_{e_{11}^{h,m*}:e_{11}^{h,m}} e_{11}^{h,m} \tag{84}$$

$$- (k^-_{e_{11}^{h,m*}:e_{11}^{h,m}} + k^{cat}_{e_{11}^{h,m*}:e_{11}^{h,m}})[E_{11}^{hms} : E_{11}^{h,m}]$$

$$\frac{d}{dt}[E_{11}^{hms} : E_{11}^{m*}] = k_{e_{11}^{h,m*}:e_{11}^{h,m*}} e_{11}^{m*} \tag{85}$$

$$- (k^-_{e_{11}^{h,m*}:e_{11}^{m*}} + k^{cat}_{e_{11}^{h,m*}:e_{11}^{m*}})[E_{11}^{hms} : E_{11}^{m*}]$$

$$\frac{d}{dt}[E_{11}^{hms} : E_2^m] = k^+_{e_{11}^{h,m*}:e_2^m} e_{11}^{h,m*} e_2^m \tag{86}$$

$$- (k^-_{hms:e_2^m \atop e_{11}} + k^{cat}_{e_{11}^{h,m*}:e_2^m}[E_{11}^{hms} : E_2^m])$$

$$\frac{d}{dt}e_{11}^h = k^{on*}_{e_{11}^h} e_{11}^h p_{111}^{avail} + k^{off*}_{e_{11}^h} e_{11}^{h,m*} - k^{on}_{e_{11}^h} e_{11}^h p_{11}^{avail} \tag{87}$$

$$+ k^{off}_{e_{11}^h} e_{11}^{h,m} - k^+_{z9:e_{11}^h} z9 e_{11}^h + (k^-_{z9:e_{11}^{h,m}} + k^{cat}_{z9:e_{11}^h})[Z_9 : E_{11}^h]$$

$$- k_{z11:e_{11}^{hp}} z11 e_{11}^h + (k^-_{z11:e_{11}^h m} + 2k^{cat}_{z11:e_{11}^h})[Z_{11:E_{11}^h}]$$

$$+ k^{cat}_{z11:e_{11}}[Z_{11} : E_{11}] + k^{cat}_{z11:e_2}[Z_{11} : E_2]$$

$$- 2k^+_{e_{11}^h:e_{11}^h} e_{11}^h e_{11}^h + (2k^-_{e_{11}^h:e_{11}^h} + k^{cat}_{e_{11}^g:e_{11}^h})[E_{11}^h : E_{11}^h]$$

$$- k^+_{e_{11}^h:e_{11}} e_{11}^h e_{11} + k^-_{e_{11}^h:e_{11}}[E_{11}^h : E_{11}]$$

$$- k^+_{e_{11}^h:e_2} e_{11}^h e_2 + k^-_{e_{11}^h:e_2}[E_{11}^h : E_2]$$

$$+ k_{flow}(e_{11}^{h,up} - e_{11}^h) - k^{AT}_{e_{11}^h} e_{11}^h[AT]$$

$$\frac{d}{dt}e_{11}^{h,m} = k^{on}_{e_{11}^h} e_{11}^h p_{11}^{avail} - k^{off}_{e_{11}^h} e_{11}^{h,m} - k^+_{z_9^m:e_{11}^{h,m}} z_9^m e_{11}^{h,m} \tag{88}$$

$$+ (k^-_{z_9^m:e_{11}^{h,m}} + k^{cat}_{z_9^m:e_{11}^{h,m}})[Z_9^m : E_{11}^{h,m}]$$

$$+ (k^-_{z_{11}^m:e_{11}^{h,m}} + 2k^{cat}_{z_{11}^m:e_{11}^{h,m}})[Z_{11}^m : E_{11}^{h,m}]$$

$$- k^+_{z_{11}^m:e_{11}^{h,m}} z_{11}^m e_{11}^{h,m} + k^{cat}_{z_{11}^m:e_{11}^{m*}}[Z_{11}^m : E_{11}^{m*}]$$

$$+ k^{cat}_{z_{11}^m:e_2^m}[Z_{11}^m : E_2^m] - k^+_{e_{11}^{h,m*}:e_{11}^{h,m}} e_{11}^{h,m*} e_{11}^{h,m}$$

$$+ (k^-_{e_{11}^{h,m*}:e_{11}^{h,m}} + k^{cat}_{e_{11}^{h,m*}:e_{11}^{h,m}})[E_{11}^{hms}:E_{11}^{h,m}] - k^{AT}_{e_{11}} e_{11}^{hm}[AT]$$

$$\frac{d}{dt}e_{11}^{h,m*} = k_{e_{11}^h}^{on*} e_{11}^h p_{111}^{avail} - k_{e_{11}^h}^{off*} e_{11}^{h,m*} \tag{89}$$

$$- k_{e_{11}^{h,m*}:e_{11}^{h,m}}^{+} e_{11}^{h,m*} e_{11}^{h,m} + k_{e_{11}^{h,m*}:e_{11}^{h,m}}^{-} [E_{11}^{hms} : E_{11}^{h,m}]$$

$$- k_{e_{11}^{h,m*}:e_{11}^{m*}}^{+} e_{11}^{h,m*} e_{11}^{m*} + k_{e_{11}^{h,m*}:e_{11}^{m}s}^{-} [E_{11}^{hms} : E_{11}^{m*}]$$

$$- k_{e_{11}^{h,m*}:e_{2}^{m}}^{+} e_{11}^{h,m*} e_{2}^{m} + k_{e_{11}^{h,m*}:e_{2}^{m}}^{-} [E_{11}^{hms} : E_{2}^{m}]$$

$$\frac{d}{dt}e_{5}^{hm} = k_{z_5^m:e_{10}^m}^{cat} [Z_5^m : E_{10}^m] + k_5^{on} e_5^m p_5^{avail} - k_5^{off} e_5^{hm} \tag{90}$$

$$- k_{e_5^{hm}:e_{10}^m}^{+} e_{10}^m e_5^{hm} + k_{e_5^{hm}:e_{10}^m}^{-} PRO^h$$

$$- k_{e_5^{hm}:e_2^m}^{+} e_2^m e_5^{hm} + k_{e_5^{hm}:e_2^m}^{-} [E_5^{hm} : E_2^m]$$

$$- k_{TFPI:e_5^{hm}}^{+} e_5^{hm} TFPI + k_{TFPI:e_5^{hm}}^{-} [TFPI : E_5^{hm}]$$

$$- k_{e_5^{hm}:APC}^{+} e_5^{hm} APC + k_{e_5^{hm}:APC}^{-} [APC : E_5^{hm}]$$

$$- k_{TFPI:e_{10}:e_5^{hm}}^{+} [TFPI : E_{10}^m] e_h^{hm}$$

$$+ k_{TFPI:e_{10}:e_5^{hm}}^{-} [E_{10}^m : TFPI : E_5^{hm}]$$

$$- k_{TFPI:e_{10}:e_5^{hm}:e_{10}^m}^{+} [TFPI : E_{10}^m] e_5^{hm}$$

$$+ k_{TFPI:e_{10}:e_5^{hm}:e_{10}^m}^{-} [TFPI : PRO_{v10}^h]$$

$$\frac{d}{dt}e_5^h = - k_{5on} e_5^h p_5^{avail} + k_5^{off} e_5^{hm} + k_{flow}(e_5^{up} - e_5^h) \tag{91}$$

$$+ (1 - f_5) N_5 dpl \cdot p - k_{e_5^h:e_2}^{+} e_5^h$$

$$+ k_{e_5^h:e_2}^{-} [E_5^h : E_2] - k_{e_5^h:APC}^{+} APC \cdot e_5^h$$

$$+ k_{e_5^h:APC}^{-} [APC : E_5^h] - k_{TFPI:e_5^h}^{+} e_5^h TFPI$$

$$+ k_{TFPI:e_5^h}^{-} [TFPI : E_5^h] - k_{TFPI:e_{10}:e_5^h}^{+} [TFPI : E_{10}] e_5^h$$

$$+ k_{TFPI:e_{10}:e_5^h}^{-} [E_{10} : TFPI : E_5^h]$$

$$- k_{TFPI:e_{10}^m:e_5^h}^{+} [TFPI : E_{10}^m] c_5^h$$

$$+ k_{TFPI:e_{10}^m:e_5^h}^{-} [E_{10}^m : TFPI : E_5^h]$$

$$\frac{d}{dt}PRO^h = k^+_{e_5^{hm}:e_{10}^m} e_{10}^m e_5^{hm} - k^-_{e_5^{hm}:e_{10}^m} PRO^h \tag{92}$$

$$- k^+_{z_2^m:PRO^h} PRO^h z_2^m + k^-_{z_2^m:PRO^h}[Z_2^m : PRO^h]$$

$$+ k^{cat}_{z_2^m:PRO^h}[Z_2^m : PRO^h]$$

$$- k^+_{TFPI:PRO^h_{v10}} PRO^h[TFPI]$$

$$+ k^-_{TFPI:PRO^h_{v10}}[TFPI : PRO^h_{v10}]$$

$$- k^+_{TFPI:PRO^h_{v5}} PRO^h[TFPI]k^-_{TFPI:PRO^h_{v5}}$$

$$[TFPI : PRO^h_{v5}]$$

$$- k^+_{PRO^h:e_2^m} PRO^h e_2^m + k^-_{PRO^h:e_2^m}[PRO^h : E_2^m]$$

$$\frac{d}{dt}[Z_2^m : PRO^h] = k^+_{z_2^m:PRO^h} PRO^h z_2^m - k^-_{z_2^m:PRO^h}[Z_2^m : PRO^h] \tag{93}$$

$$- k^{cat}_{z_2^m:PRO^h}[Z_2^m PRO^h]$$

$$\frac{d}{dt}[E_5^{hm} : E_2^m] = k^+_{e_5^{hm}:e_2^m} e_2^m e_5^{hm} - k^-_{e_5^{hm}:e_2^m}[E_5^{hm} : E_2^m] \tag{94}$$

$$- k^{cat}_{e_5^{hm}:e_2^m}[E_5^{hm} : E_2^m]$$

$$\frac{d}{dt}[E_5^h : E_2] = + k^+_{e_5^h:e_2} e_2 e_5^h - k^-_{TFPI:e_5^h}[E_5^h : E_2] \tag{95}$$

$$- k^{cat}_{e_5^h:e_2}[E_5^h : E_2] + k_{flow}([E_5^h : E_2]^{up} - [E_5^h : E_2])$$

$$\frac{d}{dt}[TFPI : E_5^{hm}] = k^+_{TFPI:e_5^{hm}} e_5^{hm} TFPI \tag{96}$$

$$- k^-_{TFPI:e_5^{hm}}[TFPI : E_5^{hm}]$$

$$- k^+_{TFPI:e_5^{hm}:e_{10}m}[TFPI : E_5^{hm}]e_{10}^m$$

$$+ k^-_{TFPI:e_5^{hm}:e_{10}m}[E_{10}^m : TFPI : E_5^{hm}]$$

$$+ k^{on}_5[TFPI : E_5^h]p_5^{avail} - k^{off}_5[TFPI : E_5^{hm}]$$

$$- k^+_{TFPI:e_{10}:e_5^{hm}}[TFPI : E_5^{hm}]e_{10}^m$$

$$+ k^-_{TFPI:e_{10}:e_5^{hm}}[TFPI : PRO^h_{v5}]$$

$$- k^+_{TFPI:e_5^{hm}:e_{10}}[TFPI:E_5^{hm}]e_{10}$$

$$+ k^-_{TFPI:e_5^{hm}:e_{10}}[E_{10}:TFPI:E_5^{hm}]$$

$$\frac{d}{dt}[APC:E_5^{hm}] = k^+_{e_5^{hm}:APC}E_5^{hm}APC - k^-_{e_5^{hm}:APC}[APC:E_5^{hm}] \tag{97}$$

$$+ k^{cat}_{e_5^{hm}:APC}[APC:E_5hm]$$

$$\frac{d}{dt}[APC:E_5^{h}] = k^+_{e_5^{h}:APC}e_5^{h}APC - k^-_{e_5^{hm}:APC}[APC:E_5^{h}] \tag{98}$$

$$- k^{cat}_{e_5^{h}:APC}[APC:E_5^{h}]$$

$$+ k_{flow}([APC:E_5^{h}]^{up} - [APC:E_5^{h}])$$

$$\frac{d}{dt}[TFPI:E_5^{h}] = k_{TFPI:e_5^{h}up}e_5^{h}TFPI - k^-_{kTFPI:e_5^{h}}[TFPI:E_5^{h}] \tag{99}$$

$$+ k_{flow}([TFPI:E_5^{h}]^{up} - [TFPI:E_5^{h}])$$

$$- k^+_{TFPI:e_5^{h}:e_{10}}[TFPI:E_5^{h}]e_{10}$$

$$+ k^-_{TFPI:e_5^{h}:e_{10}}[E_{10}:TFPI:E_5^{h}]$$

$$+ k^{on}_5[TFPI:E_5^{h}]p_5^{avail} + k^{off}_5[TFPI:E_5^{hm}]$$

$$\frac{d}{dt}[TFPI:E_{10}^{m}] = k^+_{TFPI:e_{10}^{m}}e_{10}^{m}TFPI - k^-_{TFPI:e_{10}^{m}}[TFPI:E_{10}^{m}] \tag{100}$$

$$- k^+_{TFPI:e_{10}:e_5^{h}m}[TFPI:E_{10}^{m}]e_5^{hm}$$

$$+ k^-_{TFPI:e_{10}:e_5^{hm}}[E_{10}^{m}:TFPI:E_5^{hm}]$$

$$+ k^{on}_{10}[TFPI:E_{10}]p_{10}^{avail} - k^{off}_{10}[TFPI:E_{10}^{m}]$$

$$- k^+_{TFPI:e_{10}^{m}:e_5^{hm}}[TFPI:E_{10}^{m}]e_5^{hm}$$

$$+ k^-_{TFPI:e_{10}^{m}:e_5^{hm}}[TFPI:PRO_{v10}^{h}]$$

$$- k^+_{TFPI:e_{10}^{m}:e_5^{h}}[TFPI:E_{10}^{m}]e_5^{h}$$

$$+ k^-_{TFPI:e_{10}^{m}:e_5^{h}}[E_{10}^{m}:TFPI:E_5^{h}]$$

$$\frac{d}{dt}[TFPI:PRO_{v10}^h] = k_{TFPI:PRO_{v10}^h}^+ PRO^h[TFPI] \tag{101}$$

$$- k_{TFPI:PRO_{v10}^h}^- [TFPI:PRO_{v10}^h]$$

$$+ k_{TFPI:e_{10}^m:e_5^{hm}}^+ [TFPI:E_{10}^p][TFPI:E_{10}^m]e_5^{hm}$$

$$- k_{TFPI:e_{10}^m:e_5^{hm}}^- [TFPI:PRO_{v10}^h]$$

$$\frac{d}{dt}[TFPI:PRO_{v5}^h] = k_{TFPI:PRO_{v5}^h}^+ PRO^h[TFPI] \tag{102}$$

$$- k_{TFPI:PRO_{v5}^h}^- [TFPI:PRO_{v5}^h]$$

$$+ k_{TFPI:e_{10}:e_5^{hm}}^+ [TFPI:E_5^{hm}]e_{10}^m$$

$$- k_{TFPI:e_{10}:e_5^{hm}}^- [TFPI:PRO_{v5}^h]$$

$$\frac{d}{dt}[E_{10}^m:TFPI:E_5^{hm}] = k_{TFPI:e_{10}:e_5^h m}^+ [TFPI:E_{10}^m]e_5^{hm} \tag{103}$$

$$- k_{TFPI:e_{10}:e_5^h m}^- [E_{10}^m:TFPI:E_5^{hm}]$$

$$+ k_{TFPI:e_5^{hm}:e_{10}m}^+ [TFPI:E_5^{hm}]e_{10}^m$$

$$- k_{TFPI:e_5^{hm}:e_{10}m}^- [E_{10}^m:TFPI:E_5^{hm}]$$

$$+ k_{10}^{ont} p_{10}^{avail} - k_{10}^{offt}[E_{10}^m:TFPI:E_5^{hm}]$$

$$+ k_5^{ont}[E_{10}^m:TFPI:E_5^h]p_5^{avail}$$

$$- k_5^{offt}[E_{10}^m:TFPI:E_5^{hm}]$$

$$\frac{d}{dt}[E_{10}:TFPI:E_5^h] = k_{TFPI:e_{10}:e_5^h}^+ [TFPI:E_{10}]e_5^h \tag{104}$$

$$- k_{TFPI:e_{10}:e_5^h}^- [E_{10}:TFPI:E_5^h]$$

$$+ k_{TFPI:e_5^h:e_{10}}^+ [TFPI:E_5^h]e_{10}$$

$$- k_{TFPI:e_5^h:e_{10}}^- [E_{10}:TFPI:E_5^h]$$

$$- k_5^{ont}[E_{10}:TFPI:E_5^h]p_5^{avail}$$

$$+ k_5^{offt}[E_{10}:TFPI:E_5^{hm}]$$

$$- k_{10}^{ont}[E_{10}:TFPI:E_5^h]p_{10}^{avail}$$

$$+ k_{10}^{offt}[E_{10}^m : TFPI : E_5^h]$$

$$+ k_{flow}([E_{10} : TFPI : E_5^h]^{up} - [E_{10} : TFPI : E_5^h])$$

$$\frac{d}{dt}[E_{10} : TFPI : E_5^{hm}] = k_5^{ont}[E_{10} : TFPI : E_5^h]p_5^{avail} \tag{105}$$

$$- k_5^{offt}[E_{10} : TFPI : E_5^{hm}]$$

$$- k_{10}^{ont}[E_{10} : TFPI : E_5^{hm}]p_{10}^{avail}$$

$$+ k_{10}^{offt}[E_{10}^m : TFPI : E_5^{hm}]$$

$$+ k_{TFPI:e_5^{hm}:e_{10}}^+[TFPI : E_5^{hm}]e_{10}$$

$$- k_{TFPI:e_5^{hm}:e_{10}}^-[E_{10} : TFPI : E_5^{hm}]$$

$$\frac{d}{dt}[E_{10}^m : TFPI : E_5^h] = k_{10}^{ont}[E_{10} : TFPI : E_5^h]p_{10}^{avail} \tag{106}$$

$$- k_{10}^{offt}[E_{10}^m : TFPI : E_5^h]$$

$$- k_5^{ont}[E_{10}^m : TFPI : E_5^h]p_5^{avail}$$

$$+ k_5^{offt}[E_{10}^m : TFPI : E_5^{hm}]$$

$$+ k_{TFPI:e_{10}^m:e_5^h}^+[TFPI : E_{10}^m]e_5^h$$

$$- k_{TFPI:e_{10}^m:e_5^h}^-[E_{10}^m : TFPI : E_5^h]$$

$$\frac{d}{dt}[PRO^h : E_2^m] = k_{PRO^h:e_2^m}^+ PRO^h E_2^m \tag{107}$$

$$- k_{PRO^h:e_2^m}^-[PRO^h : E_2^m] - k_{PRO^h:e_2^m}^{cat}[PRO^h : E_2^m]$$

$$\frac{d}{dt}[E_9 : AT] = - k_9^{on}p_9^{avail}[E_9 : AT] + k_9^{off}[E_9^m : AT] \tag{108}$$

$$+ k_{e9}^{AT}e_9[AT] + k_{flow}([E_9 : AT]_{up} - [E_9 : AT])$$

$$- k_9^{on}p_{91}^{avail}[E_9 : AT] + k_9^{off}[E_9^{m*} : AT]$$

$$\frac{d}{dt}[E_9^m : AT] = k_{e_9^m}^{AT}e_9^m[AT] - k_9^{off}[E_9^m : AT] \tag{109}$$

$$+ k_9^{on}p_9^{avail}[E_9 : AT]$$

$$\frac{d}{dt}[E_9^{m*} : AT] = k_{e_9^m}^{AT}e_9^{m*}[AT] - k_9^{off}[E_9^{m*} : AT] \tag{110}$$

$$+ k_9^{on}p_{91}^{avail}[E_9 : AT]$$

$$\frac{d}{dt}[E_{10}:AT] = k_{e_{10}}^{AT} e_{10}[AT] + k_{flow}([E_{10}:AT]_{up} - [E_{10}:AT]) \tag{111}$$

$$+ k_{10}^{off}[E_{10}^m:AT] - k_{10}^{on} p_{10}^{avail}[E_{10}:AT]$$

$$\frac{d}{dt}[E_{10}^m:AT] = + k_{e_{10}^m}^{AT} e_{10}^m[AT] - k_{10}^{off}[E_{10}^m:AT] \tag{112}$$

$$+ k_{10}^{on} p_{10}^{avail}[E_{10}:AT]$$

$$\frac{d}{dt}[E_2:AT] = + k_{e_2}^{off}[E_2^m:AT] - k_{e_2}^{on} p_2^{avail}[E_2:AT] \tag{113}$$

$$+ k_{e_2}^{AT} e_2 + k_{flow}([E_2:AT]_{up} - [E_2:AT])$$

$$\frac{d}{dt}[E_2^m:AT] = k_{e_2^m}^{AT} e_2^m[AT] - k_{e_2}^{off}[E_2^m:AT] \tag{114}$$

$$+ k_{e_2}^{on} p_2^{avail}[E_2:AT]$$

$$\frac{d}{dt}[E_{11}:AT] = k_{e_{11}}^{AT} e_{11}[AT] - k_{e_{11}}^{AT}[E_{11}:AT][AT] \tag{115}$$

$$+ k_{11}^{off}[E_{11}^{m*}:AT] - k_{11}^{on} p_{111}^{avail}[E_{11}:AT]$$

$$\frac{d}{dt}[AT:E_{11}:AT] = k_{e_{11}}^{AT}[E_{11}:AT][AT] \tag{116}$$

$$\frac{d}{dt}[E_{11}^{m*}:AT] = k_{e_{11}}^{AT} e_{11}^{m*}[AT] - k_{11}^{off}[E_{11}^{m*}:AT] \tag{117}$$

$$+ k_{11}^{on} p_{111}^{avail}[E_{11}:AT]$$

$$\frac{d}{dt}[E_{11}^h:AT] = k_{e_{11}}^{AT} e_{11}^h[AT] + k_{11}^{off}[E_{11}^{hm}:AT] \tag{118}$$

$$- k_{11}^{on} p_{11}^{avail}[E_{11}^h:AT]$$

$$\frac{d}{dt}[E_{11}^{hm}:AT] = k_{e_{11}}^{AT} e_{11}^{hm}[AT] - k_{11}^{off}[E_{11}^{hm}:AT] \tag{119}$$

$$+ k_{11}^{on} p_{11}^{avail}[E_{11}^h:AT]$$

$$\frac{d}{dt}[AT] = - k_{e_9}^{AT} e_9[AT] - k_{e_9^m}^{AT} e_9^m[AT] + k_{e_9^m}^{AT} e_9^{m*}[AT] \tag{120}$$

$$- k_{e_{10}}^{AT} e_{10}[AT] + k e_{10}^{m\,AT} e_{10}^m[AT] - k_{e_2}^{AT} e_2[AT]$$

$$- k_{e_2^m}^{AT} e_2^m[AT] - k_{e_{11}}^{AT} e_{11}[AT] - k_{e_{11}}^{AT}[E_{11}:AT][AT]$$

$$- k_{e_{11}}^{AT} e_{11}^{m*}[AT] - k_{e_{11}}^{AT} e_{11}^h[AT] - k_{e_{11}}^{AT} e_{11}^h[AT]$$

$$+ k_{flow}([AT]_{up} - [AT])$$

$$\frac{d}{dt}[TM_{RZ}:E_2] = k^{on}_{TM_{RZ}}e_2TM^{avail}_{RZ} - k^{off}_{TM_{RZ}}[TM_{RZ}:E_2] \tag{121}$$

$$- k^+_{pc}[TM_{RZ}:E_2][PC]$$

$$+ (k^-_{pc} + k^{cat}_{pc})[TM_{RZ}:E_2:PC]$$

$$\frac{d}{dt}[TM_{RZ}:E_2:PC] = k^+_{pc}[TM_{RZ}:E_2][PC]$$

$$- (k^-_{pc} + k^{cat}_{pc})[TM_{RZ}:E_2:PC] \tag{122}$$

Acknowledgments The work described herein was initiated during the Collaborative Workshop for Women in Mathematical Biology funded and hosted by UnitedHealth Group Optum of Minnetonka, MN and supported by University of Minnesota's Institute for Mathematics and its Applications in June 2022. Additionally, the authors and editors thank the anonymous peer reviewers for their feedback, which strengthened this work.

The authors are grateful to have participated in the workshop. We would like to thank Rebecca A. Segal, Blyrta Shtylla, Suzanne S. Sindi, and Ashlee N. Ford Versypt for organizing the workshop. We would like to thank Emma Bouck for helpful conversations regarding this work and Kenji Miyazawa for making figures and providing codes. Finally, this work was, in part, supported by the National Institutes of Health (R01 HL151984 to KL and SS), the National Science Foundation CAREER (DMS-1848221 to KL), and the National Science Foundation RTG DMS-2038056 to ACN.

References

1. F.R. Rosendaal, F.M. Helmerhorst, J.P. Vandenbroucke, Arterioscler Thromb. Vasc. Biol. **22**(2), 201 (2002). https://doi.org/10.1161/hq0202.102318
2. The Oral Contraceptive and Hemostasis Study Group, Contraception **67**(3), 173 (2003). https://doi.org/10.1016/S0010-7824(02)00476-6
3. M. Cushman, L.H. Kuller, R. Prentice, R.J. Rodabough, B.M. Psaty, R.S. Stafford, S. Sidney, F.R. Rosendaal, Women's Health Initiatives Investigators, J. Am. Med. Assoc. **292**(13), 1573 (2004)
4. A. Artero, J. Tarín, A. Cano, Semin. Thromb. Hemostasis **38**(08), 797 (2012). https://doi.org/10.1055/s-0032-1328883
5. M.Y. Abou-Ismail, D. Citla Sridhar, L. Nayak, Thromb. Res. **192**, 40 (2020). https://doi.org/10.1016/j.thromres.2020.05.008
6. R.D.T. Farmer, T.D. Preston, J. Obstet. Gynaecol. **15**(3), 195 (1995)
7. H. Jick, J.A. Kaye, C. Vasilakis-Scaramozza, S.S. Jick, Br. Med. J. **321**(7270), 1190 (2000)
8. N.R. Poulter, O. Meirik, C.L. Chang, T.M.M. Farley, M.G. Marmot, Int. J. Gynecol. Obstet. **54**(1), 81 (1996)
9. S.N. Tchaikovski, J. Rosing, Thromb. Res. **126**(1), 5 (2010)
10. J.M. Kemmeren, A. Algra, J.C.M. Meijers, B.N. Bouma, D.E. Grobbee, Thromb. Haemostasis **87**(02), 199 (2002)
11. V. Odlind, I. Milsom, I. Persson, A. Victor, Acta Obstet. Gynecol. Scand. **81**(6), 482 (2002)
12. R. Sitruk-Ware, G. Plu-Bureau, J. Menard, J. Conard, S. Kumar, J.C. Thalabard, B. Tokay, P. Bouchard, J. Clin. Endocrinol. Metab. **92**(6), 2074 (2007). https://doi.org/10.1210/jc.2007-0026

13. J.V. Johnson, J. Lowell, G.J. Badger, J. Rosing, S. Tchaikovski, M. Cushman, Obstet. Gynecol. **111**(2), 278 (2008). https://doi.org/10.1097/AOG.0b013e3181626d1b
14. A.C.P. Ferreira, M.B.A. Montes, S.A. Franceschini, M.R.T. Toloi, Contraception **64**(6), 353 (2001). https://doi.org/10.1016/S0010-7824(01)00274-8
15. J. Meijers, S. Middeldorp, W. Tekelenburg, A. van den Ende, G. Tans, M. Prins, J. Rosing, H. Büller, B. Bouma, Thromb. Haemostasis **84**(07), 9 (2000). https://doi.org/10.1055/s-0037-1613959
16. G. Tans, J. Curvers, S. Middeldorp, M.C. Thomassen, J. Meijers, M. Prins, B. Bouma, H. Büller, J. Rosing, Thromb. Haemostasis **84**(07), 15 (2000). https://doi.org/10.1055/s-0037-1613960
17. S. Middeldorp, J.C.M. Meijers, A.E. van den Ende, A. van Enk, B.N. Bouma, G. Tans, J. Rosing, M.H. Prins, H.R. Büller, Thromb. Haemostasis **84**(7), 4 (2000). https://doi.org/10.1055/s-0037-1613958
18. B.B. Gerstman, J.M. Piper, D.K. Tomita, W.J. Ferguson, B.V. Stadel, F.E. Lundin, Am. J. Epidemiol. **133**(1), 32 (1991)
19. W.O. Spitzer, M.A. Lewis, L.A. Heinemann, M. Thorogood, Br. Med. J. **312**(7023), 83 (1996)
20. A. Santamaría, J. Mateo, A. Oliver, B. Menéndez, J.C. Souto, M. Borrell, J.M. Soria, I. Tirado, J. Fontcuberta, Haematologica **86**(9), 965 (2001)
21. B.S. Andersen, J. Olsen, G.L. Nielsen, F.H. Steffensen, H.T. Sørensen, J. Baech, H. Gregersen, Thromb. Haemostasis **59**(01), 28 (1998)
22. A.L. Fogelson, K.B. Neeves, Ann. Rev. Fluid Mech. **47**, 377 (2015)
23. M.C.H. de Visser, F.R. Rosendaal, R.M. Bertina, Blood **93**(4), 1271 (1999)
24. G. Tans, A. van Hylckama Vlieg, M.C.L.G.D. Thomassen, J. Curvers, R.M. Bertina, J. Rosing, F.R. Rosendaal, Br. J. Haematol. **122**(3), 465 (2003)
25. J. Rosing, G. Tans, G.A.F. Nicolaes, M.C.L.G.D. Thomassen, R. Van Oerle, P.M.E.N. Van DerPloeg, P. Heijnen, K. Hamulyak, H.C. Hemker, Br. J. Haematol. **97**(1), 233 (1997). https://doi.org/10.1046/j.1365-2141.1997.192707.x
26. J. Curvers, M.C.L. Thomassen, G.A. Nicolaes, R. Van Oerle, K. Hamulyak, H. Coenraad HEMKER, G. Tans, J. Rosing, Br. J. Haematol. **105**(1), 88 (1999)
27. M.C.H. de Visser, A.V.H. Vlieg, G. Tans, J. Rosing, A.E.A. Dahm, P.M. Sandset, F.R. Rosendaal, R.M. Bertina, J. Thromb. Haemostasis **3**(7), 1488 (2005)
28. A.L. Kuharsky, A.L. Fogelson, Biophys. J. **80**(3), 1050 (2001). https://doi.org/10.1016/S0006-3495(01)76085-7
29. A.L. Fogelson, N. Tania, Pathophysiol. Haemostasis Thromb. **34**(2–3), 91 (2005). https://doi.org/10.1159/000089930
30. A.L. Fogelson, Y.H. Hussain, K. Leiderman, Biophys. J. **102**(1), 10 (2012). https://doi.org/10.1016/j.bpj.2011.10.048
31. K. Miyazawa, A.L. Fogelson, K. Leiderman, Biophys. J. **122**(1), 99 (2023)
32. K. Miyazawa, A.L. Fogelson, K. Leiderman, Biophys. J. **122**(1), 230 (2023)
33. K.G. Link, M.T. Stobb, J. Di Paola, K.B. Neeves, A.L. Fogelson, S.S. Sindi, K. Leiderman, PLoS ONE **13**(7), e0200917 (2018). https://doi.org/10.1371/journal.pone.0200917
34. F.M. Dekking, C. Kraaikamp, H.P. Lopuhaä, L.E. Meester, *A Modern Introduction to Probability and Statistics: Understanding Why and How*, vol. 488 (Springer, London, 2005)
35. M. Hoffman, D. Monroe, Thromb. Haemost. **85**, 958 (2001)
36. A. Fogelson, N. Tania, Pathophysiol. Haemost. Thromb. **34**, 91 (2005)
37. A. Hindmarsh, in *Scientific Computing*. IMACS Transactions on Scientific Computing, ed. by R.S. Stepleman, vol. 1 (North-Holland, Amsterdam, 1983), pp. 55–64
38. K. Mann, M. Nesheim, W. Church, P. Haley, S. Krishnaswamy, Blood **76**, 1 (1990)
39. K. Mann, E. Bovill, S. Krishnaswamy, Ann. N. Y. Acad. Sci. **614**, 63 (1991)
40. J.H. Morrissey, Thromb. Haemost. **74**, 185 (1995)
41. W. Novotny, S. Brown, J. Miletich, D. Rader, G. Broze, Blood **78**, 387 (1991)
42. H.J. Weiss, N. Engl. J. Med. **293**, 531 (1975)

43. P.N. Walsh, in *Hemostasis and Thrombosis: Basic Principles and Clinical Practice*, 3rd edn., ed. by R.W. Colman, J. Hirsh, V.J. Marder, E.W. Salzman (Lippincott, Philadelphia, 1994), pp. 629–651
44. M.E. Nesheim, D.D. Pittman, J.H. Wang, D. Slonosky, A.R. Giles, R.J. Kaufman, J. Biol. Chem. **263**, 16467 (1988)
45. S. Ahmad, R. Rawala-Sheikh, P. Walsh, J. Biol. Chem. **264**, 3244 (1989)
46. K. Mann, S. Krishnaswamy, J. Lawson, Semin. Hematol. **29**, 213 (1992)
47. F. Baglia, B. Jameson, P. Walsh, J. Biol. Chem. **270**, 6734 (1995)
48. T. Miller, D. Sinha, T. Baird, P. Walsh, Biochemistry **46**, 14450 (2007)
49. P. Tracy, L.L. Eide, E.J. Bowie, K.G. Mann, Blood **60**, 59 (1982)
50. J.A. Hubbell, L.V. McIntire, Biophys. J. **50**, 937 (1986)
51. S.T. Olson, I. Björk, J.D. Shore, in *Methods in Enzymology*, vol. 222 (Academic Press, Cambridge, 1993), pp. 525–559. https://doi.org/10.1016/0076-6879(93)22033-C
52. S. Krishnaswamy, K.C. Jones, K.G. Mann, J. Biol. Chem. **263**, 3823 (1988)
53. K.G. Mann, in *Hemostasis and Thrombosis: Basic Principles and Clinical Practice*, 3rd edn., ed. by R. Colman, J. Hirsh, V. Marder, E. Salzman (Lippincott, Philadelphia, 1994), pp. 184–199
54. J. Greengard, M. Heeb, E. Ersdal, P. Walsh, J. Griffin, Biochemistry **25**, 3884 (1986)
55. S. Butenas, K. Mann, Biochemistry **35**, 1904 (1996)
56. S.A. Limentani, B.C. Furie, B. Furie, in *Hemostasis and Thrombosis: Basic Principles and Clinical Practice*, 3rd edn., ed. by R.W. Colman, J. Hirsh, V.J. Marder, E.W. Salzman (Lippincott, Philadelphia, 1994), pp. 94–108
57. Y. Nemerson, Semin. Hematol. **29**, 170 (1992)
58. D.D. Monkovic, P.B. Tracy, J. Biol. Chem. **265**, 17132 (1990)
59. D. Hill-Eubanks, P. Lollar, J. Biol. Chem. **265**, 17854 (1990)
60. P. Lollar, G. Knutson, D. Fass, Biochemistry **24**, 8056 (1985)
61. D. Gailani, G. Broze Jr., Science **253**, 909 (1991)
62. D. Gailani, D. Ho, M.F. Sun, Q. Cheng, P.N. Walsh, Blood **97**, 3117 (2001)
63. D. Sinha, M. Marcinkiewicz, D. Navaneetham, P. Walsh, Biochemistry **46**, 9830 (2007)
64. V.T. Turitto, H.R. Baumgartner, Microvasc. Res. **17**, 38 (1979)
65. V.T. Turitto, H.J. Weiss, H.R. Baumgartner, Microvasc. Res. **19**, 352 (1980)
66. A.R.L. Gear, Can. J. Physiol. Pharmacol. **72**, 285 (1994)
67. D. Monkovic, P. Tracy, Biochemistry **29**, 1118 (1990)
68. K.G. Mann, Trends Biochem. Sci. **12**, 229 (1987)
69. R. Rawala-Sheikh, S. Ahmad, B. Ashby, P.N. Walsh, Biochemistry **29**, 2606 (1990)
70. M.E. Nesheim, R.P. Tracy, P.B. Tracy, D.S. Boskovic, K.G. Mann, Methods Enzymol. **215**, 316 (1992)
71. S.T. Olson, R. Swanson, E. Raub-Segall, T. Bedsted, M. Sadri, M. Petitou, J.P. Hérault, J.M. Herbert, I. Björk, Thromb. Haemostasis **92**(11), 929 (2004)
72. S. Solymoss, M. Tucker, P. Tracy, J. Biol. Chem. **263**, 14884 (1988)
73. J. Jesty, T.C. Wun, A. Lorenz, Biochemistry **33**(42), 12686 (1994). https://doi.org/10.1021/bi00208a020
74. S.A. Maroney, A.E. Mast, J. Thromb. Haemost. **13**, S200 (2015). https://doi.org/10.1111/jth.12897. https://onlinelibrary.wiley.com/doi/10.1111/jth.12897
75. J. Jesty, T. Wun, A. Lorenz, Biochemistry **33**, 12686 (1994)
76. J.H. Griffin, in *Williams Hematology*, ed. by E. Beutler, B. Coller, M. Lichtman, T. Kipps, U. Seligsohn (McGraw Hill, New York, 2001), pp. 1435–1447
77. S.T. Olson, I. Björk, R. Sheffer, P. Craig, J.D. Shore, J. Choay, J. Biol. Chem. **267**(18), 12528 (1992). https://doi.org/10.1016/S0021-9258(18)42309-5
78. G. Broze, J. Miletich, in *Hemostasis and Thrombosis: Basic Principles and Clinical Practice*, ed. by R.W. Colman, J. Hirsh, V.J. Marder, E.W. Salzman (Lippincott, Philadelphia, 1994), pp. 259–276

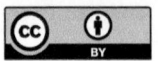

Deconstructing the Contributions of Heterogeneity to Combination Treatment of Hormone-Sensitive Breast Cancer

Samantha Linn, Jenna A. Moore-Ott, Robyn Shuttleworth, Wenjing Zhang, Morgan Craig, and Adrianne L. Jenner

1 Introduction

Breast cancer is the most common invasive malignancy in women; a woman has a one in eight chance of developing breast cancer in her lifetime [1, 2]. The treatment of breast cancer requires a multifaceted approach combining surgery, radiation, neoadjuvant, and adjuvant treatments [3]. There are five molecular subtypes of breast cancer (luminal A, luminal B, HER-2, basal, and normal)—each with a different combination of cancer cells that over- or under-express progesterone receptor (PR+/−), estrogen receptor (ER+/−), and human epidermal growth factor receptor 2 (HER2+/−) [4]. Effective treatment of these varying subtypes

S. Linn
Department of Mathematics, University of Utah, Salt Lake City, UT, USA

J. A. Moore-Ott
Department of Chemical and Biological Engineering, Princeton University, Princeton, NJ, USA

R. Shuttleworth
Department of Biology, University of Saskatchewan, Saskatoon, SK, Canada

Altos Labs, Redwood City, CA, USA

W. Zhang
Department of Mathematics and Statistics, Texas Tech University, Lubbock, TX, USA

M. Craig
Department of Mathematics and Statistics, Université de Montréal, Sainte-Justine University Hospital Azrieli Research Centre, Montréal, QC, Canada
e-mail: morgan.craig@umontreal.ca

A. L. Jenner (✉)
School of Mathematical Sciences, Queensland University of Technology, Brisbane, QLD, Australia
e-mail: adrianne.jenner@qut.edu.au

© The Author(s) 2024
A. N. Ford Versypt et al. (eds.), *Mathematical Modeling for Women's Health*,
The IMA Volumes in Mathematics and its Applications 166,
https://doi.org/10.1007/978-3-031-58516-6_5

of breast cancer requires a deep understanding of heterogeneity in their responses to the different treatment types; unfortunately, there is still no completely curative treatment for any subtype.

Combination therapy, i.e., combining two or more therapeutic agents, is a cornerstone of cancer therapy [5]. The major goal of combination therapy in oncology is to enhance the therapeutic efficacy of a single anti-cancer drug through co-administration with a synergistic or additive drug that targets key pathways [5]. For example, metformin, an agent used to treat type 2 diabetes, was found to increase the susceptibility of a p53 breast cancer cell line to therapeutic molecule tumor necrosis factor-related apoptosis inducing ligand (TRAIL) [6]. To that end, a combination therapy in breast cancer with recognized potential is palbociclib combined with fulvestrant [7, 8], which was approved in early 2016 by the FDA to treat hormone receptor-positive breast cancer [9].

Palbociclib (brand name Ibrance) is an orally available, highly selective inhibitor of cyclin-dependent kinase 4 and 6 (CDK4 and CDK6) [10–12]. CDK4/6 are critical mediators of the cellular transition into the S phase and are crucial for the initiation, growth, and survival of many cancer types [13]. As such, pharmacological inhibitors of CDK4/6 are rapidly becoming a new standard of care for patients with advanced hormone receptor-positive breast cancer. Palbociclib is an inhibitor of CDK4/6 and thus forces cells to stay in the G1 phase in lieu of undergoing cell division (Fig. 1). Importantly, palbociclib does not induce apoptosis but instead halts cellular division. According to the US National Institutes of Health (NCT03007979), patients have historically been on a 21-day-on, 7-day-off palbociclib schedule, though there were concerns that the off days were the reason behind worse patient outcomes [14]. A 5-day-on, 2-day-off schedule has thus far shown better health outcomes for the treatment of ER+ breast cancer [14, 15], though this study is still ongoing.

Fulvestrant is a novel endocrine therapy for breast cancer that binds, blocks, and degrades the estrogen receptor, leading to complete inhibition of estrogen signaling through the ER [16, 17]. Through extensive preclinical and clinical trials, fulvestrant has demonstrated improved clinical efficacy compared to established endocrine agents [17]. Fulvestrant has been combined with several different classes of therapeutics, in particular, CDK4/6 inhibitors [16]. The PALOMA-3 study investigated fulvestrant with palbociclib or placebo in both pre- and postmenopausal patients who had progressed on previous endocrine treatment [18, 19]. The trial demonstrated a substantial increase in progression-free survival from 4.6 months to 9.5 months in the placebo compared to palbociclib arms [16]. The FLIPPER trial was a phase II study comparing fulvestrant and palbociclib with fulvestrant and placebo in the first-line metastatic setting [20].

While it can be challenging to fully capture the effects of heterogeneity on treatment outcomes experimentally and clinically, mathematical modeling is well-placed to provide insight into how cancer treatments are affected by multiple scales of heterogeneity. Previously, groups have used deterministic mathematical models to examine the combined treatment of breast cancer using palbociclib and AZD9496 [21]. For example, He et al. [22] used a mathematical model that captured the cell cycle and signaling pathways in response to endocrine therapy and CDK4/6 inhibition. Their model successfully predicted the combined effects of estrogen deprivation and palbociclib and was used to explore combination scheduling. Mathematical modeling studies can also be extended to a virtual or in silico clinical trial setting to account for variations in patient characteristics and comprehensively explore dosing regimens in ways that are clinically unfeasible [23–26]. There are many examples of virtual clinical trials employed in cancer therapies [27–31] to this end, and this approach continues to gain traction within pharmaceutical and other applications [32, 33].

As heterogeneity can impact combination strategies aimed at CDK4/6 inhibition at multiple levels, we developed a simple mathematical model of two unique ER+ breast cancer cell types and their responses to combination treatment with palbociclib and fulvestrant to understand how different sources of variation impact this therapeutic approach. We examined *in situ* how co-culturing of heterogeneous cell types, specifically two commonly used breast cancer cell lines exhibiting different degrees of aggressivity, affects their responses to treatment. Using our model, we next explored how interindividual variability in PK within a virtual breast cancer patient cohort affects treatment outcomes. Lastly, we used our integrated framework to establish how therapeutic scheduling determines treatment responses, providing insight into effective regimens using this combination treatment.

2 Methods

2.1 Mathematical Model of Breast Cancer Co-cultures and Combination Therapy

In this section, we detail the development of a mathematical model to capture the action of palbociclib and fulvestrant on a heterogeneous population of breast cancer cells. This type of model of the effects of the drugs on the target tissue in the body is known as a pharmacodynamic (PD) model. To capture and understand how intrinsic cell characteristics affect combination palbociclib and fulvestrant treatment, we considered co-cultures of MCF7 and T47D cell lines—two commonly used breast cancer cell lines that display different sensitivities to each drug, with T47D thought to be more aggressive (i.e., exhibit stronger/faster growth) than MCF7. A schematic overview of our model is provided in Fig. 1.

2.1.1 Palbociclib's Impact on Cell Growth

We first constructed a mathematical PD model describing the growth of a cell population under treatment by palbociclib, a drug whose effects were assumed to inhibit the cell cycle. To model these effects, we adopted a general inhibitory effects model given by

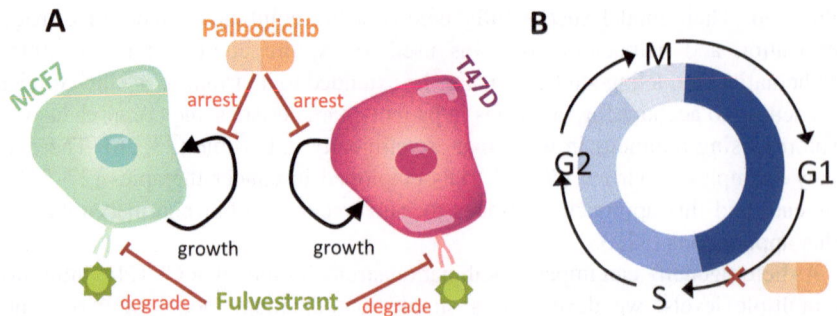

Fig. 1 Schematic summarizing the mathematical model of combination therapy on breast cancer cells. (**a**) Our model consists of pharmacokinetics and pharmacodynamics (PK/PD) of two drugs (palbociclib and fulvestrant), each with different mechanisms of action. Palbociclib targets and arrests the cell in the cell cycle, and fulvestrant degrades the estrogen receptor on cells, essentially causing cell death. To examine the impact of heterogeneity on tumor composition prior to, during, and at the end of treatment with this combination, we considered heterogeneous tumors composed of less aggressive and more aggressive cells. We parameterized the model's parameters to *in vitro* data from two cell lines: MCF7 and T47D. (**b**) Schematic overview of the mechanism of action of palbociclib on cell cycle arrest—inhibiting the cell cycle transition from G1 to S phase

$$E_i = E_{0,i} - \frac{E_{0,i} I_{max,i} [P]^{h_i}}{[P]^{h_i} + [IC_{50,i}]^{h_i}}, \tag{1}$$

where E_i with $i = M, T$ specifies the effect on cell line MCF7 and T47D, respectively, P is the concentration of palbociclib at the tumor site, $E_{0,i}$ denotes the baseline effect of the drug palbociclib on cell type i, $I_{max,i}$ represents the maximal effect of the drug at high concentrations, h_i is the Hill coefficient measuring the slope of the inhibitory curve for cell type i, and $IC_{50,i}$ represents the drug concentration eliciting 50% of the maximal inhibition. This model formulation is regularly used to capture the effect of a drug on inhibiting a cell population [34].

As palbociclib arrests cells in the cell cycle, the general growth inhibition model for a population of cells of type i, $C_i(t)$, inhibited by palbociclib is given by

$$\frac{dC_i}{dt} = \lambda(C_i) E_i C_i,$$

where $\lambda(C_i)$ is a function describing cell population growth in the absence of treatment (see Fig. 1). As *in vitro* tumor growth is constrained by the availability of nutrients, space, etc., we modeled cell growth using the logistic growth law

$$\lambda(C_i) = r_i \left(1 - \frac{C_i}{K}\right), \tag{2}$$

where r_i is the cell line-specific proportionality constant and K is the total cell population carrying capacity in a given space. We chose logistic growth as this provided the most accurate fit to cell count measurements; however, Gompertzian tumor growth also provided a close (but less accurate) fit to our data (Figs. 10 and 11). Thus, the complete model of monoculture growth under treatment with palbociclib is given by

$$\frac{dC_i}{dt} = C_i r_i \left(1 - \frac{C_i}{K}\right) \left(E_{0,i} - \frac{E_{0,i} I_{max,i}[P]^{h_i}}{[P]^{h_i} + [I_{50,i}]^{h_i}}\right).$$

Typically, tumors are not homogeneous in nature and are comprised of a variety of different cell types. We accounted for this phenotypic heterogeneity by modeling both MCF7 and T47D cell types within a single tumor environment in co-culture. As mentioned above, while both MCF7 and T47D are ER+, they differ in their responses to treatment. We, therefore, considered each to have separate parameters, with co-culture growth rates affected by the available space in the domain. In addition, since the effect parameters will depend on the drug being applied, we now update our PD variables to be drug specific. In other words, since parameters are cell line and drug specific, the specific combination is represented in their subscript as $X_{CELL^{drug}}$, where $drug$ is denoted by either p for palbociclib or f for fulvestrant and the cell line is denoted either by M for MCF7 or by T for T47D.

For simplicity, and owing to the absence of data, we did not consider switching between tolerant and resistant types [35]. We assumed that the carrying capacity and growth of each cell type is affected by the presence of the other cell type in the dish, thus modifying our logistic growth model. Therefore, to account for the impact of variable growth between each cell type, we included cell-specific carrying capacities in our model of cell growth in Eq. (2). Our final model describing the change in population of two, indirectly interacting, cell types (MCF7 and T47D) is given by

$$\frac{dC_M}{dt} = C_M r_M \left(1 - \frac{C_M + C_T}{K_M \phi_M + K_T \phi_T}\right) \left(E_{0,Mp} - \frac{E_{0,Mp} I_{max,Mp}[P]^{h_{Mp}}}{[P]^{h_{Mp}} + [I_{50,Mp}]^{h_{Mp}}}\right) \tag{3}$$

$$\frac{dC_T}{dt} = C_T r_T \left(1 - \frac{C_M + C_T}{K_M \phi_M + K_T \phi_T}\right) \left(E_{0,Tp} - \frac{E_{0,Tp} I_{max,Tp}[P]^{h_{Tp}}}{[P]^{h_{Tp}} + [I_{50,Tp}]^{h_{Tp}}}\right), \tag{4}$$

where ϕ_i is the volume fraction of cell type i for either type MCF7 or T47D, respectively, in the domain and is calculated by $\phi_i = C_i / (C_M + C_T)$ with $\phi_M + \phi_T = 1$. The global carrying capacity in the domain is given by $K = K_M \phi_M + K_T \phi_T$, where K_i is the individual carrying capacity for each cell type, MCF7 and T47D, respectively.

2.1.2 Modeling the Effects of Fulvestrant on a Heterogenous Tumor

As fulvestrant degrades cells, we modeled its effect on the rates of decay for both the MCF7 and T47D cells. We updated the ordinary differential equation (ODE) system for the effect of palbociclib in Eqs. (3) and (4) to include a decay term for both cell populations, d_i, that is affected by the concentration of fulvestrant (F), using a modified version of the effect function in Eq. (1):

$$\frac{dC_M}{dt} = C_M r_M \left(1 - \frac{C_M + C_T}{K_M \phi_M + K_T \phi_T} \right) \left(E_{0,MP} - \frac{E_{0,MP} I_{max,MP} [P]^{h_{MP}}}{[P]^{h_{MP}} + [I_{50,MP}]^{h_{MP}}} \right)$$
$$- C_M d_M \left(\frac{E_{0,Mf} I_{max,Mf} [F]^{h_{Mf}}}{[F]^{h_{Mf}} + [IC_{50,Mf}]^{h_{Mf}}} \right) \tag{5}$$

$$\frac{dC_T}{dt} = C_T r_T \left(1 - \frac{C_M + C_T}{K_M \phi_M + K_T \phi_T} \right) \left(E_{0,TP} - \frac{E_{0,TP} I_{max,TP} [P]^{h_{TP}}}{[P]^{h_{TP}} + [I_{50,TP}]^{h_{TP}}} \right)$$
$$- C_T d_T \left(\frac{E_{0,Tf} I_{max,Tf} [F]^{h_{Tf}}}{[F]^{h_{Tf}} + [IC_{50,Tf}]^{h_{Tf}}} \right), \tag{6}$$

where $E_{0,if}$ is the basal effect of fulvestrant on cell type i, $I_{max,if}$ is the maximum effect of fulvestrant, h_{if} is the Hill coefficient for fulvestrant, and $IC_{50,if}$ is the half-effect of fulvestrant, given i represents either M or T for cell type MCF7 or T47D, respectively. Note that this modified effect function for fulvestrant aims to capture the death rate increase that results from fulvestrant concentration F increase. To determine the concentration of palbociclib and fulvestrant after administration, we introduced pharmacokinetic (PK) models parameterized from clinical PK studies for both drugs.

2.1.3 Palbociclib and Fulvestrant Pharmacokinetic Models

We used a linear two-compartment PK model with first-order absorption and absorption lag to model the dynamics of orally administered palbociclib,

$$\frac{dM_0}{dt} = -k_a M_0, \tag{7}$$

$$\frac{dM_1}{dt} = k_a M_0 - M_1 \left(\frac{k_e + k_{el}}{V_C} \right) + M_2 \left(\frac{k_e}{V_P} \right), \tag{8}$$

$$\frac{dM_2}{dt} = -M_2 \left(\frac{k_e}{V_P} \right) + M_1 \left(\frac{k_e}{V_C} \right), \tag{9}$$

where M_0, M_1, and M_2 are the palbociclib concentrations pre-absorption, in plasma, and in peripheral tissue, respectively. Furthermore, k_a is the rate of absorption into plasma, k_{el} is the rate of linear elimination, k_e is the exchange rate between plasma and tissue, and V_C and V_P are the apparent plasma and peripheral tissue volumes, respectively. The concentration of palbociclib at the tumor site is then calculated by $P(t) = M_1(t)/V_C$ in Eqs. (5) and (6).

Based on data from 38 postmenopausal women with advanced breast cancer who received 250 mg doses of extended-release fulvestrant (in a single 5 mL intramuscular (IM) injection or two 2.5 mL IM injections [36]), a two-compartment PK model with zero-order administration and linear elimination was developed. The fulvestrant PK model is given by

$$\frac{dF_1}{dt} = In - k_{el}F_1 + k_{21}F_2 - k_{12}F_1, \tag{10}$$

$$\frac{dF_2}{dt} = -k_{21}F_2 + k_{12}F_1, \tag{11}$$

$$In = \frac{D}{T_{k0}V}, \tag{12}$$

where F_1 and F_2 denote fulvestrant concentrations in the plasma and tissues, respectively, In represents the administered dose (here taken to be an IM administration), k_{el} is the rate of linear elimination, k_{12} and k_{21} are transit rates between the plasma and tissue compartments, D represents the IM dose, T_{k0} is the time for absorption, and V is the volume of distribution. The concentration of fulvestrant at the tumor site is set as $F(t) = F_1(t)$ in Eqs. (5) and (6).

2.2 Parameter Estimation

2.2.1 Estimating Tumor Growth Parameters

Cell counting was performed in breast cancer cell lines MCF7 and T47D by Vijayaraghavan et al. [37]. Cells were plated in six-well plates and treated with indicated agents for 10 days. The medium was replaced every other day over the course of the experiment. Cells were then collected and counted using BioRad TC20 Automated Cell Counter on days 0, 3, 6, and 10 (see data in Figs. 10 and 11). We estimated parameters governing cell growth by setting all drug concentrations to zero in our model, Eqs. (5) and (6), and fitting the proliferation rate r_i and carrying capacity K_i to cell type i count data. Fitting was performed in MATLAB using the nonlinear least-squares fitting function lsqnonlin; the trust-region-reflective algorithm with 1000 maximum function evaluations was chosen. The model was solved using ode45.

2.2.2 Estimating Drug Effect Parameters from Cell Viability Assays

Cell viability measurements for MCF7 and T47D with palbociclib were measured by Vijayaraghavan et al. [37]. For these dose–response studies, cells were plated on a 96-well plate and treated with increasing concentrations $(0.01–12\,\mu M)$ of palbociclib for 1, 2, 4, 6, or 8 days. The medium was replaced with drug-containing medium every other day. At the completion of drug treatment, cultures were continued in drug-free medium until day 12 after which they were stained with 0.5% crystal violet solution. Values were normalized to those of their no treatment controls. We assumed that after 8 days of drug exposure, the drug effects were saturated. We fit the 8-day data (see Fig. 12) and estimated the values of E_{0,i^p}, I_{max,i^p}, and IC_{50,i^p} for each cell type i in Eqs. (5) and (6) by minimizing the least-squares error between the data and the inhibitory growth model using lsqnonlin in MATLAB. We additionally estimated the 95% confidence intervals for the parameters using the Jacobian returned from the lsqnonlin fit. All fitted parameters and their bounds are given in Fig. 12 for both MCF7 and T47D cell lines.

In similar experiments, Nukatsuka et al. [38] measured MCF7 cell growth under varying fulvestrant concentrations. Measurements were calculated as means and standard deviation of cell growth relative to that of the control for three independent experiments. Lewis-Wambi et al. [39] measured DNA (μg/well) from T47D cells after treatment with fulvestrant. Cells were seeded in 24-well dishes and after 24h were treated with varying drug concentrations for 7 days. At the conclusion of the experiment, cells were harvested, and proliferation was assessed as cellular DNA mass (μg/well). We assumed this as a proxy for cell viability relative to control. As with the palbociclib experiments from Vijayaraghavan et al. [37], we estimated the PD parameters in Eqs. (5) and (6) by minimizing the least-squares error between the data and the inhibitory growth model using lsqnonlin in MATLAB (see Fig. 12).

2.2.3 Estimating Pharmacokinetic Parameters

We used a nonlinear mixed effects model in Monolix to estimate parameters of the fulvestrant PK model in Eqs. (10)–(12). As the data reported in Robertson et al. [36] were pooled, we extracted the reported mean and lower and upper bounds to estimate interindividual variability (IIV). We then fit the model in Eqs. (10)–(12) to this data assuming lognormal distributions on parameters (Fig. 13) subject to IIV according to

$$\rho_i = \theta_j \exp(\eta_{ji}), \qquad \eta_{ji} \sim N(0, \omega_j^2),$$

where ρ_i is the value of a given model parameter (e.g., k_{el}, k_{12}, etc.) for subject i, θ_j is the population mean, and η_{ji} represents the deviation from the mean (i.e., IIV) for the i-th individual. Estimated model parameters are presented in Tables 1 and 2.

Table 1 Estimated parameter values for the palbociclib population pharmacokinetic model in Eqs. (7)–(9)

Fixed effects			
Parameter	Units	Mean	Variance
k_a	1/hour	0.617	0.484
k_{el}/F	L/hour	48	0.0451
k_e/F	L/hour	4.49	0.61
V_C/F	L	1520	0.0185
V_P/F	L	2780	–

Table 2 Estimated parameter values for the fixed effects of the fulvestrant population pharmacokinetic model in Eqs. (10)–(12) and other terms used in Monolix

Parameter	Units	Mean
Fixed effects		
T_{k0}	hour	6.74
V	L	5.61
k_{el}	1/hour	0.284
k_{12}	1/hour	15.2
k_{21}	1/hour	3.01
Standard deviation of random effects		
ω_{Tk0}	–	0.276
ω_V	–	0.554
ω_{kel}	–	0.0311
ω_{k12}	–	0.0263
ω_{k21}	–	0.068
Correlations		
ρ_{Tk0-V}	–	0.976

The model in Eqs. (7)–(9) is based on the clinical and theoretical work of Yu et al. [40], which described data from 26 advanced breast cancer patients who received palbociclib and letrozole on a 3-weeks-on, 1-week-off treatment regimen. The palbociclib PK model parameters were taken from Yu et al. and were used to simulate patient populations. We assumed lognormal distributions on parameters subject to interindividual variability using ρ_i from Eq. (2.2.3). Parameter values for fulvestrant were taken directly from Robertson et al. [36].

2.3 Generating Heterogeneous Pharmacokinetics and Pharmacodynamics

2.3.1 Pharmacokinetic Parameters

We investigated palbociclib and fulvestrant individually to quantify each of their contributions to the effects of PK IIV on tumor growth. For palbociclib, we sampled V_C, V_P, k_{el}, k_e, and k_a from lognormal distributions according to the parameters in Table 1 to produce a virtual patient population. Similarly, for fulvestrant, we sampled T_{k0}, V, k_{el}, k_{12}, and k_{21} from lognormal distributions according to the best-fit nonlinear mixed effects model determined by our parameter fitting (see Table 2) to generate virtual patients. In the case of each drug, by simulating the full model (with all other components' parameters set to their average values), we selected only those virtual patients whose predicted trajectories were realistic (as confirmed by visual predictive check of their concentration time courses) before accepting them into our cohort. This left 500 virtual patients in the case of palbociclib and 438 for fulvestrant.

2.3.2 Pharmacodynamic Parameters

To investigate the effect of heterogeneity of the PD of palbociclib and fulvestrant, we generated 400 sets of parameter values by sampling E_0, I_{max}, h, and IC_{50} for each cell type–drug combination from the ranges established during parameter fitting (Fig. 14). As each of these four parameters is drug and cell type specific, this gave 16 parameters to sample:

$$\hat{p} = [E_{0,Mp}, I_{max,Mp}, h_{Mp}, IC_{50,Mp}, E_{0,Tp}, I_{max,Tp}, h_{Tp}, IC_{50,Tp}, \dots$$
$$E_{0,Mf}, I_{max,Mf}, h_{Mf}, IC_{50,Mf}, E_{0,Tf}, I_{max,Tf}, h_{Tf}, IC_{50,Tf}]. \qquad (13)$$

Parameters were sampled from a multivariate normal distribution with mean μ set to the fitted values in Table 3 for $\hat{p}$ and standard deviation σ determined from the confidence intervals (CIs) returned for the fitted parameters and the formula

$$\frac{CI - \mu}{1.96} = \sigma,$$

where 1.96 was chosen to return values in the 95% confidence interval. Any samples resulting in negative parameter values were discarded. The resulting distributions of parameters are provided in Fig. 14.

3 Results

3.1 Shorter Treatment Cycle Reduces Aggressive Cell Viabilities as Compared to Conventional Schedule

We first set out to predict whether a shortened treatment cycle (i.e., 5 days on of palbociclib followed by 2 days of rest, repeated for 28 days) was a viable strategy as compared to a conventional (21 days on of palbociclib followed by 7 days of rest) schedule. Both protocols included combination therapy with 125 mg of fulvestrant on days 1 and 15. For this, we simulated the complete model with mean values for both the palbociclib and fulvestrant PK models (Tables 1 and 2 for Eqs. (7)–(12))

Table 3 Fitting parameter values obtained by fits in Figs. 10 and 11 to the pharmacodynamics model in Eqs. (5) and (6)

Cell line	Variable	Description	Units	Fit	Bounds	
					Upper	Lower
MCF7	r_M	Growth rate	1/day	0.6083	0.5390	0.6776
	k_M	Carrying capacity	cells	7.61×10^6	3.98×10^6	11×10^6
	$E_{0,MP}$	Palbo. initial no-drug effect	–	98.9	96.1	101.7
	$I_{max,MP}$	Palbo. max inhibition	–	0.879	0.86	0.9
	h_{MP}	Palbo. hill coefficient	–	1.65	1.38	1.91
	$IC_{50,MP}$	Palbo. half-effect	μM	0.67	0.61	0.73
	$E_{0,Mf}$	Fulv. initial no-drug effect	–	96.3	95.4	97.2
	$I_{max,Mf}$	Fulv. max inhibition	–	0.81	0.81	0.82
	h_{Mf}	Fulv. hill coefficient	–	1.85	1.75	1.95
	$IC_{50,Mf}$	Fulv. half-effect	μM	4.4×10^{-4}	4.3×10^{-4}	4.5×10^{-4}
T47D	r_T	Growth rate	1/day	0.6726	0.6512	0.6941
	k_T	Carrying capacity	cells	5.27×10^6	4.6×10^6	5.9×10^6
	$E_{0,TP}$	Palbo. initial no-drug effect	–	96.2	87.5	96.2
	$I_{max,TP}$	Palbo. max inhibition	–	0.95	0.79	1.11
	h_{TP}	Palbo. hill coefficient	–	1.02	0.53	1.52
	$IC_{50,TP}$	Palbo. half-effect	μM	1.13	0.64	1.62
	$E_{0,Tf}$	Fulv. initial no-drug effect	–	100	82.5	117.5
	$I_{max,Tf}$	Fulv. max inhibition	–	0.84	0.74	0.95
	h_{Tf}	Fulv. hill coefficient	–	1.85	0.2	1.32
	$IC_{50,Tf}$	Fulv. half-effect	μM	1.2×10^{-4}	-1×10^{-4}	2×10^{-4}

and PD effects model (Table 3 for Eqs. (5) and (6)). We considered only the case where the two cell types were present in equal fractions (i.e., $\phi_i = 0.5$) with a total cell count of $C_{M,i} + C_{T,i} = 7 \times 10^4$ cells. We called this an "average patient." Left untreated over the course of 28 days, unsurprisingly both cell lines were predicted to grow to the global carrying capacity of $K = K_M \phi_M + K_T \phi_T$ (see Fig. 15).

We then introduced treatment to this average patient. We first simulated 125 mg of palbociclib daily for 21 days followed by a period of rest for 7 days, with 125 mg of fulvestrant on days 1 and 15, consistent with current treatment schedules [40] (Fig. 2). Our model predicted the resulting viabilities at the end of treatment to be 0.39 (MCF7) and 0.51 (T47D). Here, cell viability was determined by comparing treatment outcomes to the untreated scenario for the same parameters, i.e., the viability of each cell line was calculated by comparing the total cells at the end of treatment to the total number of cells under no treatment. Repeating this strategy for a treatment course of 125 mg of palbociclib for 5 days followed by 2 days of rest repeated over a period of 28 days, with 125 mg of fulvestrant administered on days 1 and 15, we found viabilities after 28 days of 0.38 (MCF7) and 0.47 (T47D), respectively (Fig. 2). Notably, this change in treatment schedule was predicted to somewhat lower the viability of T47D, which is the more aggressive cell type.

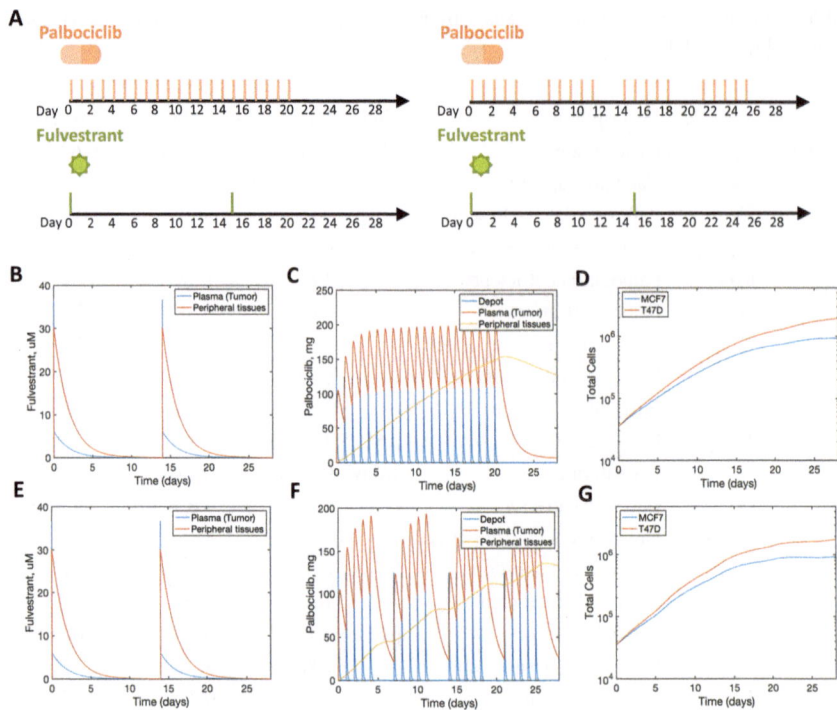

Fig. 2 Comparison of alternate protocols for combination therapy. (**a**) Two established protocols are considered for combination palbociclib and fulvestrant treatment denoted by this schematic: (left) conventional treatment with 21 days on of palbociclib followed by 7 days of rest, and (right) shortened treatment with 5 days on of palbociclib followed by 2 days of rest, repeated for 28 days. Both protocols include combination therapy with 125 mg of fulvestrant on days 1 and 15. (**b**)–(**d**) Tumor growth dynamics on conventional treatment. (**e**)–(**g**) Tumor growth dynamics on shortened treatment. (**b**) and (**e**) Fulvestrant pharmacokinetic model (Eqs. (10)–(12)). (**c**) and (**f**) Palbociclib pharmacokinetic model (Eqs. (7)–(9)). (**d**) and (**g**) Tumor response to treatment by pharmacodynamic model (Eqs. (5) and (6)). For (**d**) by comparing the total number of MCF7 and T47D cells at the end of treatment to the trial that did not receive treatment (Fig. 14), the cell viability was calculated as 0.39 and 0.51 for MCF7 and T47D cells, respectively. For (**g**) by comparing the total number of MCF7 and T47D cells at the end of treatment to the trial that did not receive treatment (Fig. 15), the cell viability was calculated as 0.38 and 0.47 for MCF7 and T47D cells, respectively. Figure 16 shows plots of the corresponding effect function values over time

3.2 Initial Tumor Composition Has Little Impact on Treatment Outcomes

Given that our model predicted a slight reduction in T47D viability under shortened schedules for an average patient, we next interrogated how various levels of heterogeneity (e.g., intrinsic to the tumor population, PK, PD, and treatment scheduling) would affect outcomes. First, we explored the effects of the initial tumor

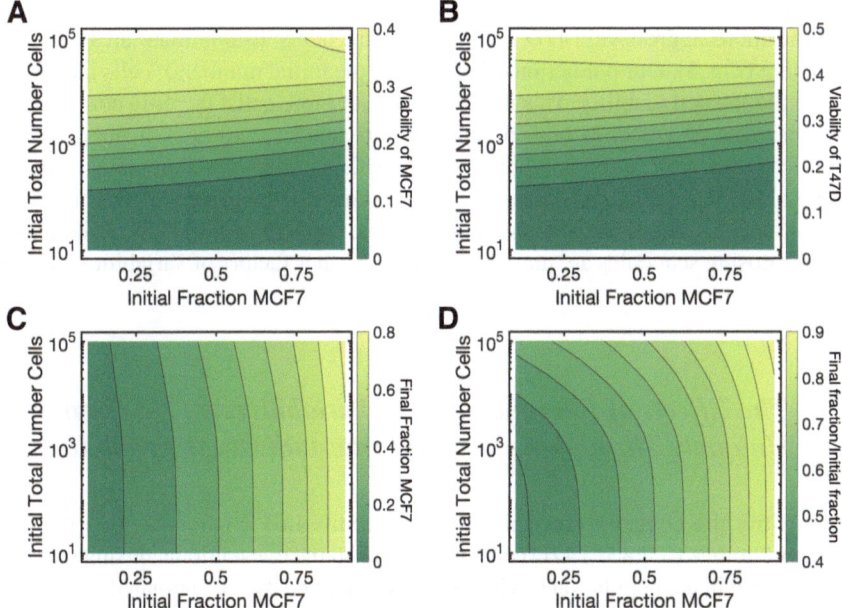

Fig. 3 Results for varying initial tumor composition and total initial cell count, conventional treatment (i.e., 21 days on followed by 7 days off for palbociclib, Fig. 2a). Initial fraction of MCF7 cell line (ϕ_M) and the total number of cells ($C_M + C_T$) are varied over $0 < \phi_M < 1$ and $101 < C_M + C_T < 105$. (**a**) Viability of the MCF7 line for conventional treatment over varied ϕ_M and $C_M + C_T$. Viability is calculated by comparing the total number of MCF7 cells with treatment compared to the total number of MCF7 cells without treatment after 28 days; both trials have the same initial conditions and only differ in whether treatment is administered. (**b**) Viability of the T47D line for conventional treatment over varied ϕ_M and $C_M + C_T$. Viability for T47D is larger than that of MCF7. (**c**) The final fraction of MCF7 cell line (ϕ_M) after the 28 days of treatment. (**d**) The final fraction of MCF7 cell line (ϕ_M) after the 28 days of treatment compared to the initial fraction. Note the differences in the color bars between panels

composition and initial total cancer cell count on the outcomes of different treatment regimens (Figs. 3 and 17).

To isolate the effect of the initial tumor composition, we set all model parameters in both the palbociclib and fulvestrant PK models and their PD models to be their mean values (Tables 1, 2, and 3), as in the previous section. We then varied the initial fraction of MCF7 cells (ϕ_M), i.e., the less aggressive cell type, from 0 to 1 and the total initial cell count ($C_M + C_T$) from 10^1 to 10^5 cells. To accurately capture the effect of changing these parameters, we chose values of ϕ_M and $C_M + C_T$ in these ranges evenly spaced apart to give 440 unique parameter combinations.

We found that the cell viability—defined as stated earlier by comparing the total cells at the end of treatment to the total number of cells under no treatment—and final fraction (ϕ_M after 28 days of treatment) over these varying initial conditions showed decreased T47D viability for the shortened treatment (i.e., shortened vs.

conventional, see Figs. 3 and 17). At lower initial cell counts, our model predicted that the more aggressive T47D cells were more likely to dominate at the end of treatment (Fig. 3). Our predictions show that as the initial number of cells increased, so too did the cell viability. This implies that with more cells, the drug combination becomes less effective. In both regimens, there appeared to be a switching point for the initial cell fraction above which MCF7 cells can dominate (~0.75). Though T47D viability did decrease with the shortened treatment, there was not an exceptional difference between the shortened and conventional treatment, indicating that the dosing schedules are more dependent on other factors of variability, i.e., PK and/or PD.

3.3 The Effects of Pharmacokinetic Variability Are Determined Uniquely Through Fulvestrant Interindividual Variability

To quantify the effects of interindividual variability in PK parameters on the outcomes of both MCF7 and T47D cells, we simulated the conventional dosing regimen of each drug in the population of virtual patients defined by our estimated population PK models (see Figs. 4 and 18) using the methods described in Sect. 2.3.1.

For palbociclib, our results suggest that variability in the PK parameters has a negligible influence on tumor growth outcomes for both cell types (Figs. 4 and 19). Interestingly, examining the palbociclib parameters by classifying patients as either responders or non-responders based on their predicted terminal T47D cell count,

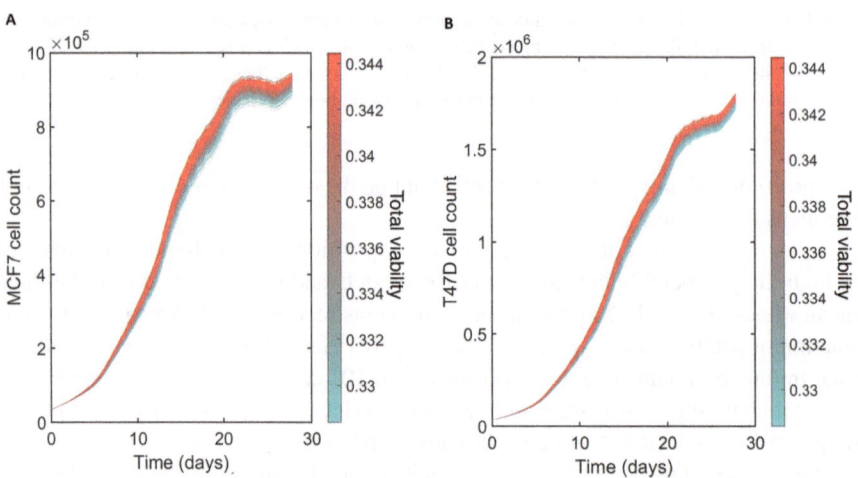

Fig. 4 Predicted outcomes on conventional regimen in palbociclib virtual patient cohort. (**a**) MCF7 and (**b**) T47D cell counts over the course of the conventional treatment regimen with variation in palbociclib pharmacokinetic parameters summarized in Fig. 18. Color bar: viability of T47D cells

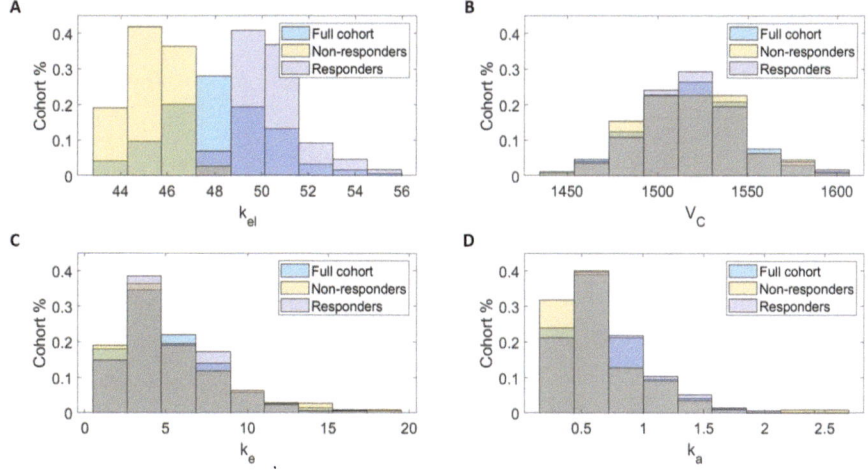

Fig. 5 Some palbociclib pharmacokinetic parameters differ between responders and non-responders. Virtual patients were classified as responders or non-responders based on the predicted terminal T47D cell count of each virtual patient. Upon performing a two-sided Kolmogorov–Smirnov test for each parameter between these two subcohorts, significant differences were found in (**a**) the elimination rate (k_{el}) and (**d**) the absorption rate (k_a); no significant differences were observed in (**b**) the central volume (V_C) and (**c**) the intercompartmental clearance rate (k_e)

we see a clear distinction in the cohort's value for k_{el}, which is high for those with high terminal concentrations and low for those without (Fig. 5). Upon performing a two-sided Kolmogorov–Smirnov test for each parameter between the responder and non-responder subcohorts, we found significant differences in the elimination rate (k_{el}) and the absorption rate (k_a) (Fig. 5).

In contrast, our results suggest that fulvestrant PK variability has a significant impact on tumor growth outcomes for both cell types (Fig. 6a, b). Distributions of fulvestrant PK parameters in the virtual patient cohort are provided in Fig. 20. We observed that virtual patients who sustained high concentrations of fulvestrant over the treatment period have significantly and consistently lower tumor growth as compared to virtual patients who more rapidly cleared the drug (Fig. 6c).

Given the clear relationship between terminal fulvestrant concentrations and outcomes in our virtual patients, we defined virtual patients with "high concentration" to be those with terminal fulvestrant concentrations above $-1.4 \log(\mu M)$ and those with "low concentration" as those with concentrations below $-3.97 \log(\mu M)$ (Fig. 7). Using a two-sided Kolmogorov–Smirnoff test to test for statistically significant differences in distributions between these two subcohorts, we found significant differences in all fulvestrant PK parameters between these groups. This suggests that not only is fulvestrant the key driver of PK heterogeneity (as compared to palbociclib), but that differences in final tumor viability were related to higher t_{k0}, V_D, k_{el}, and k_{21} values and lower k_{12} values than those virtual patients who were predicted to have strong responses to fulvestrant treatments (Fig. 7).

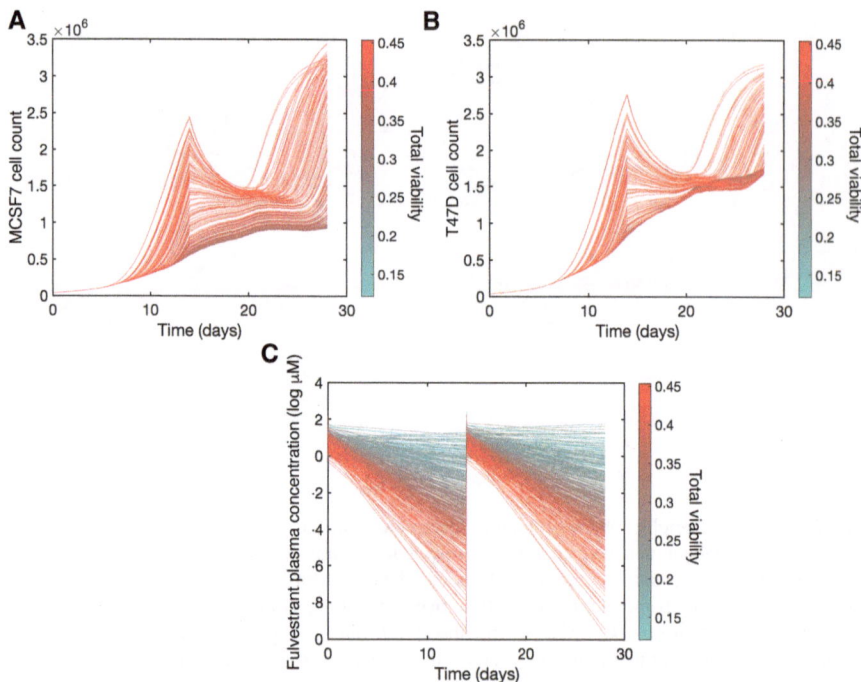

Fig. 6 Spaghetti plots for fulvestrant virtual patients. Predicted dynamics for 438 patients in fulvestrant virtual patient cohort after treatment with 125 mg of fulvestrant on days 1 and 15. (**a**) MCF7 cells, (**b**) T47D cells, and (**c**) fulvestrant concentrations. In all, color bar indicates total tumor viability

3.4 Variability in Each Drug's Maximal Effect Drives Heterogeneity in Outcomes

To next explore the effect of PD variability on treatment outcomes, we fixed all model parameters to be that of an average patient (see Tables 1, 2, and 3) except the PD parameters in Eqs. (5) and (6) noted in $\hat{p}$ from Eq. (13). We generated 400 parameter sets within this range as described in Sect. 2.3.2. We then examined whether we could discern a relationship between each individual's response to treatment and their inherent PD response. For each virtual patient, the shortened treatment protocol was simulated, and the corresponding MCF7 and TD47 cell counts were recorded (Fig. 8a) along with the total number of tumor cells (Fig. 8b). We observed large variance in counts of both cell types across the cohort, but our model did not predict tumor eradication for any patient (Fig. 21).

To examine the correlation between the final numbers of MCF7 and T47D cells at the end of the treatment, we next plotted the total number of MCF7 and T47D cells at the end of treatment (Fig. 8c) for each generated parameter set in our ensemble. Our results clearly show that as the final amount of MCF7 cells decreased, there

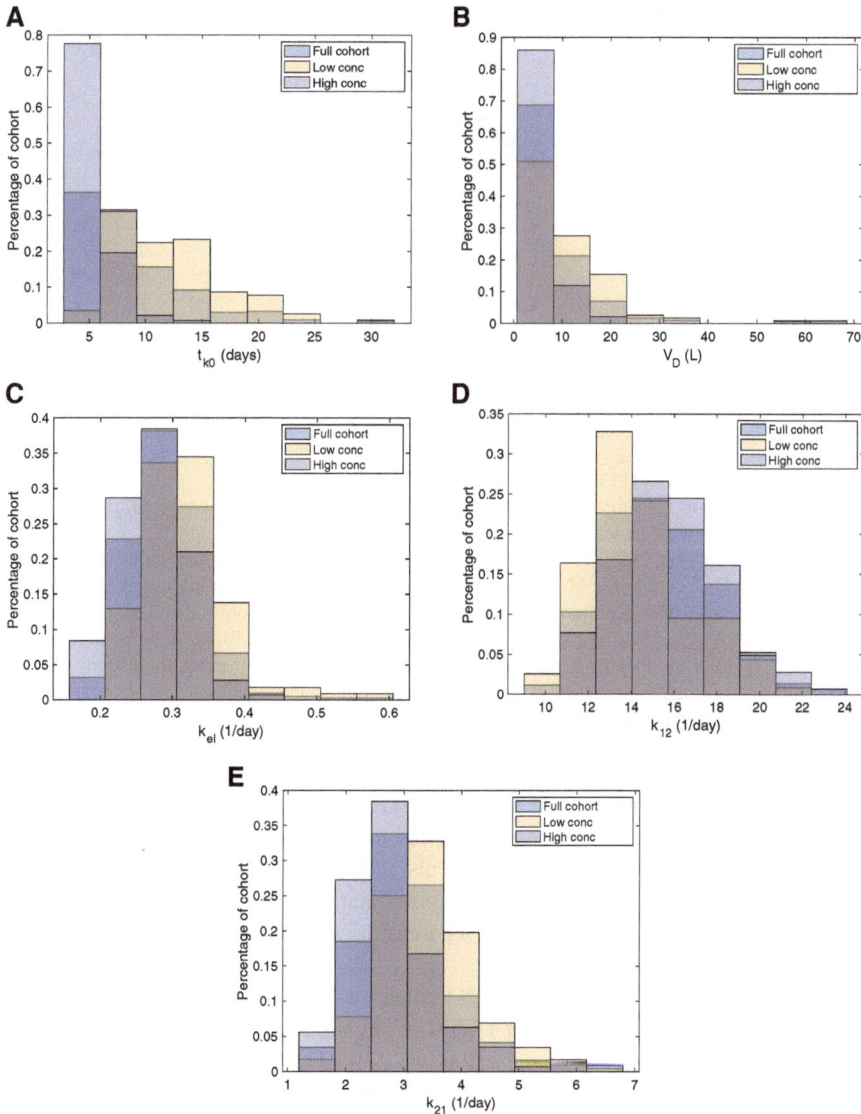

Fig. 7 Fulvestrant pharmacokinetic parameters differ between virtual patients with high terminal fulvestrant concentrations and those with low terminal concentrations. We classified virtual patients as "high concentration" or "low concentration" based on the predicted terminal fulvestrant concentration of each virtual patient (Fig. 6c). Using a Kolmogorov–Smirnoff test for differences in distributions, significant differences were found in (**a**) absorption delay (t_{k0}), (**b**) central volume of distribution (V_D), (**c**) rate of elimination (k_{el}), (**d**) rate of transfer from central to peripheral compartment (k_{12}), and (**e**) rate of transfer from peripheral to central compartment (k_{21}). In legends, "high conc" corresponds to those virtual patients with high terminal fulvestrant concentrations above $-1.4 \log(\mu M)$ and "low conc" to those with low terminal fulvestrant concentrations below $-3.97 \log(\mu M)$

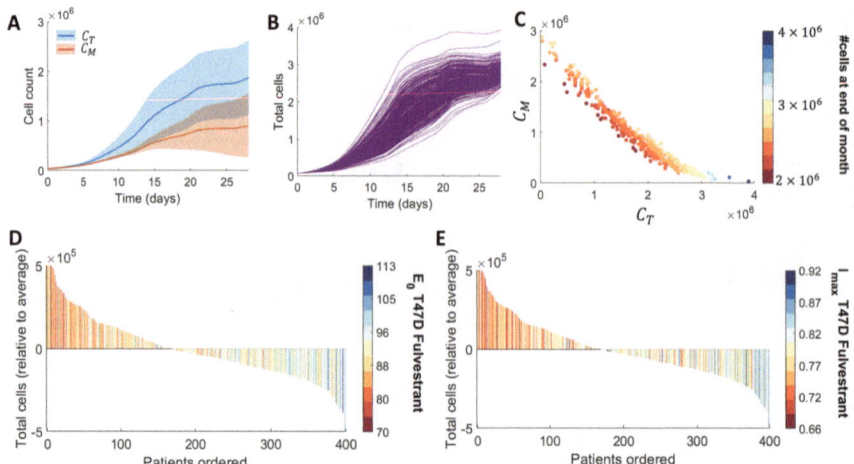

Fig. 8 Virtual cohort investigation into the effect of pharmacodynamics on combination treatment. 400 virtual patients were generated with varying pharmacodynamic parameters (see $\hat{p}$ in Eq. (13)). The shortened 5-day-on, 2-day-off palbociclib regimen combined with two fulvestrant dosages was considered. (**a**) Cell counts for MCF7 (C_M) and T47D (C_T) cells over time plotted as mean and standard deviation of patient cohort. (**b**) Individual patient trajectories for total cell count $C_M + C_T$. (**c**) A scatter plot of the final number of each cell type after 28 days of treatment, colored by the corresponding total number of cells after treatment. (**d**)–(**e**) Waterfall plots for patient specific $E_{0,Tf}$ and $I_{max,Tf}$ against the total cells relative to the cohort average. Color bar corresponds to the value of each patient's parameter normalized to a range between 0 and 1

was a corresponding increase in the final T47D cell count and vice versa. In other words, final MCF7 and T47D were predicted to have an inverse linear relationship when PD variability was considered. We also correlated the final MCF7 and T47D cell counts with the final tumor size and found that the largest tumors are those that are predominately made up of T47D, whereas the smallest tumors are a mixture of both cell types.

We then examined which PD characteristics were the major drivers of final tumor size. We found that E_0 and I_{max} for T47D (i.e., $E_{0,Tf}, I_{max,Tf}$) were most correlated with the final tumor size (Fig. 8d, e), as we predicted that small values of either parameter corresponded to large tumor sizes relative to the average (Fig. 22). All other parameters were not found to contribute significantly to the final tumor size.

3.5 Examining the Long-Term Effect of Variation on the Combination Protocol

Finally, with our understanding of the effects of cell-intrinsic, PK, and PD variability on combination palbociclib and fulvestrant therapy, we explored alternative treatment regimens to study whether we could improve upon the conventional

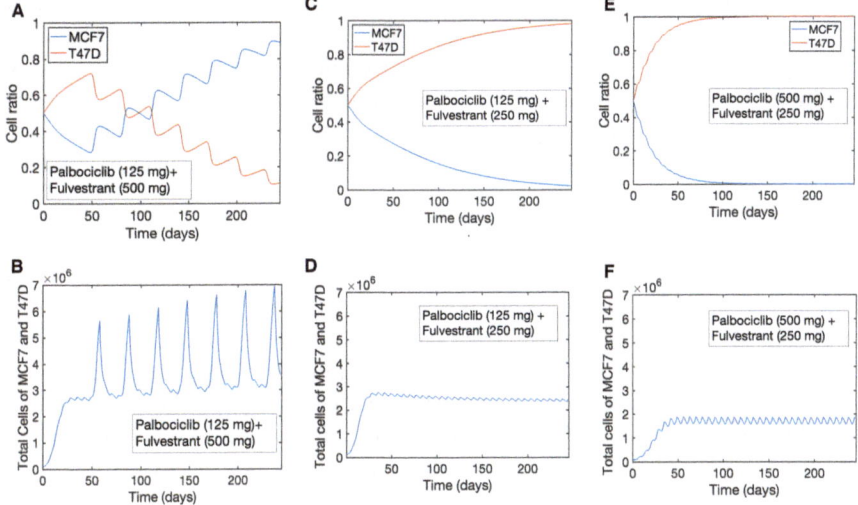

Fig. 9 Investigating long-term dynamics of Regimens 1, 2, and 3. (**a**), (**c**), and (**e**) cancer cell ratios after treatment with Regimens 1, 2, and 3, respectively. Blue curves: ratio of cancer cell line MCF7 given by $C_M/(C_M + C_T)$, and red curves: ratio of cancer cell T47D given by $C_T/(C_M + C_T)$. (**b**), (**d**), and (**f**) comparison of the total cell load of both cell lines after treatment with Regimens 1, 2, and 3, respectively

and investigational schedules using virtual clinical trials of three different dosing regimens. On the current standard-of-care conventional 3-weeks-on/1-week-off dosing schedule of palbociclib, numerous patients have been reported to develop grade 3 or higher degree of neutropenia [15]. This adverse event could result in dose reduction or treatment discontinuation [15]. Furthermore, it has been hypothesized that the one week off-drug in the conventional combination schedule could potentially lead to an increase in the retinoblastoma tumor suppressor protein (Rb) [15].

Therefore, we explored alternative schedules with the aim of reducing the time off-drug (as compared to the conventional regimen) and limiting dose intensity to minimize off-target effects (Regimen 1). Additionally, clinical reports suggest that fulvestrant is most likely to cause acute liver injury [41, 42]. To reduce the risk of hepatotoxicity, we virtually reduce the dose level of fulvestrant and test the effectiveness of this dual-agent combination therapy in Regimens 2 and 3. Thus, we compared the following three schedules (Fig. 9):

Regimen 1   125 mg of oral palbociclib administered once daily for 5 consecutive days, followed by 2 days off, plus 500 mg of intramuscular (IM) fulvestrant administered every 14 days for the first three injections and then every 28 days.

Regimen 2   125 mg of oral palbociclib administered once daily for 5 consecutive
                  days, followed by 2 days off, plus 250 mg IM of fulvestrant adminis-
                  tered every 7 days for the first five injections and then every 14 days.
Regimen 3   500 mg of oral palbociclib administered once daily for 5 consecutive
                  days, followed by 2 days off, plus 250 mg IM of fulvestrant adminis-
                  tered every 7 days for the first five injections and then every 14 days.

Simulations of Regimen 1 suggest that this schedule leads to the competitive
exclusion of aggressive T47D cells. Selective killing of the therapy sensitive
cells removes competitive restriction of MCF7 cells (Fig. 9a). The troughs of the
fluctuating total cell loads come down to 3×10^7, while the peaks still reach a high
level (Fig. 9b). In contrast, by reducing the dose of fulvestrant, Regimen 2 did not
lead to competitive exclusion of the T47D cells (Fig. 9c) but resulted in an overall
significant decrease in the total number of tumor cells (Fig. 9d). We further found
that Regimen 3 did not lead to competitive exclusion of the T47D cells (Fig. 9e),
and our model predictions suggest that the total number of cancer cells from both
lines would continue to decrease (Fig. 9f).

Overall, we found that increasing the dose level of palbociclib within acceptable
toxicity levels could achieve a lower level of total cancer cell load. Importantly,
based on the simulations of our three dosing regimens, it is possible that higher
doses/concentrations of fulvestrant could cause competitive exclusion of the T47D
cell line. As a result, the relative strength of the less aggressive MCF7 cells may in
fact inhibit the efficacy of the palbociclib-fulvestrant combination therapy.

4   Discussion

Heterogeneity is a key factor in cancer therapeutic planning, particularly when
considering combination therapies that may have overlapping and interacting factors
driving treatment responses. The interest in establishing different treatment regimens for palbociclib plus fulvestrant for the treatment of hormone-sensitive breast
cancers gives rise to a number of questions relating to optimal scheduling. These
include various scales of heterogeneity and their impact on combination palbociclib
and fulvestrant, i.e., cell-intrinsic, PK, and PD heterogeneities. Understanding the
contributions of each of these elements to tumor responses helps to establish new,
and perhaps more potent and less toxic, therapeutic regimens. In this work, we used
a simple model of interacting cells to quantify these contributions to help guide
preclinical studies of palbociclib plus fulvestrant.

Considering a tumor composed of lesser and more aggressive cells (i.e., MCF7
and T47D cell lines), each type sensitive to a different degree to each drug,
we predicted the overall tumor cell population and composition after treatment
under variable initial fractions. Our model's predictions showed that the initial cell
fractions have little impact on the final tumor composition after treatment on either

the shortened (i.e., 5 days on of palbociclib followed by 2 days of rest, repeated for 28 days) or conventional (21 days on of palbociclib followed by 7 days of rest) schedule. This is encouraging, in the sense that it suggests that it is primarily PK/PD variability controlling outcomes, and these can be more easily modulated to provide better outcomes.

When considering both PK and PD heterogeneity through the generation of virtual patients, we found that palbociclib PK variability alone had little impact on outcomes, whereas the PK of fulvestrant (as a cytotoxic agent) was a strong determinant of final tumor compositions. This is perhaps expected, as palbociclib acts to freeze the cell cycle rather than to induce apoptosis. Our results further show that palbociclib and fulvestrant are truly synergistic when given in combination, with each being less effective on its own.

Lastly, we used our investigations of the impact of various scales of heterogeneity to propose three alternative regimens to conventional and shortened. These regimens were designed to account for the undesired side effects of each drug through dose fractionation. Our model predictions suggest that it is possible that fulvestrant could cause competitive exclusion of the MCF7 (i.e., less aggressive) cells composing the tumors in our study. Indeed, our results showed that the more aggressive T47D cells act to inhibit the efficacy of the palbociclib-fulvestrant combination therapy, acting similarly to drug tolerant cells despite us not considering resistance in our study. Moreover, within acceptable toxicity levels, increasing the dose level of palbociclib could achieve better outcomes with respect to final tumor size.

In our model, we implemented a logistic growth function to model tumor growth. While Gompertzian tumor growth returned a similar Akaike information criterion and was able to capture the data (Fig. 11), we do not anticipate a large difference in our predictions between the two growth models. This is largely due to the fact both exhibit sigmodal style growth to a carrying capacity.

There are limitations to our approach. The lack of robust clinical data measuring this combination therapy presents a limitation in the reliability of our predictions. Given this is an exploratory study focused primarily on the effects of different sources of heterogeneity, we believe our results support further experiments into combination therapy that may be used in the future to validate our model predictions. Our parameterization could be further validated with *in vivo* experiments. Future work could also look to refine the number of parameters in the model and introduce simpler terms to model the effect of the drugs on the cancer cell population or simplify the cancer growth function.

Though we considered heterogeneity in tumor composition, we did not include the mammary stem cell cascade [43, 44], nor does our model include the actions of the immune system. We opted for the simpler model studied herein to provide a straightforward initial (more *in vitro*-focused) conceptualization of the impact of many scales of heterogeneity affecting treatment outcomes under combination palbociclib and fulvestrant; future studies will include key *in vivo* factors impacting therapies. As seen in Fig. 12a, there is a loss of fidelity in the MCF7 fits for high fulvestrant concentrations. This is a byproduct of the standard nonlinear effect function in Eq. (1) used here. As can be seen in the data, after an initial plateau beginning around 10^{-2} µM, there is a second dip in the observed viability around 1 µM. Unfortunately, our effect function is unable to capture this second decrease, resulting in a higher predicted maximal effect than what is suggested in the data. To confirm the asymptotic behavior of MCF7 cells under treatment with fulvestrant, it would be necessary to have more viability data above 1µM. Nonetheless, as our maximal effect is potentially slightly higher than in the data, our predictions are more conservative with respect to the overall treatment response to fulvestrant. We also did not consider the ways in which each degree of heterogeneity interacts with one another, opting instead to study each individually. This can be incorporated in subsequent iterations of our work.

Our study provides a roadmap for the continued study of CDK4/6 inhibitors and combination therapies in anti-cancer treatments more broadly. Despite using a simple model of tumor growth, our model's predictions showed important, perhaps unexpected behaviors, including how competition between less and more aggressive cells in a heterogeneous tumor impacts treatment scheduling. Ultimately, this work demonstrates the importance of merging mathematical modeling within preclinical studies to improve drug development considerations.

Supplementary Information

See Figs. 10–22.

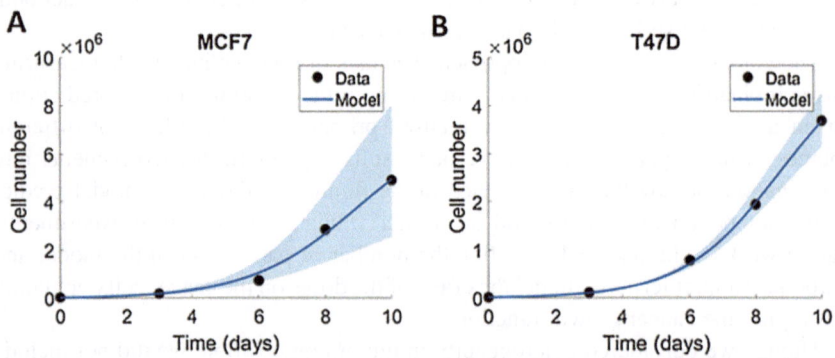

Fig. 10 Fit of logistic growth to cell count measurements for MCF7 and T47D from Vijayaraghavan et al. [37]. A logistic growth curve (Eq. (2)) was fit to the cell count measurements to obtain a cell growth rate r_i and a cell carrying capacity K_i, where i represents either M or T for cell type MCF7 or T47D, respectively. The fit is given as a solid curve with a shaded 95% confidence interval. The data are represented as solid points. The resulting parameters are in Table 3

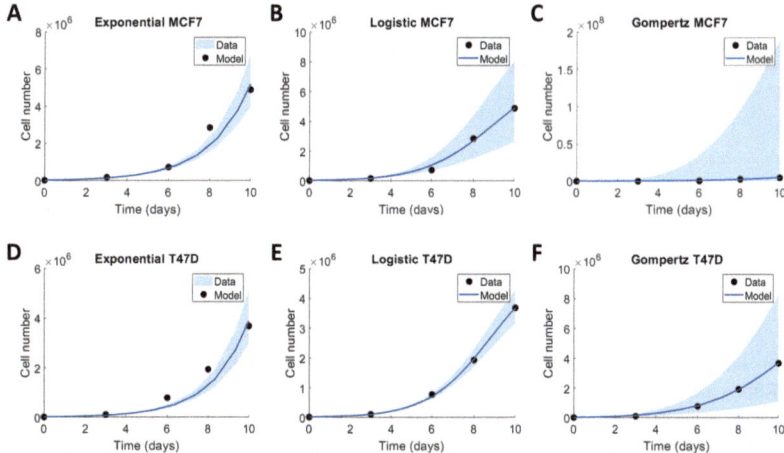

Fig. 11 Comparative analysis of model selection to MCF7 and T47D cell count data. We compared the least-squares fit for (**a, d**) exponential growth, (**b, e**) logistic growth, and (**c, f**) Gompertzian growth. We calculated the corrected Akaike information criterion for each figure, which returned (**a**) 134.3, (**b**) 131.1, (**c**) 134.5, (**d**) 131.7, (**e**) 116.5, and (**f**) 115.0. We also considered the confidence intervals plotted for each model fit. Given that the Gompertzian growth has wider confidence intervals compared to logistic growth, we concluded that logistic growth was a good model choice for tumor growth

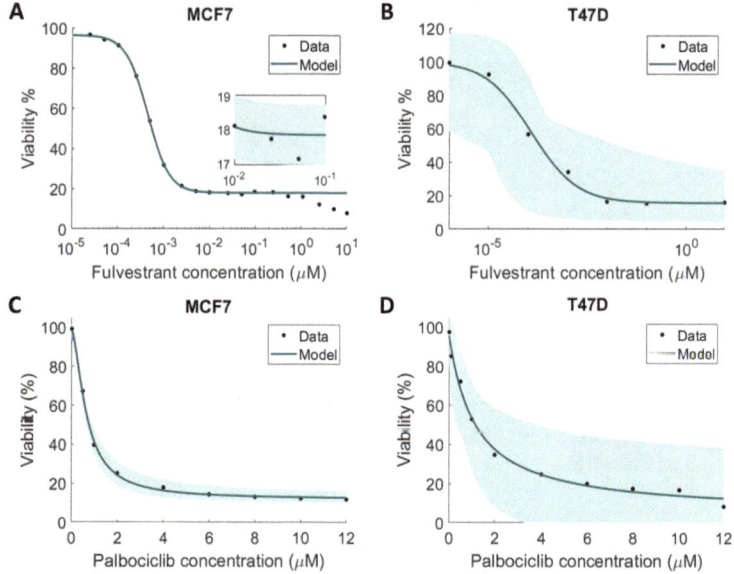

Fig. 12 Fit of drug effect parameters in the pharmacodynamics model in Eqs. (5) and (6) to cell viability measurements for fulvestrant and palbociclib on MCF7 and T47D. Cell viability measurements for (**a**) fulvestrant on MCF7 cells [38] and (**b**) fulvestrant on T47D cells [39]. Cell viability measurements for (**c**) palbociclib on MCF7 and (**b**) palbociclib on T47D by Vijayaraghavan et al. [37]. The resulting parameters are in Table 3. In (**a**) the inset zooms in on the confidence intervals surrounding the data fit

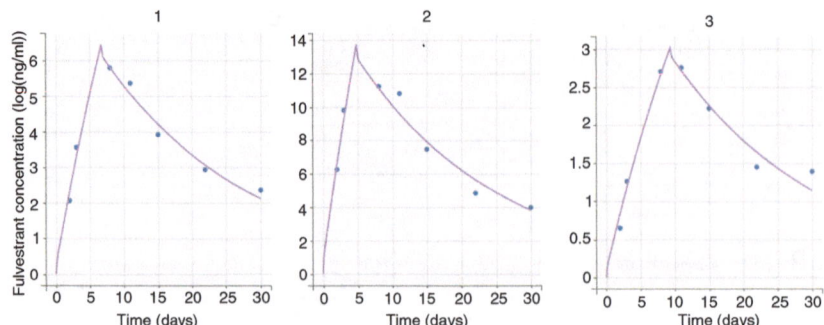

Fig. 13 Fulvestrant pharmacokinetic parameter estimation fits. Population pharmacokinetic data from Robertson et al. [36] was pooled to extract the mean and lower and upper bounds of the data. Pharmacokinetic parameters were estimated for Eqs. (10)–(12) to these data assuming lognormal distributions on parameters subject to interindividual variability using a standard nonlinear mixed effects model in Monolix

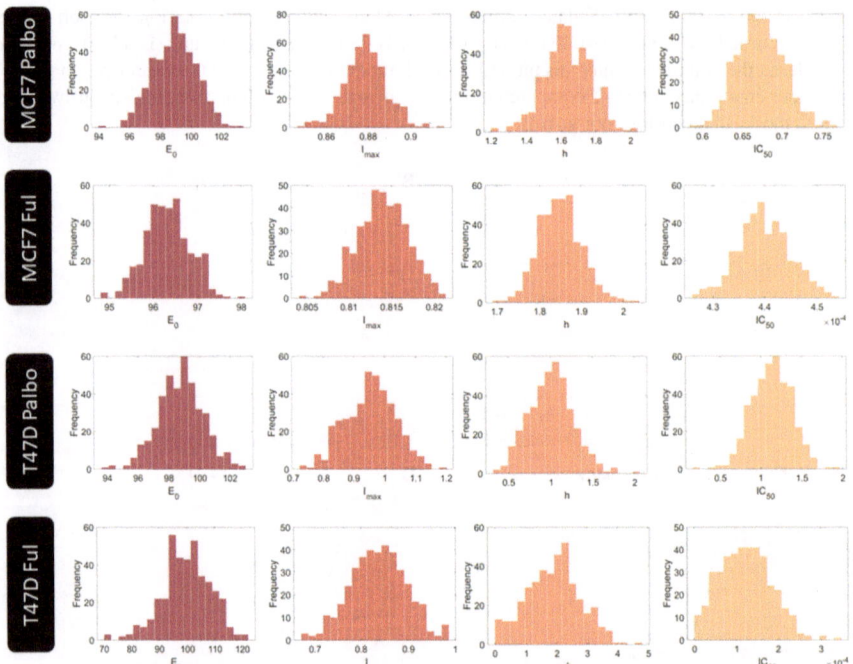

Fig. 14 Virtual patient parameter values for investigation varying the drug effect pharmacodynamic parameters. Parameters relating the effect of fulvestrant and palbociclib on MCF7 and T47D were sampled from normal distributions as described in Sect. 2.3.2 to obtain 400 unique parameter combinations corresponding to 400 virtual patients

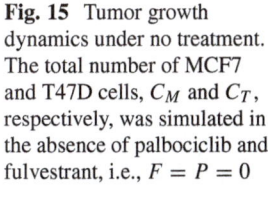

Fig. 15 Tumor growth dynamics under no treatment. The total number of MCF7 and T47D cells, C_M and C_T, respectively, was simulated in the absence of palbociclib and fulvestrant, i.e., $F = P = 0$

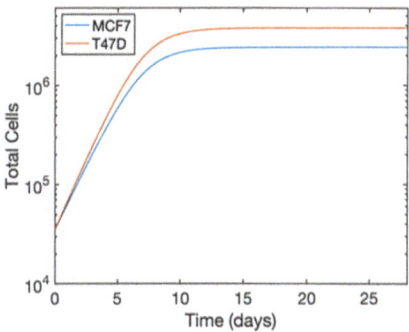

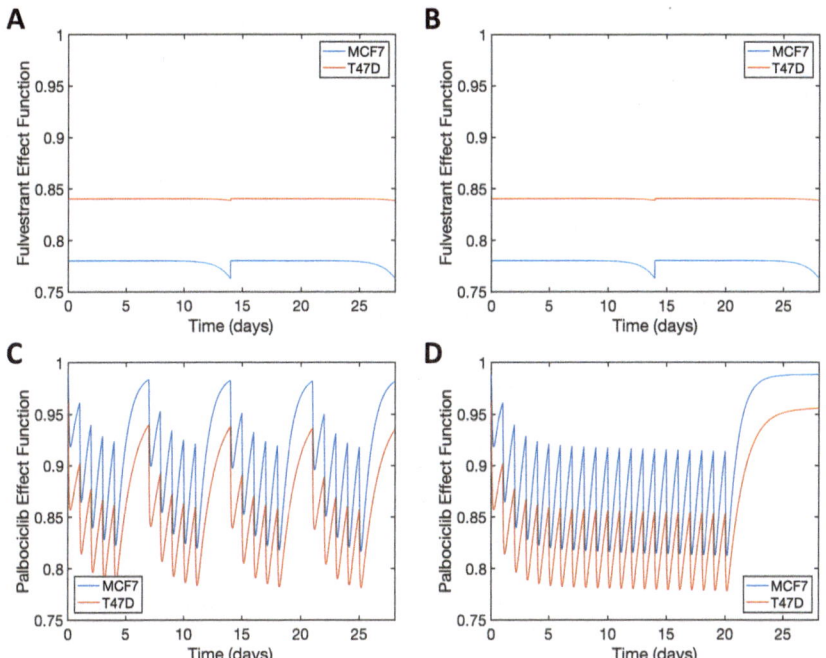

Fig. 16 The effect functions corresponding to the simulation in Fig. 2. The effect function Eq. (1) for fulvestrant (**a**)–(**b**) and palbociclib (**c**)–(**d**) for the two dosage protocols considered in Fig. 2: (**a**), (**c**) shortened treatment with palbociclib 5 days on and 2 days off and (**b**), (**d**) conventional treatment with palbociclib 21 days on and 7 days off. Both protocols include combination therapy with 125 mg of fulvestrant on days 1 and 15

Analysis of the Two Cell Line Model Dynamics

To understand the dynamics of the system of ODEs used in our study, we carried out a linear steady-state analysis to determine long-term stability of a simplified version

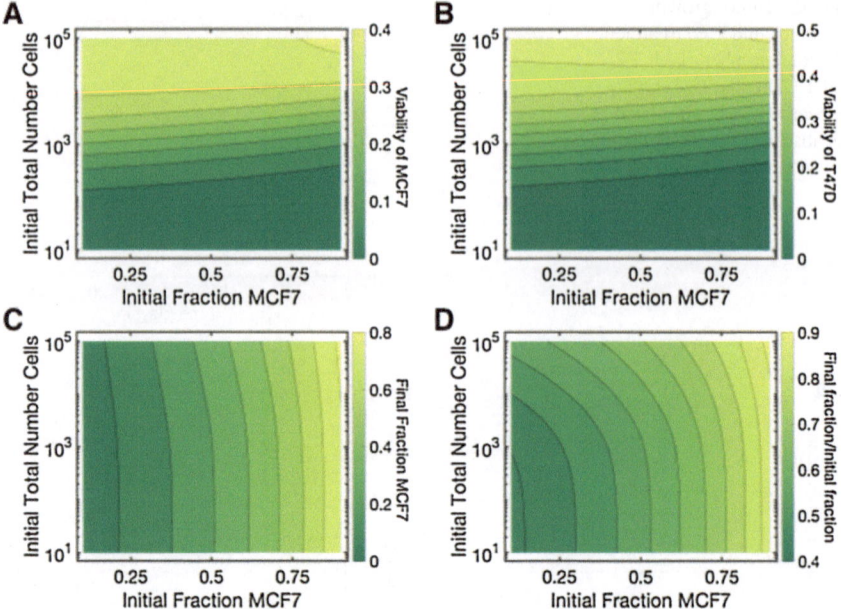

Fig. 17 Results for varying initial tumor composition and total initial cell count, shortened treatment (i.e., 5 days on followed by 2 days off, repeated for 28 days, for palbociclib, Fig. 2a). Initial fraction of MCF7 cell line (ϕ_M) and the total number of cells ($C_M + C_T$) are varied over $0 < \phi_M < 1$ and $101 < C_M + C_T < 105$. (**a**) Viability of the MCF7 line for shortened treatment over varied ϕ_M and $C_M + C_T$. Viability is calculated by comparing the total number of MCF7 cells with treatment compared to the total number of MCF7 cells without treatment after 28 days; both trials have the same initial conditions and only differ in whether treatment is administered. The overall viability of MCF7 is less for the shortened treatment, compared to the conventional treatment. (**b**) Viability of the T47D line for shortened treatment over varied ϕ_M and $C_M + C_T$. The overall viability of T47D is less for the shortened treatment, compared to the conventional treatment. (**c**) The final fraction of MCF7 cell line (ϕ_M) after the 28 days of treatment. (**d**) The final fraction of MCF7 cell line (ϕ_M) after the 28 days of treatment compared to the initial fraction. At lower initial total cell numbers, T47D has a greater propensity to overtake the cancer tumor and take up a greater fraction of the total tumor. Note the differences in the color bars between panels

of the system. We took the one-drug model in Eqs. (3) and (4) and considered only the effects of palbociclib on MCF7 and T47D cells.

We assumed both cell lines share the same carrying capacity due to them coexisting in the same spatial location and hence having the same spatial limitations, that is, $K = K_M = K_T$. For convenience for this analysis, we also assume that all PD parameters are equivalent for the two cell types, i.e., $I_{50,MP} = I_{50,TP} = IC_{50}$, $h_{MP} = h_{TP} = h$, $I_{max,MP} = I_{max,TP} = I_{max}$, and $E_0 = E_{0,MP} = E_{0,TP}$, etc. For convenience, we denote $r_M E_0 = \tilde{r}_M$ and $r_T E_0 = \tilde{r}_T$.

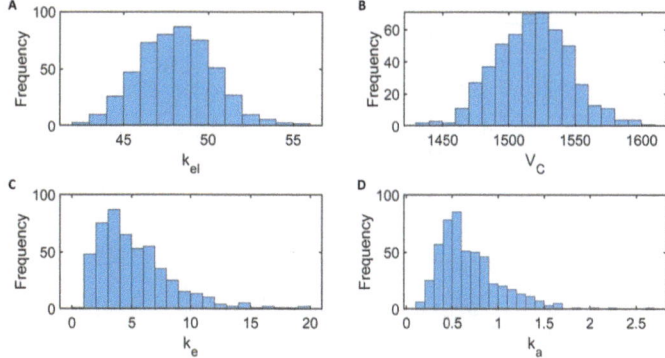

Fig. 18 Distributions of palbociclib virtual patient pharmacokinetic parameters: (**a**) elimination rate (k_{el}), (**b**) central volume (V_C), (**c**) intercompartmental clearance rate (k_e), and (**d**) absorption rate of palbociclib (k_a) constructed from 500 virtual patients sampled as described in Sect. 2.3.1. Experimental data suggests negligible variation in the parameter V_P. This parameter is thus taken to be constant, hence the lack of distribution. The virtual patient population randomly generated to produce these distributions is maintained in Fig. 5

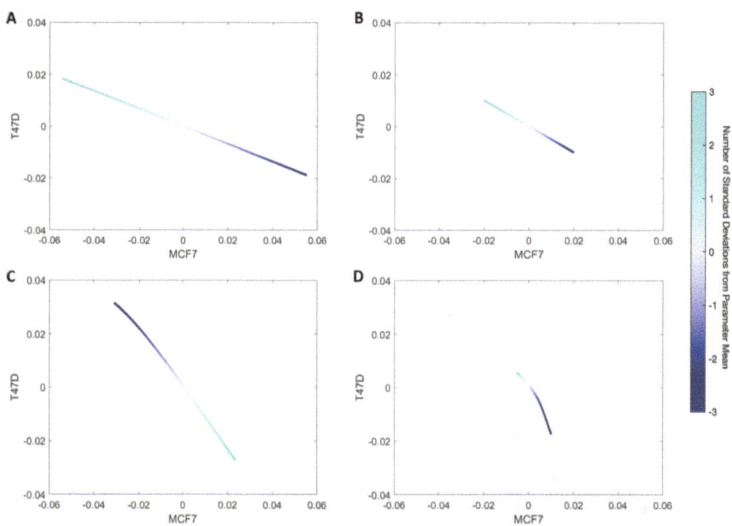

Fig. 19 Parameter sensitivity of the palbociclib pharmacokinetic parameters: (**a**) elimination rate (k_{el}), (**b**) central volume (V_C), (**c**) intercompartmental clearance rate (k_e), and (**d**) absorption rate of palbociclib (k_a). Curves convey changes in cell counts resulting from varying each parameter uniformly within three standard deviations of its mean (Table 1). Numerical values on the axes correspond to the fractional deviation of an end-of-treatment cell count from its mean scaled by the fractional deviation of one standard deviation of a parameter from that parameter mean. Qualitatively, the distribution of a parameter non-negligibly influences final cell count if the curves have coordinate values far from zero and close to one. Thus, our results show that all the parameter distributions have a negligible influence on the final cell counts

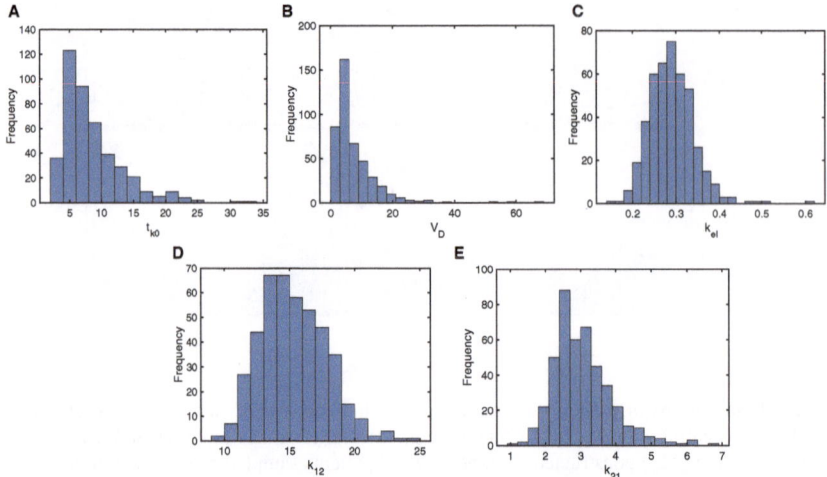

Fig. 20 Distributions of fulvestrant virtual patient pharmacokinetic parameters: (**a**) absorption delay (t_{k0}), (**b**) central volume of distribution (V_D), (**c**) rate of elimination (k_{el}), and (**d**)–(**e**) rates of transit between central and peripheral compartments (k_{12} and k_{21}). Distributions describe the 438 virtual patients in the fulvestrant virtual patient cohort, generated as described in Sect. 2.3.1

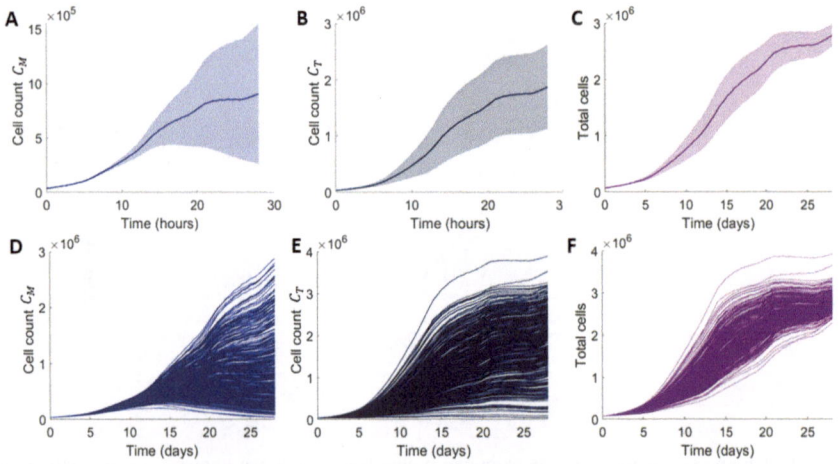

Fig. 21 Results of simulating the virtual cohort with varying pharmacodynamics. (**a**)–(**c**) the mean and standard deviation for the 400 virtual patients' cohort simulated with varying effect parameters (Fig. 12). (**d**)–(**f**) the corresponding individual patient trajectories

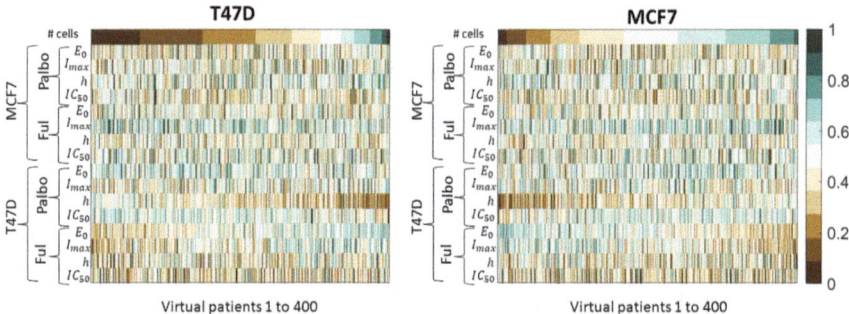

Fig. 22 Results of simulating the virtual cohort with varying pharmacodynamics. Each column corresponds to a virtual patient where the patients are ordered by the number of cells (left: T47D and right: MCF7) at the end of the treatment. The rows correspond to the parameter value where the column is the normalized value for that parameter. The color bar corresponds to the value of the parameter normalized between 0 and 1

The model has three isolated equilibrium points

$$(C_M^*, C_T^*) = (0, 0), \quad (C_M^*, C_T^*) = (0, K), \quad (C_M^*, C_T^*) = (K, 0)$$

and a line containing infinite number of equilibrium points:

$$C_M + C_T - K = 0.$$

Linear stability analysis shows that $(C_M^*, C_T^*) = (0, 0)$ has two positive eigenvalues

$$\lambda_{01} = \left[IC_{50}^h - (I_{max} - 1) P^h \right] \tilde{r}_T K > 0 \text{ and}$$

$$\lambda_{02} = \left[IC_{50}^h - (I_{max} - 1) P^h \right] \tilde{r}_M K > 0,$$

with corresponding eigenvectors

$$v_{01} = [1, 0]^T, \quad v_{02} = [0, 1]^T.$$

The equilibrium $(C_M^*, C_T^*) = (0, K)$ has eigenvalues:

$$\lambda_{11} = \left[(I_{max} - 1) P^h - IC_{50}^h \right] \tilde{r}_T K < 0 \text{ and } \lambda_{12} = 0,$$

with corresponding eigenvectors

$$v_{11} = [0, 1]^T, \quad v_{12} = [-1, 1]^T.$$

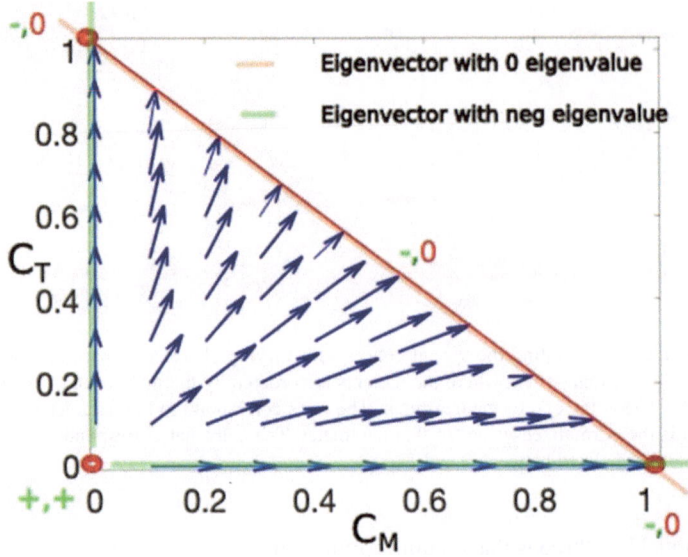

Fig. 23 Simulated vector field. Three equilibria $(0, 0)$, $(0, K)$, and $(K, 0)$ are denoted in red circles. The line with infinite equilibria, $C_M + C_T - K = 0$, is plotted in red. The eigenvector corresponding to the negative eigenvalues (i.e., $\lambda_{11}, \lambda_{22}$, and λ_2) is denoted by the green lines. The eigenvector corresponding to the zero eigenvalue is represented as a pink line and collapses with the equilibrium line, $C_M + C_T - K = 0$. As a result, the vector field (blue arrows) points toward the red equilibrium line. Moreover, there is no movement in the red line

Lastly, the equilibrium $(C_M^*, C_T^*) = (K, 0)$ has two eigenvalues

$$\lambda_{21} = 0 \quad \text{and} \quad \lambda_{22} = \left[(I_{max} - 1) P^h - IC_{50}^h \right] < 0 \tilde{r}_M K < 0,$$

with corresponding eigenvectors

$$v_{21} = [-1, 1]^T, \quad v_{22} = [1, 0]^T.$$

The line of infinite equilibria $C_M + C_T - K = 0$ has two eigenvalues

$$\lambda_1 = 0, \quad \text{and} \quad \lambda_2 = \left[(I_{max} - 1) P^h - IC_{50}^h \right] (K - C_M) \tilde{r}_M$$

$$+ \tilde{r}_T C_T \left[(I_{max} - 1) P^h - IC_{50}^h \right] < 0,$$

with corresponding eigenvectors

$$v_1 = [-1, 1]^T, \quad v_2 = \left[\frac{(K - C_T)(P^h (I_{max} - 1) - IC_{50}^h) \tilde{r}_M}{C_T \tilde{r}_T (P^h (I_{max} - 1) - IC_{50}^h)}, 1 \right].$$

Flow along nonzero eigenvalues is much faster than flow along zero eigenvalues. Therefore, in the fast timescale, cell trajectories are repelled from equilibrium $(C_M^*, C_T^*) = (0, 0)$ and quickly converge to the neighborhood of the eigenvector associated with zero eigenvalue. Moreover, the eigenvector associated with zero eigenvalue $v_{12} = v_{21} = v_1 = [-1, 1]^T$ collides with the equilibrium line $C_M + C_T - K = 0$, which is the slow manifold. Therefore, in the slow timescales, the flow on the slow manifold has no movement. It implies that the cancer population will eventually converge to the carrying capacity K but will have a different proportion depending on the initial fractions. A simulated vector field under the assumption $K_M = K_T = K$ is shown in Fig. 23.

Acknowledgments The work described herein was initiated during the Collaborative Workshop for Women in Mathematical Biology funded and hosted by UnitedHealth Group Optum of Minnetonka, MN and supported by University of Minnesota's Institute for Mathematics and its Applications in June 2022. Additionally, the authors and editors thank the anonymous peer reviewers for their feedback, which strengthened this work.

The authors would like to thank the organizers, sponsors, and participants of the workshop for bringing them together and providing support. SL was supported by the National Science Foundation (No. 2139322). WZ acknowledges the generous support from Simons Foundation Collaboration Grants for Mathematicians (award number: A21-0013-001). MC was funded by a Natural Sciences and Engineering Research Council of Canada Discovery Grant (RGPIN-2018-04546), the Fondation du CHU Sainte-Justine, and a Fonds de recherche du Québec-Santé J1 Research Scholar award. ALJ would like to thank the Queensland University of Technology Early Career Researcher Scheme.

References

1. C. Núñez, J.L. Capelo, G. Igrejas, A. Alfonso, L.M. Botana, C. Lodeiro, Biomaterials **97**, 34 (2016). https://doi.org/10.1016/j.biomaterials.2016.04.027
2. E. Warner, N. Engl. J. Med. **365**(11), 1025 (2011). https://doi.org/10.1056/NEJMcp1101540
3. F.A. Fisusi, E.O. Akala, Pharm. Nanotechnol. **7**(1), 3 (2019). https://doi.org/10.2174/2211738507666190122111224
4. M. Pujani, H. Jain, V. Chauhan, C. Agarwal, K. Singh, M. Singh, Breast Dis. **39**(2), 61 (2020). https://doi.org/10.3233/bd-200442
5. R.B. Mokhtari, T.S. Homayouni, N. Baluch, E. Morgatskaya, S. Kumar, B. Das, H. Yeger, Oncotarget **8**(23), 38022 (2017). https://doi.org/10.18632/oncotarget.16723
6. M.T. Do, H.G. Kim, J.H. Choi, H.G. Jeong, Free Radicals Biol. Med. **74**, 21 (2014). https://doi.org/10.1016/j.freeradbiomed.2014.06.010
7. N. Masuda, K. Inoue, R. Nakamura, Y. Rai, H. Mukai, S. Ohno, F. Hara, Y. Mori, S. Hashigaki, Y. Muramatsu, et al., Int. J. Clin. Oncol. **24**(3), 262 (2018). https://doi.org/10.1007/s10147-018-1359-3
8. S. Verma, C.H. Bartlett, P. Schnell, A.M. DeMichele, S. Loi, J. Ro, M. Colleoni, H. Iwata, N. Harbeck, M. Cristofanilli, et al., Oncologist **21**(10), 1165 (2016). https://doi.org/10.1634/theoncologist.2016 0097
9. A.J. Walker, S. Wedam, L. Amiri-Kordestani, E. Bloomquist, S. Tang, R. Sridhara, W. Chen, T.R. Palmby, J. Fourie Zirkelbach, W. Fu, et al., Clin. Cancer Res. **22**(20), 4968 (2016). https://doi.org/10.1158/1078-0432.Ccr-16-0493

10. T. Traina, K. Cadoo, A. Gucalp, Breast Cancer Targets Ther. (2014). https://doi.org/10.2147/bctt.S46725
11. N.C. Turner, J. Ro, F. André, S. Loi, S. Verma, H. Iwata, N. Harbeck, S. Loibl, C. Huang Bartlett, K. Zhang, et al., N. Engl. J. Med. **373**(3), 209 (2015). https://doi.org/10.1056/NEJMoa1505270
12. F. Serra, P. Lapidari, E. Quaquarini, B. Tagliaferri, F. Sottotetti, R. Palumbo, Drugs Context **8**, 1 (2019). https://doi.org/10.7573/dic.212579
13. S. Goel, J.S. Bergholz, J.J. Zhao, Nat. Rev. Cancer **22**(6), 356 (2022). https://doi.org/10.1038/s41568-022-00456-3
14. Washington University School of Medicine and Pfizer. Alternative dosing schedule of palbociclib in metastatic hormone receptor positive breast cancer. https://ClinicalTrials.gov/show/NCT03007979 (2017)
15. J. Krishnamurthy, J. Luo, R. Suresh, F. Ademuyiwa, C. Rigden, T. Rearden, K. Clifton, K. Weilbaecher, A. Frith, A. Roshal, et al., npj Breast Cancer **8**(1) (2022). https://doi.org/10.1038/s41523-022-00399-w
16. M.R. Nathan, P. Schmid, Oncol. Ther. **5**(1), 17 (2017). https://doi.org/10.1007/s40487-017-0046-2
17. S. Johnston, K. Cheung, Curr. Med. Chem. **17**(10), 902 (2010). https://doi.org/10.2174/092986710790820633
18. S. Loibl, N.C. Turner, J. Ro, M. Cristofanilli, H. Iwata, S.A. Im, N. Masuda, S. Loi, F. André, N. Harbeck, et al., Oncologist **22**(9), 1028 (2017). https://doi.org/10.1634/theoncologist.2017-0072
19. H. Iwata, S.A. Im, N. Masuda, Y.H. Im, K. Inoue, Y. Rai, R. Nakamura, J.H. Kim, J.T. Hoffman, K. Zhang, et al., J. Glob. Oncol. **3**(4), 289 (2017). https://doi.org/10.1200/JGO.2016.008318
20. J. Albanell, M.T. Martînez, M. Ramos, M. O'Connor, L. de la Cruz-Merino, A. Santaballa, N. Martînez-Jañez, F. Moreno, I. Fernández, J. Alarcón, et al., Eur. J. Cancer **161**, 26 (2022). https://doi.org/10.1016/j.ejca.2021.11.010
21. H.C. Wei, AIMS Math. **5**(4), 3446 (2020). https://doi.org/10.3934/math.2020223
22. W. He, D.M. Demas, I.P. Conde, A.N. Shajahan-Haq, W.T. Baumann, J. R. Soc. Interface **17**(169) (2020). https://doi.org/10.1098/rsif.2020.0339
23. F. Pappalardo, G. Russo, F.M. Tshinanu, M. Viceconti, Briefings Bioinf. **20**(5), 1699 (2019). https://doi.org/10.1093/bib/bby043
24. T.M. Polasek, A. Rostami-Hodjegan, Clin. Pharmacol. Ther. **107**(4), 742 (2020). https://doi.org/10.1002/cpt.1778
25. J. Scott, Lancet Oncol. **13**(3) (2012). https://doi.org/10.1016/s1470-2045(12)70098-0
26. S. Alfonso, A.L. Jenner, M. Craig, Chaos **30**(12), 123128 (2020). https://doi.org/10.1063/5.0019556. http://aip.scitation.org/doi/10.1063/5.0019556
27. E. Kim, V.W. Rebecca, K.S.M. Smalley, A.R.A. Anderson, Euro. J. Cancer **67**, 213 (2016). https://doi.org/10.1016/j.ejca.2016.07.024
28. M.U. Zahid, A.S.R. Mohamed, J.J. Caudell, L.B. Harrison, C.D. Fuller, E.G. Moros, H. Enderling, J. Pers. Med. **11**(11) (2021). https://doi.org/10.3390/jpm11111124
29. H. Wang, R.J. Sové, M. Jafarnejad, S. Rahmeh, E.M. Jaffee, V. Stearns, E.T.R. Torres, R.M. Connolly, A.S. Popel, Front. Bioeng. Biotechnol. **8** (2020). https://doi.org/10.3389/fbioe.2020.00141
30. O. Cardinal, C. Burlot, Y. Fu, P. Crosley, M. Hitt, M. Craig, A.L. Jenner, Comput. Syst. Oncol. **2**, e1035 (2022). https://doi.org/10.1101/2022.03.29.486309
31. A.L. Jenner, T. Cassidy, K. Belaid, M.C. Bourgeois-Daigneault, M. Craig, J. Immuno. Ther. Cancer **9**(2), e001387 (2021). https://doi.org/10.1136/jitc-2020-001387. https://jitc.bmj.com/lookup/doi/10.1136/jitc-2020-001387
32. W. Xiong, M. Friese-Hamim, A. Johne, C. Stroh, M. Klevesath, G.S. Falchook, D.S. Hong, P. Girard, S. El Bawab, CPT Pharmacometrics Syst. Pharmacol. **10**(5), 428 (2021). https://doi.org/10.1002/psp4.12602

33. E.L. Bradshaw, M.E. Spilker, R. Zang, L. Bansal, H. He, R.D.O. Jones, K. Le, M. Penney, E. Schuck, B. Topp, et al., CPT: Pharmacometrics Syst. Pharmacol. **8**(11), 777 (2019). https://doi.org/10.1002/psp4.12463

34. P. Crosley, A. Farkkila, A.L. Jenner, C. Burlot, O. Cardinal, K.G. Potts, K. Agopsowicz, M. Pihlajoki, M. Heikinheimo, M. Craig, et al., Int. J. Mol. Sci. **22**(9), 4699 (2021). https://doi.org/10.3390/ijms22094699

35. M. Craig, K. Kaveh, A. Woosley, A.S. Brown, D. Goldman, E. Eton, R.M. Mehta, A. Dhawan, K. Arai, M.M. Rahman, et al., PLoS Comput. Biol. **15**(8), 1 (2019). https://doi.org/10.1371/journal.pcbi.1007278. https://doi.org/10.1371/journal.pcbi.1007278

36. J.F.R. Robertson, M.P. Harrison, Cancer Chemother. Pharmacol. **52**(4), 346 (2003). https://doi.org/10.1007/s00280-003-0643-7

37. S. Vijayaraghavan, C. Karakas, I. Doostan, X. Chen, T. Bui, M. Yi, A.S. Raghavendra, Y. Zhao, S.I. Bashour, N.K. Ibrahim, et al., Nat. Commun. **8**(1) (2017). https://doi.org/10.1038/ncomms15916

38. M. Nukatsuka, H. Saito, S. Noguchi, T. Takechi, In Vivo **33**(5), 1439 (2019). https://doi.org/10.21873/invivo.11622

39. J.S. Lewis-Wambi, H. Kim, R. Curpan, R. Grigg, M.A. Sarker, V.C. Jordan, Mol. Pharmacol. **80**(4), 610 (2011)

40. Y. Yu, W. Sun, Y. Liu, D. Wang, J. Clin. Pharmacol. **62**(3), 376 (2021). https://doi.org/10.1002/jcph.1971

41. National Institute of Diabetes and Digestive and Kidney Diseases. Livertox: Clinical and research information on drug-induced liver injury [internet]. https://www.ncbi.nlm.nih.gov/books/NBK548072/ (2018)

42. A. Schlotman, A. Stater, K. Schuler, J. Heideman, V. Abramson, Case Rep. Oncol. **13**(1), 304 (2020). https://doi.org/10.1159/000506442

43. J. Le Sauteur-Robitaille, Z.S. Yu, M. Craig, AIMS Math. **6**(10), 10861 (2021). https://doi.org/10.3934/math.2021631

44. P. Tharmapalan, M. Mahendralingam, H.K. Berman, R. Khokha, EMBO J. **38**(14) (2019). https://doi.org/10.15252/embj.2018100852

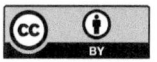

Towards a Mathematical Understanding of Ventilator-Induced Lung Injury in Preterm Rat Pups

Rayanne A. Luke, Gess Kelly, Melissa Stoner, Jordana Esplin O'Brien, Sharon R. Lubkin, and Laura Ellwein Fix

1   Introduction

Infants born at less than 28 weeks gestation or less than 1000 g in weight are considered extremely preterm and are prone to a multitude of breathing issues

R. A. Luke (✉)
Department of Applied Mathematics and Statistics, Johns Hopkins University, Baltimore, MD, USA

Current address: Department of Mathematical Sciences, George Mason University, Fairfax, VA, USA
e-mail: rluke@gmu.edu

G. Kelly
Martin A. Fisher School of Physics, Brandeis University, Waltham, MA, USA

Current address: Ab Initio, Lexington, MA, USA
e-mail: gkelly@abinitio.com

M. Stoner
Department of Mathematical Sciences, Salisbury University, Salisbury, MD, USA
e-mail: mastoner@salisbury.edu

J. E. O'Brien
School of Mathematical Sciences, Rochester Institute of Technology, Rochester, NY, USA
e-mail: jeo8857@g.rit.edu

S. R. Lubkin
Department of Mathematics, North Carolina State University, Raleigh, NC, USA
e-mail: lubkin@ncsu.edu

L. E. Fix
Department of Mathematics and Applied Mathematics, Virginia Commonwealth University, Richmond, VA, USA
e-mail: lellwein@vcu.edu

A. N. Ford Versypt et al. (eds.), *Mathematical Modeling for Women's Health*,
The IMA Volumes in Mathematics and its Applications 166,
https://doi.org/10.1007/978-3-031-58516-6_6

associated with bronchopulmonary dysplasia (BPD) and lifelong co-morbidities [1]. BPD may also be linked to maternal infection during pregnancy [2], such as chorioamnionitis, which can lead to preterm birth and lung distress [3]. Respiratory therapies, such as non-invasive pressure support and surfactant replacement, are less effective in the extremely preterm demographic, resulting in the use of invasive mechanical ventilation applied as a last resort. However, such treatment can cause various forms of trauma, such as hyperoxia or endotracheal tube injury, leading to ventilator-induced lung injury (VILI) that can be exacerbated by infection. Inflammation from VILI and co-infections changes lung structure in ways that are hypothesized to stiffen them by increasing their resistance (opposition to movement) or decreasing compliance (ability to change the volume with applied pressure). However, it is not clear which particular histological changes are the proximate agents of inflammation-induced stiffening. The challenges associated with clinical studies in this and other fragile demographics necessitate computational and animal experiments to understand these mechanisms.

Mandell et al. [2] addressed this question using a neonatal rat model of chorioamnionitis to investigate what altered lung mechanics underlie the pressure-volume (PV) responses to mechanical ventilation. Maternal infection was simulated by prenatal exposure to an endotoxin (ETX). Their data included rat pup pressures, volumes, and histology images acquired at birth (D0) and day 7 (D7) of life, with and without ETX, and under either protective (SAFE) ventilation or two levels of injurious ventilation. Metrics of respiratory mechanics were determined from a forced oscillation technique and by applying standard image analysis techniques to the histology. Their results suggested that infection-related inflammation was correlated with a stronger VILI response, presenting in part as progressive airspace enlargement and increased compliance with increased inspiratory pressure. Interestingly, this is counter to what happens in adults. In the preterm pup demographic, this was speculated to result from insufficient collagen due to incomplete development of lung tissue. However, the group receiving ETX and the highest level of injurious ventilation exhibited decreased compliance and increased stiffness, possibly resulting from alveolar flooding, reducing lung capacity. This counter-intuitive result (see Fig. 2 of [2] for visualization) suggests additional analyses are needed to help elucidate mechanisms.

The histology was evaluated in Mandell et al. [2] with radial alveolar counts (RAC) and mean linear intercept (MLI). These techniques, while well-established, are subject to sampling bias [4, 5], can be labor-intensive, and may quantify only some aspects of the lung structure. Hence, there may be some benefit to the development of additional histological metrics. In particular, we expect that septal thinness and crimp (tortuosity) would be significantly correlated with parenchymal compliance; thickened and/or straight septa would be expected to be stiffer. Localized pockets of inflammation, in the form of locally thickened tissue, would be expected to pre-stress the parenchyma, rendering it stiffer. Bespoke image processing techniques could quantify such features.

Adverse respiratory system mechanics have traditionally been assessed with the classical single-compartment hydraulic model, dating back at least as far as the

1950s [6, 7]. This model consists of a pressure drop across the lung represented as a combination of a resistive component proportional to the rate of change of volume and an elastic component proportional to volume. Such models have grown in complexity over the years with the addition of new features such as lung heterogeneity, chest wall and other compartments, nonlinear components, respiratory muscle driving functions, ventilation, and even gas composition and cardiovascular dynamics [8–12] but have been overlooked in favor of simpler models in the context of data fitting [13–15]. The newer constant phase model used by Mandell et al. [2, 16] fits impedance measurements from a forced oscillation technique to determine a different set of mechanics metrics describing airways resistance, energy dissipation, and energy storage in the lungs [17]. However, it is also only a single-compartment model and a linear characterization of respiratory dynamics. Despite their growing prevalence and complexity, respiratory models are still primarily applied to adult physiology, in part due to the scarcity of relevant data and associated parameterization for children and infants. Applying parameter inference to a compartmental model specifically designed and evaluated against typical PV data obtained in a clinical setting may help assess the mechanics of a challenging demographic such as the extremely preterm infant.

In the current study, we explored new alternate parameter inference techniques via nonlinear modeling and image analysis applied to respiratory mechanics data from a neonatal demographic that is difficult to obtain and analyze. The experimental data described above obtained from ventilated rat pups in Mandell et al. [2] was used because it serves as an accepted surrogate of human data, is current, is amenable to our techniques, and generates an interesting open question. The primary objective was to determine if the proposed parameter inference techniques can uncover new information about breathing mechanics in the context of additional model complexity and for preterm infants. Secondary to the main objective was the opportunity to address the particular open questions from Mandell et al. [2] regarding mechanisms underlying changes in breathing mechanics related to inflammation and VILI.

We proposed a reduced compartmental model of lung pressure and volume that can capture the observed PV dynamics in the data for simulated experimental mechanical ventilation (Sect. 2.2). We then applied global and local sensitivity analyses to remove non-influential model components and obtain a minimal model (Sect. 2.3). Next, we performed parameter estimation on this minimal model (Sect. 2.4) with a gradient-based optimization algorithm and compared key parameter values for different groups. Concurrently, we developed and applied novel image analysis procedures for quantifying injury-related parameters extracted from histological images (Sect. 2.5). Metrics extracted from the images were statistically analyzed and their connections with the model parameters from this study and previously obtained biomechanical and biochemical data [2] were investigated. Results for all analyses are given in Sect. 3. Finally, we discussed our results in Sect. 4, including correlations that inspire new hypotheses that may improve understanding of the driving mechanisms of VILI response.

2   Methods

2.1   Experimental Data

The experimental groups from Table 1 of [2] were divided into pups from mothers injected with ETX and pups from healthy controls injected with saline (SAL). These were further divided into a group ventilated immediately upon birth at D0 and a group ventilated at D7. A protective level and two harmful levels (maximum pressures of 20 cm H_2O and 24 cm H_2O, denoted P20 and P24, respectively) of ventilation were administered. See Fig. 1 for a flowchart describing the experimental ventilation procedure.

2.2   Reduced Compartmental Model of Pressure-Volume Dynamics

We adapted a "reduced model" from a previous compartmental model developed by Ellwein Fix et al. [18] to simulate breathing mechanics in a preterm human infant.

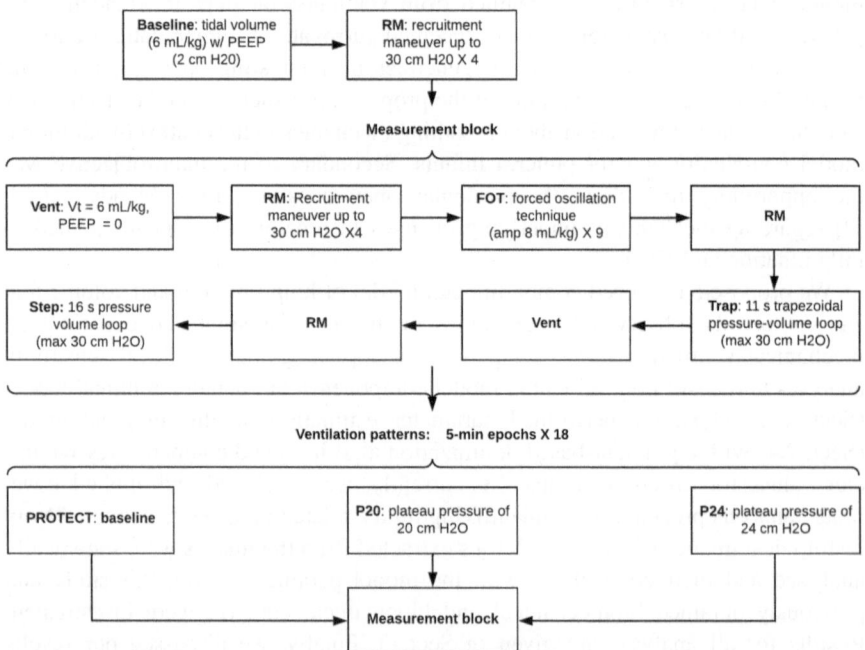

Fig. 1 Flowchart of the experimental procedure of ventilation. The experiment includes several recruitment maneuvers within the measurement block, which would add complexity that may be challenging to model if we chose to extend our model to simulate the entire experimental sequence

We retained compartments describing the airways with nonlinear resistance and the lungs with a nonlinear compliance in series with a parallel viscoelastic resistance and compliance. The diaphragm driving pressure from the previous model was replaced by mechanical ventilation. The original model [18] included components that could differentiate between types of airways and levels of chest wall compliance; given the limited experimental pressure and volume data characterizing the lungs directly, it was expected that the impact of these components would be unobservable. The extremely high chest wall compliance observed in preterm rat pups and human neonates was expected to contribute a nearly negligible amount to total compliance, further supporting the choice to remove the related compartment from the model [2, 19]. A schematic of the reduced model is given in Fig. 2. Tables 1, 2 give definitions for all variables and parameters, respectively.

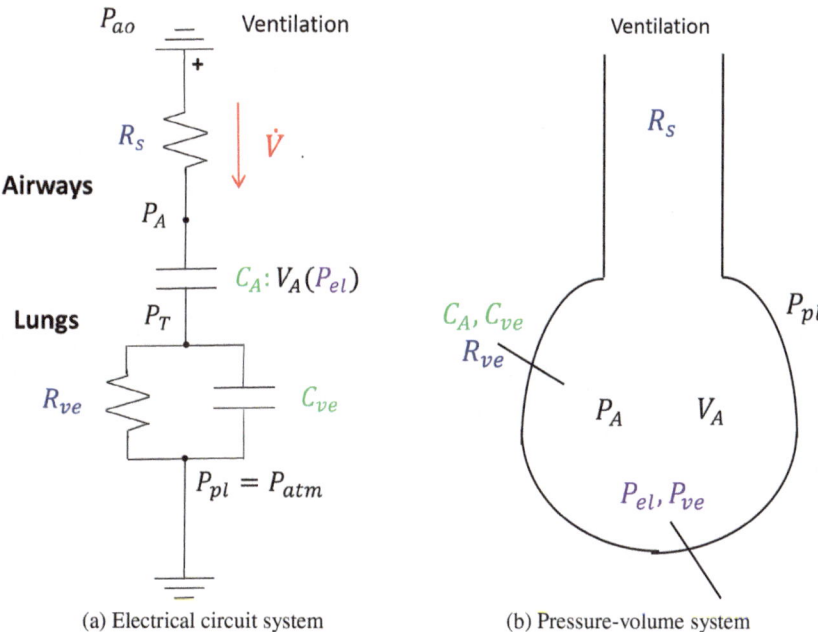

(a) Electrical circuit system   (b) Pressure-volume system

Fig. 2 Schematic of compartmental model shown as (**a**) an electrical circuit analog and (**b**) a pressure-volume analog. Simulated mechanical ventilation pressure drives the model at the airway opening (P_{ao}). Fixed pressures include alveolar (P_A) and lung tissue (P_T). Pressures across compliant boundaries include lung elastic (P_{el}) and lung viscoelastic (P_{ve}), such that transmural pressure across the lung $P_l = P_{el} + P_{ve}$. Volumetric air flow $\dot{V}$ is positive in the direction of the arrows across the small airways with resistance R_s and across the lungs with viscoelastic resistance R_{ve}, thereby increasing lung volume with positive ventilation. Compliances are denoted by C_i, where i matches the subscript for the corresponding zone. V_A denotes the volume of the airway lung compartment. Pleural pressure (P_{pl}) is modeled as equal to atmospheric pressure (P_{atm}) since chest wall effects and diaphragm breathing are considered negligible

Table 1 Variables (states and constitutive quantities) used in the reduced and minimal models. (-) denotes a dimensionless quantity

Variable	Physiological description (units)	Source
R_s	Small airway resistance (cm $H_2O \cdot s \cdot mL^{-1}$)	Eq. (7)
C_A	Lung compliance (mL $\cdot$ (cm $H_2O)^{-1}$)	
$\dot{V}$	Airflow rate (mL $\cdot s^{-1}$)	
t	Time (s)	a
F_{rec}	Recruitment function, contribution of (de)recruitment of alveoli to compliance (-)	Eq. (10)
Pressures (cm H_2O)		
P_{ao}	Airway opening pressure	Eq. (11), a
P_A	Lung alveolar pressure	
P_T	Lung tissue pressure	
P_{el}	Lung elastic recoil pressure	
P_{ve}	Viscoelastic pressure	
$P_{l,\text{dyn}}$	Dynamic pulmonary pressure	Eq. (1)
Volumes (mL)		
V_A	Volume of airway lung compartment	Eq. (8)
V_{el}	Volume due to lung unit structure aggregate elasticity	Eq. (9)

a Adapted from data [2]

2.2.1 State Equations

State variables include pressures $P(t)$ and volumes $V(t)$ within each compartment. Airflow through the cumulative airways is equal to the change in volume across lung compartments and is represented by both $\dot{V}(t)$ and dV/dt. The basic mass balance, Ohm's law, and Kirchoff's law are used with the circuit/hydraulic analog of flow such that resistances $R = \Delta P/\dot{V}$ and compliances $C = \partial V/\partial P$. Pressures denoted by P_{ao}, P_A, and P_T represent airway opening, alveolar, and (lung) tissue pressures, respectively. The dynamic transmural pressure across the lung $P_{l,\text{dyn}}$ is the sum of elastic recoil pressure and the viscoelastic pressure:

$$P_{l,\text{dyn}} = P_{el} + P_{ve}, \tag{1}$$

which are described respectively by

$$P_{el} = P_A - P_T, \tag{2}$$

$$P_{ve} = P_T - P_{atm}. \tag{3}$$

Differentiating P_{el} with respect to time and applying the chain rule gives

$$\frac{dP_{el}}{dt} = \frac{\partial P_{el}}{\partial V_A} \frac{dV}{dt} = \frac{1}{C_A(V_A)} \frac{dV}{dt}, \tag{4}$$

Table 2 Parameters used in the reduced and minimal models. Nominal values are those used for simulation (Sect. 3.1) and as initial guesses in optimization (Sect. 2.4); ranges are the bounds used in Morris screening (Sect. 2.3.1). R_s is included here as a parameter due to results from sensitivity analyses. The fixed parameters are not varied via sensitivity analyses (Sect. 2.3). (-) denotes a dimensionless quantity

Parameter	Physiological description (units)	Notes	Nominal value [range]
IC	Inspiratory capacity volume (mL)	[a]	0.5 [0.233, 1.11][c]
C_{ve}	Viscoelastic compliance (mL · (cm H$_2$O)$^{-1}$)	[a]	0.0159 [0.0044, 0.0769][c]
k	Characterizes slope, aggregate lung elasticity ((cm H$_2$O)$^{-1}$)	[b]	0.1 [0.001,0.2]
K_s	Small airway resistance low pressure coefficient (-)	[b]	15 (0,20]
γ	Maximum recruitable fraction of lung (-)	[b]	0.9 (0,1]
β	Baseline fraction of lung recruited at $P_{el} = 0$ (-)	[b]	0.5 (0,1]
Pressures (cm H$_2$O)			
c_{FR}	Mean opening pressure (recruitment curve)	[b]	17 [4, 25]
c_{FD}	Mean closing pressure (derecruitment curve)	[b]	6 [0, 20]
d_{FR}	Variance in opening pressure (recruitment curve)	[b]	4 [1, 10]
d_{FD}	Variance in closing pressure (derecruitment curve)	[b]	0.1 [0.01, 5]
Resistances (cm H$_2$O · s · mL^{-1})			
$R_{s,d}$	Change in small airway resistance	[b]	0.3 [0.225, 0.375]
$R_{s,m}$	Minimum small airway resistance	[b]	0.1 [0.075, 0.125]
R_s	Small airway resistance	[a]	0.276 [0.0002, 4.231]
R_{ve}	Lung viscoelastic resistance	[a]	12 [2.27, 73.5]
Fixed parameters for ventilation			
P_{atm}	Atmospheric pressure (cm H$_2$O)		0
$P_{\max}$	Maximum recruitment maneuver pressure (cm H$_2$O)	[a]	30
n_{steps}	Number of stair-step pressure increases (-)	[a]	7
τ	Total time of ventilation procedure (s)	[a]	16
t_d	Time delay in recruitment maneuver (s)	[a]	0.25
t_w	Stair-step width (s)	[a]	0.05

[a] Nominal values taken directly from or motivated by biomechanical metrics computed from the data [2]; in particular, R_s, R_{ve}, and total compliance compare to R_N, G, and $1/H$
[b] Morris screening bounds hand-tuned to produce realistic pressure volume curves
[c] Ranges widened during sensitivity analysis to [0.1, 1.5] for IC and [0.001, 0.1] for C_{ve}; see Sect. 2.3

where lung elastic compliance is calculated via symbolic computation as $C_A(V_A) = \frac{dV_A}{dP_{el}}$. Likewise, for the viscoelastic component, we can state

$$\frac{dP_{ve}}{dt} = \frac{\partial P_{ve}}{\partial V_{Cve}} \frac{dV_{Cve}}{dt} = \frac{1}{C_{ve}} \frac{dV_{Cve}}{dt}, \tag{5}$$

where C_{ve} is a constant viscoelastic compliance. Since $\frac{dV}{dt} = \frac{dV_{Rve}}{dt} + \frac{dV_{Cve}}{dt}$, i.e., the sum of the volume changes (flows) over R_{ve} and C_{ve}, Eq. (5) can be rewritten as

$$\frac{dP_{ve}}{dt} = \frac{1}{C_{ve}}\left(\frac{dV}{dt} - \frac{dV_{Rve}}{dt}\right) \tag{6a}$$

$$= \frac{1}{C_{ve}}\left(\frac{dV}{dt} - \frac{P_{ve}}{R_{ve}}\right). \tag{6b}$$

2.2.2 Constitutive Equations

The previous representation of upper airway dependence on airflow from [18] does not apply in our minimal model due to the quasi-static ventilation scheme and resulting data. We restrict the airway model to the resistance of the small airways such as bronchioles, which are presumed to decrease resistance with increased alveolar volume V_A. The equation for the resistance of the small airways R_s reflects this inverse relationship:

$$R_s = R_{s,d}\exp\left[\frac{-K_s V_A}{\text{IC}}\right] + R_{s,m}, \tag{7}$$

where $R_{s,d}$ is the change in small airway resistance, K_s is the small airway resistance low pressure coefficient, IC is the inspiratory capacity volume, and $R_{s,m}$ is the minimum small airway resistance. Parameter $K_s > 0$. As $V_A \to 0$, $R_s \to R_{s,m} + R_{d,m}$; likewise, as $V_A \to \text{IC}$, $R_s \to R_{s,m}$.

The volume of the lung compartment V_A is given by the product of the volume due to aggregate elasticity of the lung unit structure V_{el} and the fraction of recruited alveoli F_{rec}. Both quantities are functions of lung elastic recoil pressure:

$$V_A = V_{el}(P_{el})F_{rec}(P_{el}). \tag{8}$$

We model V_{el} as in prior studies [18, 20, 21] with a saturated exponential function

$$V_{el} = \text{IC}[1 - \exp(-kP_{el})], \tag{9}$$

where k is a parameter that characterizes slope and aggregate lung elasticity. The alveolar (de)recruitment function F_{rec} changes for recruitment and derecruitment of alveoli. Following [21], we assume that the form of the models for recruitment and derecruitment are identical except that the closing alveolar pressures are lower than the opening pressures. Thus, we model (de)recruitment by the piecewise function

$$F_{\text{rec}} = \begin{cases} \beta + (\gamma - \beta) \left\{ \dfrac{1 - \exp(-P_{el}/d_{F_R})}{1 + \exp[-(P_{el} - c_{F_R})/d_{F_R}]} \right\}, & 0 \le t < \tau/2 \\[2ex] \beta + (\gamma - \beta) \left\{ \dfrac{1 - \exp(-P_{el}/d_{F_D})}{1 + \exp[-(P_{el} - c_{F_D})/d_{F_D}]} \right\}, & \tau/2 \le t \le \tau \end{cases}, \tag{10}$$

where β is the baseline fraction of lung recruited at $P_{el} = 0$, γ is the maximum recruitable fraction of lung, d_{F_R} is the variance in opening pressure, c_{F_R} is the mean opening pressure, d_{F_D} is the variance in closing pressure, c_{F_D} is the mean closing pressure, and τ is the total time of the ventilation procedure. Here, the parameters c_F and d_F are allowed to vary between inspiration and expiration for one breathing loop for a single rat. The subscripts R and D denote recruitment and derecruitment, respectively.

The data in this study come from a quasi-static stepwise ventilation [2], thus the driving pressure is a simulated protocol applied at the airway opening P_{ao}. The stepwise ventilation is a 16-second procedure, whereby the airway opening pressure P_{ao} is increased from zero (or nearly zero) to 30 cm H_2O in a stair-step fashion: the maneuver comprises seven increases in pressure (7 "steps up") with roughly equal duration and comparable decreases from maximum pressure back down to zero. We approximate the stair-step function with the sum of narrow tanh functions and include the possibility for a slight delay t_d in order to better match the experiments:

$$P_{ao}(t) = \frac{1}{2} \frac{P_{\max}}{n_{\text{steps}}} \text{switch}(\tau/2 - t) \sum_{j=0}^{N} \left[1 + \tanh\left(\frac{t - [j\tau/(2n_{\text{steps}}) + t_d]}{t_w} \right) \right], \tag{11}$$

where $P_{\max}$ is the maximum recruitment maneuver pressure, n_{steps} is the number of stair-step pressure increases, τ is the total time of the ventilation procedure, t_d is the time delay in the recruitment maneuver, and t_w is the stair-step width. The narrowness of the stair-step is enforced by using $t_w = 0.05$ s; the value is selected to best match experimental ventilation pressure curves [2]. Here, $N = \lfloor t/(\tau/2n_{\text{steps}}) \rfloor$, $P_{\max} = 30$ cm H_2O, $n_{\text{steps}} = 7$, $\tau = 16$ s, $t_d = 0.25$ s, and switch(x) is essentially the sign function that returns 1 or -1 depending on the sign of the argument, but takes the value 1 at 0:

$$\text{switch}(x) = \begin{cases} \text{sgn}(x), & x \ne 0 \\ 1, & x = 0 \end{cases}. \tag{12}$$

The switch function facilitates the step direction: "up" until $t = 8$ s and then "down" until $t = 16$ s.

2.2.3 Model Formulation

Recalling the relationship between resistance, change in pressure, and time derivative of volume, we may write

$$\frac{dV}{dt} = \frac{P_{ao} - P_{l,\text{dyn}}}{R_s}.$$ (13)

We also recall that $P_{l,\text{dyn}}$ is calculated via Eq. (1). Substitution into Eqs. (4) and (6b) yields the system for viscoelastic and elastic pressure, P_{ve} and P_{el}, which is given by

$$\frac{dP_{el}}{dt} = \frac{1}{C_A(V_A)\,R_s}(P_{ao} - P_{el} - P_{ve}),$$ (14)

$$\frac{dP_{ve}}{dt} = \frac{1}{C_{ve}}\left(\frac{P_{ao} - P_{el} - P_{ve}}{R_s} - \frac{P_{ve}}{R_{ve}}\right).$$ (15)

Equations (14)–(15) are solved on $t \in [0, 16]$ s because this time frame represents one mechanical PV ventilation maneuver. The initial conditions used are $P_{el}(0) = 0.954$ cm H_2O and $P_{ve}(0) = 0$ cm H_2O following [18]. Then, V_{el} and V_A are calculated via Eqs. (8) and (9).

The system in Eqs. (14)–(15) is solved using `ode15s` in MATLAB. We use default ODE solver and optimization tolerances of 10^{-3} for the relative error tolerance and 10^{-6} for the absolute error tolerance.

2.3   Sensitivity Analysis

The reduced model includes components that we hypothesize could reflect the physiology of breathing mechanics as evident in PV data, especially nonlinearities. Determination of the impact of model parameters on model output and estimation of reasonable parameter values requires a mathematical understanding of the practical identifiability of the underlying parameters in the context of available data [22]. Parameters that are identified as insensitive or correlated with other sensitive parameters are commonly handled either by keeping them at a reasonable nominal value for all data sets or by removing them from the model altogether. We perform two analyses for this purpose: (1) the Morris global screening method, a one-at-a-time method that calculates a set of randomized but structured finite difference derivatives over the full parameter space versus a set of key scalar outputs; and (2) a coarse univariate local sensitivity analysis, whereby each parameter is perturbed by a factor of two starting with nominal parameters describing an 'average' animal subject and the magnitude of change in scalar outputs is categorized by effect size. Sensitivity analyses are applied to the parameter set:

$$\boldsymbol{p} = \{K_s, \text{IC}, C_{ve}, R_{ve}, R_{s,d}, R_{s,m}, k, c_{F_R}, c_{F_D}, d_{F_R}, d_{F_D}, \beta, \gamma\}.$$ (16)

Note that the ventilation setting parameters are not included, as they are taken directly from the experimental settings.

2.3.1 Morris Effects Analysis

As a global sensitivity analysis, we use the technique of Morris elementary effects, or Morris screening analysis [23]. In contrast to local sensitivity analysis, which studies the effect of individual parameter perturbations, global sensitivity analysis explores the effect of combinations of parameters that are perturbed across the feasible parameter space [24]. The method aims to compute "elementary effects" that approximate derivatives of model output changes with respect to parameter perturbations [25, 26]. These are then combined in order to rank parameter sensitivities.

Let $f(t; p_1, \ldots, p_n)$ be the model output with respect to parameters $\boldsymbol{p} = (p_1, \ldots, p_n)$. Let $\boldsymbol{e}_i$ denote the ith unit vector. Then the ith Morris elementary effect is given by

$$\text{EE}_i = \frac{f(t; \boldsymbol{p} + \boldsymbol{e}_i \delta) - f(t; \boldsymbol{p})}{\delta}. \tag{17}$$

Here, $\delta = \ell/2/(\ell - 1)$ describes a step in the parameter space such that $\boldsymbol{p} + \boldsymbol{e}_i \delta$ is still within the allowable bounds for the parameter. We choose $\ell = 60$, which gives $\delta \approx 0.51$. This choice allows for a symmetric sampling distribution [23]. We also normalize the parameters to the interval $[0, 1]$ for the sampling following Colebank and Chesler [24].

The summary statistics are computed after conducting N random initializations for each EE_i. The arithmetic mean associated with each parameter is given by

$$\mu_i = \frac{1}{N} \sum_{j=1}^{N} \text{EE}_i^j, \tag{18}$$

and the corresponding sample variance is found by

$$\sigma_i^2 = \frac{1}{N-1} \sum_{j=1}^{N} (\text{EE}_i^j - \mu_i)^2. \tag{19}$$

To study the magnitude of the elementary effects while avoiding potential cancellation issues, we use the mean absolute elementary effect μ_i^* in our analysis following [27]:

$$\mu_i^* = \frac{1}{N} \sum_{j=1}^{N} \left| \text{EE}_i^j \right|. \tag{20}$$

To compare and rank the relative effects of the parameters on the model output, we use the combined statistic termed the Morris ranking:

Table 3 Scalar metrics used by the Morris screening analysis

Metric (units)	Equation	Physiological description	Nominal value
WOB (mL · cm H_2O)	(22)	Work of breathing	2.49
v_{max} (mL)	max(V_A)	Maximum lung volume	0.428
Slopes (mL/(cm H_2O))			
sl_R 0–10	(23)	Recruitment slope, $P_1 = 0$, $P_2 = 10$ cm H_2O	0.0157
sl_D 0–10	(24)	Derecruitment slope, $P_1 = 0$, $P_2 = 10$ cm H_2O	0.0311
sl_R 10–30	(23)	Recruitment slope, $P_1 = 10$, $P_2 = 30$ cm H_2O	0.0131
sl_D 10–30	(24)	Derecruitment slope, $P_1 = 10$, $P_2 = 30$ cm H_2O	0.00621

$$M_i = \sqrt{\mu_i^{*2} + \sigma_i^2}. \tag{21}$$

We vary thirteen free parameters (the set listed in Eq. (16)) for our Morris sensitivity analysis; descriptions and their nominal values are given in Table 2. We choose six scalar metrics as key characteristics of the PV curves that were observed in the data to differ across treatment and ventilation groups. These are work of breathing (WOB) [mL · cm H_2O], maximum lung volume v_{max} [mL], and four slopes of the PV curve [mL/(cm H_2O)]: (1) recruitment and (2) derecruitment curves between pressures of 0 and 10 cm H_2O and (3) recruitment and (4) derecruitment curves between pressures of 10 and 30 cm H_2O. These four slopes are denoted by sl_R 0–10, sl_D 0–10, sl_R 10–30, and sl_D 10–30, respectively. WOB is defined as

$$\text{WOB} = \int_{P_{l,\text{dyn}}} \left[V_{A,D}(P_{l,\text{dyn}}) - V_{A,R}(P_{l,\text{dyn}}) \right] dP_{l,\text{dyn}}, \tag{22}$$

where the D and R subscripts on V_A denote the derecruitment and recruitment portions of the lung volume, respectively. This integration is calculated numerically with a standard trapezoid rule in MATLAB. The second metric v_{max} is the maximum of V_A, the lung volume, over the PV loop. The slope sl_R between pressures of P_1 and P_2 cm H_2O is calculated by

$$sl_R = \frac{V_{A,R}(P_2) - V_{A,R}(P_1)}{P_2 - P_1}, \tag{23}$$

and the slope sl_D between pressures of P_1 and P_2 cm H_2O is calculated by

$$sl_D = \frac{V_{A,D}(P_2) - V_{A,D}(P_1)}{P_2 - P_1}. \tag{24}$$

The metrics are summarized in Table 3.

2.3.2 Local Sensitivity Analysis

We perform a coarse univariate sensitivity analysis in which we change one parameter at a time by a fixed percent of its nominal value and note the percent change in the scalar output, while holding all other parameters constant at their nominal values. In particular we look at the effect of doubling and halving each of the following parameters $R_{s,m}$, $R_{s,d}$, K_s, R_{ve}, c_{F_R}, c_{F_D}, d_{F_R}, d_{F_D}, and k on the six scalar effects defined in Sect. 2.3.1. Parameters γ and β nominally represent fractions of recruitable lung such that $0 < \gamma, \beta \leq 1$, so their low and high values are chosen explicitly to be $[0.85, 0.95]$ and $[0.1, 0.9]$, respectively. These also represent reasonable possible values for these fractions with respect to the physiology. Finally, for C_{ve} we use 70% as opposed to 50% to ensure simulation output remains physiologically viable. The nominal values are listed in Table 2.

2.4 Optimization

For the parameters identified as sensitive for at least one metric (Table 3), we determined optimal values by minimizing the sum of the squared differences between the quasi-static points of the PV curves given in the data and the quasi-static points of our simulated PV curves. Parameters that were identified as non-sensitive across all six metrics used in the Morris screening were kept constant at the nominal values reported in Table 2. Optimization was performed using the constrained optimization algorithm lsqnonlin in MATLAB using the Levenberg-Marquardt option and an ODE solver relative tolerance of 10^{-12}. The step tolerance for the optimization algorithm was set equal to the square root of the relative tolerance of the ODE solver. The nominal values from our sensitivity analyses were used as the initial guesses for k, c_{F_R}, c_{F_D}, β, and γ.

We determined optimal parameters for each rat in the data by minimizing the least-squares objective function J:

$$J = \frac{1}{2} \sum [V_A(P_{l,\mathrm{dyn}_{\mathrm{data}}}) - V_{A_{\mathrm{data}}}(P_{l,\mathrm{dyn}_{\mathrm{data}}})]^2. \tag{25}$$

Since pulmonary system elastance and tissue damping coefficient for individual rats were reported in the data as obtained from the forced oscillation impedance fit to the constant phase model ([2, 16], denoted as H and G), we based our initial iterates for C_{ve} and R_{ve} on these data. Constant upper and lower bounds were imposed on the algorithm. Initially, we used the same bounds that were imposed in the Morris screening for C_{ve}, which were estimated from pulmonary system elastance in [2]. However, we were unable to achieve acceptable fits without decreasing the lower bound for C_{ve} (equivalent to increasing the upper bound on $1/C_{ve}$, which is how we later present our results). Since the viscoelastic compliance, C_{ve}, in our model does

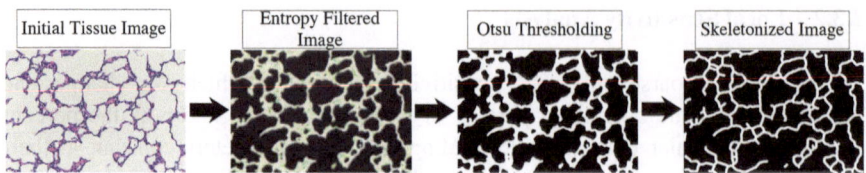

Fig. 3 Schematic of image processing procedure. Raw RGB image is entropy filtered and converted to grayscale, before thresholding (Otsu method) to create a binary image where tissue is white and lumens are black. The binary image is quantitatively analyzed by a variety of additional methods, including skeletonization. Here, the skeletonized image is dilated for visibility

not actually encapsulate the whole system compliance/elastance, it is unsurprising that we had to adjust the bounds.

For some rat data sets, the optimization appeared to stagnate or converge to a local minimum that creates a non-physiological simulated output. In these cases, the initial iterate was modified so the optimization algorithm converged appropriately. We calculated the mean optimal values for each rat group along with the variance. The average values and the variances are reported in Table 5.

2.5 Image Analysis

In order to quantify morphological differences relevant to biomechanical differences among treatment and age groups, we develop a procedure to analyze lung histology images using known techniques. The implementation uses the MATLAB Image Processing Toolbox. RGB images from a previous study [2], of uniform pixel dimensions and magnification, are passed through an entropy filter, using a filter neighborhood equivalent to a disk of radius 5 microns, which minimizes the impact of irrelevant small-scale details such as individual cells. Filtered images are converted to grayscale and binarized with Otsu thresholding. Binarized images are then analyzed by skeletonization, erosion, connected region identification, and other methods to yield image metrics. The basic procedure is illustrated in Fig. 3. We do not identify or segment specific non-alveolar structures in the images, such as blood vessels.

2.5.1 Metrics of Lumens

Lumen Count

A raw count c of lumens in each image will include both lumens falling completely within the image and also lumens clipped by the edge of the image (Fig. 4). Considering them together would skew the lumen metrics (count, area, and others),

(a) (b)

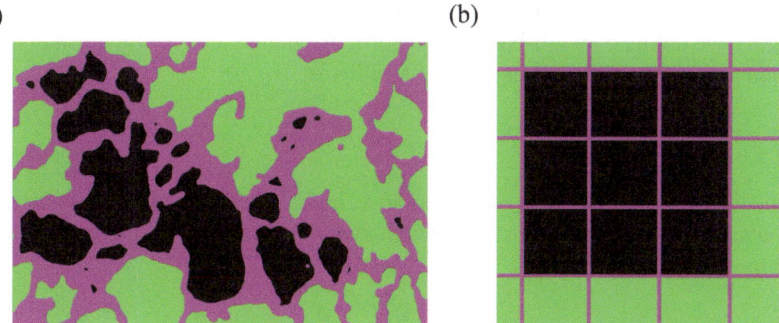

Fig. 4 The images include lumens cropped by the image boundaries. The lumen count is used as an independent metric and in calculating other image analysis metrics. (**a**) When segmenting lumens to count them, it is crucial to make a distinction between complete lumens (black) and those clipped by the edge of the image (green). (**b**) A simple estimate of the corrected image analysis metrics in the presence of the clipped lumens begins with assuming a regular grid of square lumens and deriving the correction terms for the lumen count

as well as other tissue metrics, such as tortuosity, that are derived from lumen counts. Therefore, we use a corrected lumen count, on the assumption that, on average, each edge lumen is missing half its true area.

Consider a grid of equal square lumens, each of side length l, in a square image of side length L (Fig. 4). On average, the image length and width each hold $n = L/l$ lumens, each of area l^2, so the whole image holds $N = n^2 = (L/l)^2$ lumens, each of area l^2, for a total area of L^2. The image has a perimeter of $4L$, and the image edge cuts $4n$ lumens. The edge lumens are assumed to be, on average, cut in half. Thus, if N is the true lumen count, and c is the raw lumen count, on average, $c = N + 2n = N + 2\sqrt{N}$. Solving for N gives the corrected lumen count, N, based on the raw lumen count c:

$$N = c + 2 - 2\sqrt{c + 1}. \tag{26}$$

Lumen Areas

For each lumen found, we measure its area A_i. Again, the image contains a significant fraction of lumens that are cut by the image edge, so aggregate area metrics have to account for the estimated true areas of clipped lumens. Again considering a regular grid of squares, the total lumen area of the image is $L^2 = Nl^2 = c\bar{A}_{\text{meas}}$ where $\bar{A}_{\text{meas}} = [(N - 2n)l^2 + 4nl^2/2)]/c$. Thus, the true mean area per lumen $\bar{A}$, for the square grid model, would be $\bar{A} = l^2 = \bar{A}_{\text{meas}}c/N = \bar{A}_{\text{meas}}c/(c + 2 - 2\sqrt{c + 1})$.

For comparison with the existing lumen metric mean linear intercept (MLI, a length), we define a length-based area metric for image lumens, by normalizing the lumen areas A_i to an equivalent disk of radius R_i

$$R_{\text{equiv},i} = \sqrt{A_i/\pi}. \tag{27}$$

Since the set of lumens with measured areas includes clipped lumens, when finding the average R_{equiv}, we adjust the calculation of the mean by the expected edge counts:

$$\bar{R}_{\text{equiv}} = \sqrt{c/N} \frac{1}{N} \sum_{i=1}^{N} \sqrt{A_i/\pi}. \tag{28}$$

All analyses of R_{equiv} use this corrected quantity, $\bar{R}_{\text{equiv}}$, but we drop the bar for convenience.

We define another area metric using the cumulative distribution of areas A_i:

$$A_{\text{cumul}}(n) = \sum_{i=1}^{n} A_i. \tag{29}$$

Then the total area of lumens $A_{\text{lum}} = A_{\text{cumul}}(c)$. We define R_{mid} as the equivalent radius $R_{\text{equiv},j}$ such that $A_{\text{cumul}}(j) \approx \frac{1}{2} A_{\text{cumul}}(c)$ as closely as possible.

2.5.2 Metrics of Tissue

Tissue Area Fraction

We calculate the total tissue area A_{tiss} by counting tissue pixels in the binarized image. Tissue area fraction ϕ is then (tissue pixel count)/(total pixel count).

Tissue Length

For each image, tissue length or total perimeter Λ is calculated from the skeletonized image as the total length of tissue centerlines. We expect an allometric relation between Λ and the number of lumens N. Following the extremely simplified square grid models of Sect. 2.5.1 and Fig. 4, an image of a tissue grid has total area $A = L^2 = Nl^2$ and the total length (perimeter) is $\Lambda = 2Nl = 2(L/l)L = 2nL = 2L\sqrt{N}$. Thus, for a general image of lung parenchyma, we expect that the total tissue length Λ is proportional to $\sqrt{N}$.

Tortuosity

At the length scale of whole alveoli, septal tortuosity results from adjacent alveoli pushing into the septum. Collagen tortuosity is at a much smaller length scale. We quantify tortuosity at an intermediate length scale, smaller than a single alveolus, but substantially larger than the polymer length scale. For a single geometric object of area A and perimeter Λ, we define its dimensionless tortuosity ratio T as

$$T = \frac{\Lambda^2}{4\pi A}. \tag{30}$$

A circle has $T = \frac{(2\pi r)^2}{4\pi(\pi r^2)} = 1$, the minimum possible value. For a square, $T = 4/\pi \approx 1.3$.

To extend this metric to a non-simply-connected object or a partitioning of the space, we propose the following. Suppose an image contains N lumens with total perimeter length Λ and total of lumen areas A_{lum}, which in our images may be close to the total area of the image. Because of the shared boundaries between adjacent lumens, we need to count the total perimeter twice. We generalize the tortuosity ratio T for a single object to an image tortuosity metric

$$\tilde{T} = \frac{\bar{\Lambda}^2}{4\pi \bar{A}} = \frac{(2\Lambda_{\text{tot}}/N)^2}{4\pi A_{\text{lum}}/N} = \frac{\Lambda^2}{N\pi A_{\text{lum}}}, \tag{31}$$

where N is the estimated true lumen count from Eq. (26). Note that Λ does not need an adjustment for intersecting the edge.

For a grid of squares, it is easy to confirm that $\tilde{T} = \frac{4}{\pi} \approx 1.3$, and for regular hexagonal packing $\tilde{T} = \frac{2\sqrt{3}}{\pi} \approx 1.1$, which is the lower bound, approaching the minimum of 1 for a circle. We expect a greater tortuosity ratio $\tilde{T}$ to correspond to higher tissue compliance since the lumens would expand by straightening their crimps.

Septal Width and Its Distribution

Our most basic width metric is the mean width:

$$\bar{w} = A_{\text{tiss}}/\Lambda, \tag{32}$$

where A_{tiss} is the total area of septa and Λ is the total length of septa. We can infer more about the histology, and potentially the biomechanics, if we go beyond the mean of width to look at its distribution.

Consider a 2D image containing a network of septa with a fixed total tissue area A_{tiss} or area fraction ϕ and total tissue length Λ. If the tissue is uniformly thick, i.e., the tissue area is spread evenly along its length, we expect that network to have a

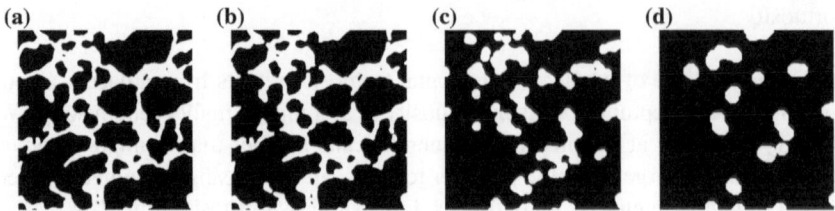

Fig. 5 Example of applying the pruning algorithm with a disk of radius r to a representative lung tissue image, where r increases from left to right. Eventually using a large enough disk, no tissue remains. (**a**) The binarized tissue image before pruning. The structuring element used in (**b**) is a disk with radius 5 μm, in (**c**) 10 μm, in (**d**) 15 μm. Hence (**b**), (**c**), and (**d**) show all tissue in (**a**) thicker than 10, 20, and 30 μm, respectively

certain stiffness. If that same area is distributed unevenly over the same total length, i.e., in a mixture of thick and thin septa, we would expect that network to be more compliant, as the thin portions would stretch more than the thick portions under the same load. We expect uneven width from a variety of pathological conditions, including fibrosis and inflammation.

To obtain the distribution of septa of width w, we follow a previous study [28] quantifying vascular networks. We define a pruning algorithm as follows:

1. Starting with the binarized image, erode the image using a disk of radius $w/2$.
2. Dilate the eroded image by the same disk.
3. Measure the area of tissue remaining in the image.
4. Repeat 1–3 until no tissue remains.

This serves to remove any image details finer than the gauge w (Fig. 5). For a large enough w, no pixels remain after the pruning. We can consider the functions $A_{\text{tiss}}(w)$ or $\phi(w)$, which show the distribution of structures of width greater than w.

2.6   Statistical Analysis

Statistical analyses were performed in MATLAB and JMP (SAS Institute, Cary, NC) and using the FactoMineR package [29] in R [30]. Analyses included distributions of image metrics and their fits to multivariate models, as well as correlations with previously reported metrics of biomechanics and cytokine concentrations [2] and with fitted model parameters. Multivariate statistical models for metric distributions were evaluated by significance (strictly $p < 0.05$, but generally much lower) and by small-sample corrected Akaike information criterion (AICc). Optimized parameter values are compiled by mean and variance.

3 Results

3.1 Model Solutions

A representative dimensional solution of the reduced model output airway lung compartment volume V_A against dynamic pressure $P_{l,\text{dyn}}$ is shown in Fig. 6. This solution uses nominal values for a D0 control (SAL) group rat under SAFE ventilation, also used for the later sensitivity analyses (Sect. 2.3), and is plotted against the corresponding data for a single rat from this group. The effect of the stair-step ventilation pressure applied at the airway opening P_{ao} on model solutions can be seen in Fig. 6b, as the curve exhibits cusps at quasi-static points corresponding to the edge of a stair. The expiration curve is higher than the inspiration curve, which is consistent with the lung air sacs requiring a higher pressure on average to open than to close for a given volume.

3.2 Morris Screening

We implement Morris screening using 100 randomized initializations of model parameters. A small number of outlier results are excluded: any model imple-mentation with an integration tolerance failure, an infeasible metric output (e.g., a negative WOB), or a drastic outlier EE_i^j (defined as more than 1000 times the median magnitude) that suggests an infeasible combination of parameter values. These excluded results comprise less than 1% of the original simulations.

The absolute value of the mean μ_i^* from Eq. (20) is plotted on a log-log scale against the sample variance σ_i^2 for the thirteen parameters studied over the six

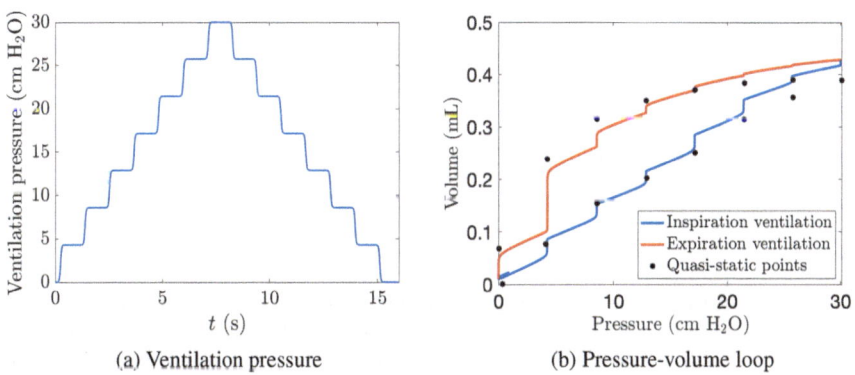

(a) Ventilation pressure   (b) Pressure-volume loop

Fig. 6 (**a**) Simulated stair-step ventilation pressure applied at the airway opening. (**b**) Simulated pressure-volume loop for nominal parameter values (see Tables 1, 2). The experimental quasi-static points from a pressure-volume loop [2] are shown for comparison

metrics analyzed (Fig. 7). Parameters with both large μ_i^* and σ_i^2 are considered influential because they have both nonlinear and interacting effects with other parameters. Conversely, parameters with both small μ_i^* and σ_i^2 are considered non-influential. For example, inspiratory capacity IC is consistently the most influential parameter, and change in small airway resistance $R_{s,d}$ is consistently a non-influential parameter across all six metrics. This categorization follows [23].

The Morris statistic rankings per elementary effect by Eq. (21) are shown in Fig. 8 on a log scale. Note that the Morris ranking plot for the recruitment slope between pressures of 0 and 10 cm H_2O does not show values for c_{F_D} and d_{F_D} because the analysis determined those rankings to be zero. Parameter rankings falling above the mean ranking, shown by the dashed horizontal line, are deemed sensitive and are indicated in red [31, 32]. Eight out of the thirteen reduced model parameters form the union of the sensitive parameters over all six effects (Table 4), which we call the "Morris sensitive set" and summarize as

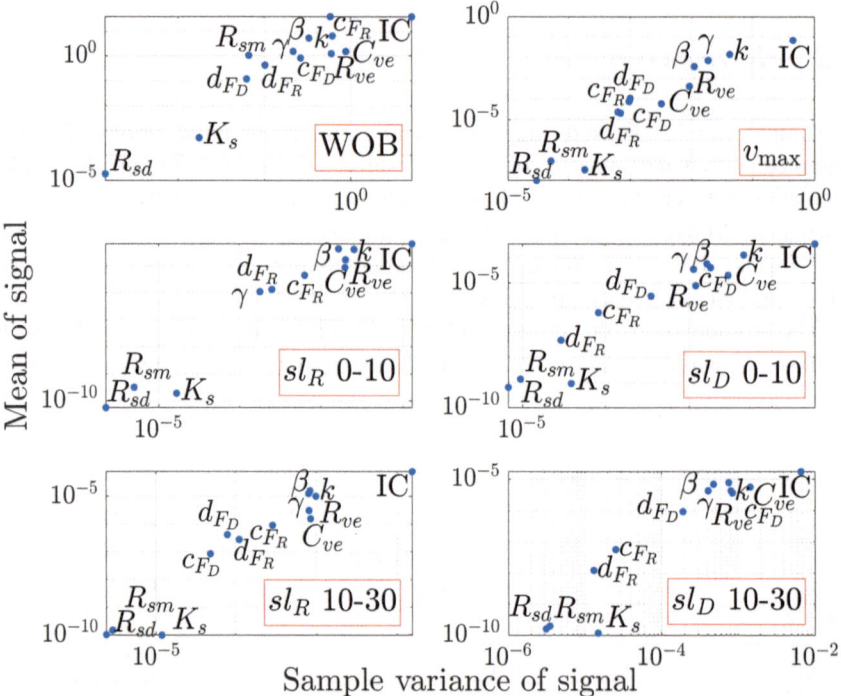

Fig. 7 Absolute value of the mean elementary effect μ_i^* of signal (Eq. 20) is plotted against sample variance of signal for the results of the Morris effects analysis. The six metrics analyzed are work of breathing (WOB), maximum lung compartment volume (v_{max}), and recruitment or inspiratory (R) and derecruitment or expiratory (D) slopes between pressures 0 and 10 cm H_2O and 10 and 30 cm H_2O

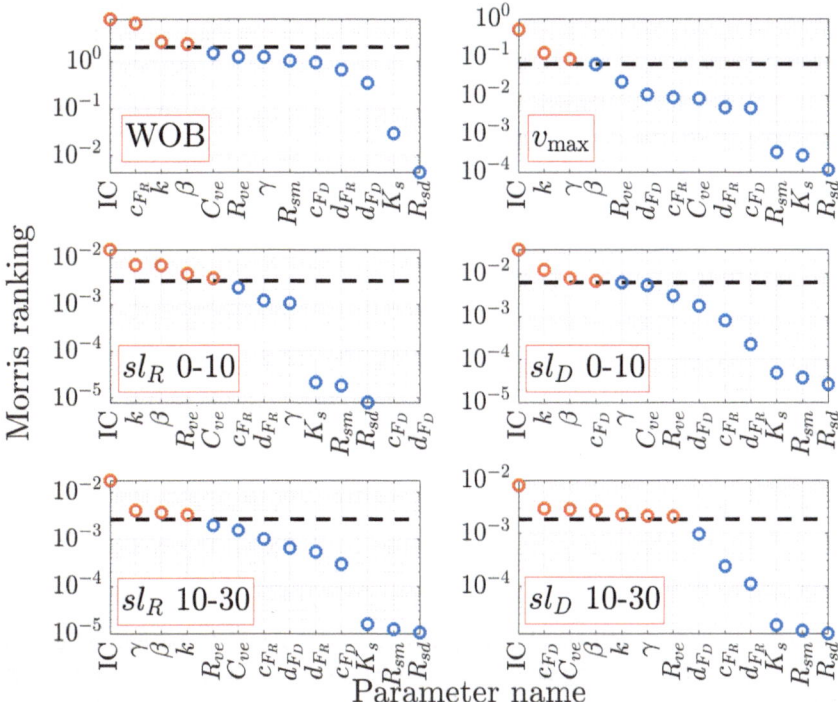

Fig. 8 Ranking from the Morris effects analysis. The black dashed line shows the mean ranking. Parameters with rankings above the mean are colored in red, and those with ranking below the mean are colored in blue. The six metrics analyzed are work of breathing (WOB), maximum lung compartment volume (v_{max}), and recruitment or inspiratory (R) and derecruitment or expiratory (D) slopes between pressures 0 and 10 cm H_2O and 10 and 30 cm H_2O

$$p_{\text{Mor}} = \{IC, C_{ve}, R_{ve}, k, c_{F_R}, c_{F_D}, \beta, \gamma\}. \tag{33}$$

Inspiratory capacity was evaluated as the most sensitive parameter for all six metrics. This is not surprising since IC has a direct effect on V_A through Eqs. (8)–(9), which comprise one axis of a PV curve. The parameter k, representing lung elasticity and contributing to the saturation of the function V_{el}, is sensitive across all six metrics. For the four slope metrics, the Morris rankings of the mean and variance in recruitment or decruitment pressure seem to correspond to the part of the PV curve studied. For example, c_{F_R} and d_{F_R} have higher Morris rankings than the same for derecruitment (c_{F_D} and d_{F_D}) for sl_R from 0 to 10 cm H_2O. This holds for all c_{F_R} and c_{F_D} pairs and for all but one pair of d_{F_R} and d_{F_D}; in this case, the values are nearly identical.

Certain parameter groupings were found to be either all part of p_{Mor} (Eq. 33) or all non-influential. The pressure range parameters d_{F_R} and d_{F_D} describing the

Table 4 Sensitive parameters as determined by the Morris screening. The middle six columns indicate the parameters that had ranking larger than the mean for each of the metrics. The far right column indicates the union of the rankings above the mean to form the Morris sensitive set

Parameter	Sensitive by Morris analysis						Union of sensitive parameters
	WOB	v_{max}	sl_R 0–10	sl_D 0–10	sl_R 10–30	sl_D 10–30	
K_s							
IC	X	X	X	X	X	X	X
C_{ve}			X			X	X
R_{ve}			X			X	X
$R_{s,d}$							
$R_{s,m}$							
k	X	X	X	X	X	X	X
c_{FR}	X						X
c_{FD}				X		X	X
d_{FR}							
d_{FD}							
β	X		X	X	X	X	X
γ		X			X	X	X

variance of the opening or closing pressures were found to be non-influential; however, the mean opening and closing pressures, c_{FR} and c_{FD}, and baseline and maximum recruitment fractions, β and γ, respectively, were found to be influential. We note that d_{FR} and d_{FD} characterize the heterogeneity of the lung by allowing for a transition to full (de)recruitment [18, 21]; since these are not sensitive, this suggests that our model cannot identify this alveolar variation in the opening or closing of lung units. In contrast, c_{FR} and c_{FD} represent average information. This suggests that a limitation of our model is that it must treat alveolar recruitment as a homogeneous action across the lung, although we recognize that in actual lungs, recruitment occurs in specific locations. The parameters K_s, $R_{s,d}$, and $R_{s,m}$ govern the nonlinear small airway resistance, R_s, and are not part of p_{Mor} (Eq. 33). Note that the function for R_s is also dependent upon inspiratory capacity IC, but IC governs the variable volume V_A as well. This suggests that the reduced model can be minimized further to eliminate the nonlinearity of our small airway resistance so that R_s is set to a constant value: $R_s = R_{s,const}$. For the optimization that follows in Sect. 3.4, we use this "minimal" version of our reduced model. We also create our vector of free parameters to vary in the optimization as p_{Mor} (Eq. 33) without IC, which is given by

$$p_{free} = \{C_{ve}, R_{ve}, k, c_{FR}, c_{FD}, \beta, \gamma\}. \tag{34}$$

While IC was determined to be sensitive by the Morris screening, subject-specific metrics were directly measured by Mandell et al. [2], so we use each animal's IC data for our analysis.

3.3 Local Sensitivity Analysis

Results from the univariate local sensitivity analysis are shown in the tornado plots in Fig. 9. The parameters in a given plot are organized from the largest increase to the largest decrease in metric upon an increase in parameter, with parameters with the least effect on the metric in the center of the plot. For most parameters, there is either a positive relationship across all metrics or a negative relationship across all metrics.

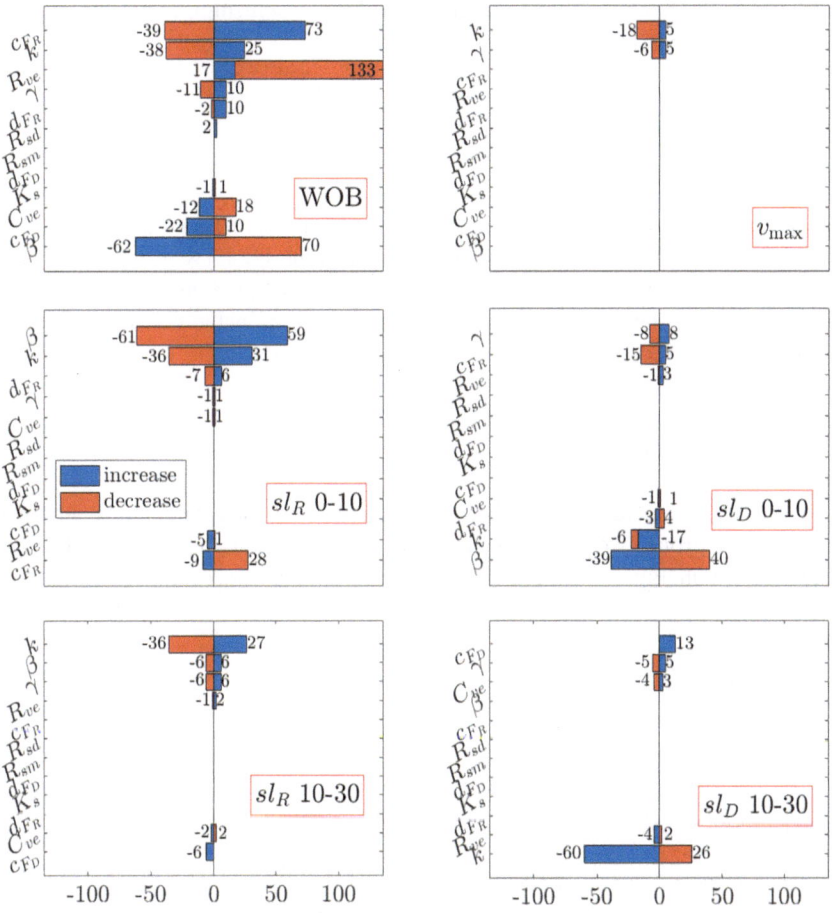

Fig. 9 Local sensitivity analysis of the twelve parameters discussed in Sect. 2.3.2 and their impact on six scalar metrics defined in Table 3. The blue bar for each parameter represents the percent change in the scalar metric given an increase (typically 200%) in the parameter, while the orange bar represents the percent change in the scalar output given a decrease (typically 50%) in the parameter. Parameters are organized from the largest increase to the largest decrease in metric upon an increase in parameter

Of note are the parameters that do not have a strictly positive or negative relationship. WOB increases when viscoelastic resistance R_{ve} is both increased and decreased. While WOB increases when change in airway resistance R_{sd} increases, there is no change to the metric when R_{sd} is decreased. Both positive and negative changes to k cause a decrease in sl_R 10–30. When decreased, mean closing pressure c_{F_D} causes no change in sl_D 0–10 and sl_D 10–30; however, increasing c_{F_D} increases sl_D 10–30 and decreases sl_D 0–10.

We see that k has the largest impact on sl_D 0–10, sl_D 10–30, and $v_{\max}$, while β has the largest impact on sl_R 0–10 and sl_R 10–30. Multiple parameters affect WOB; c_{F_R}, R_{ve}, and β have the largest impacts. Encouragingly, the findings of local sensitivity analysis (Fig. 9) confirm the Morris screening conclusions of the least influential parameters.

3.4   Optimization Results

Optimized PV curves are plotted against the data in Fig. 10 for one rat in the D0 SAL group under SAFE ventilation and one rat in the D7 ETX group under high-pressure ventilation (P24). Using the nominal parameters, the initial guesses for the optimization are shown as dashed lines to indicate reasonable convergence to the optimal solution representing observed rat lung dynamics.

3.4.1   Mean Values

After fitting parameters to the data for each rat and then calculating summary statistics per group, we obtained the mean optimized parameter values for each group shown in Fig. 11 (blue circles) with normalized maximum and minimum

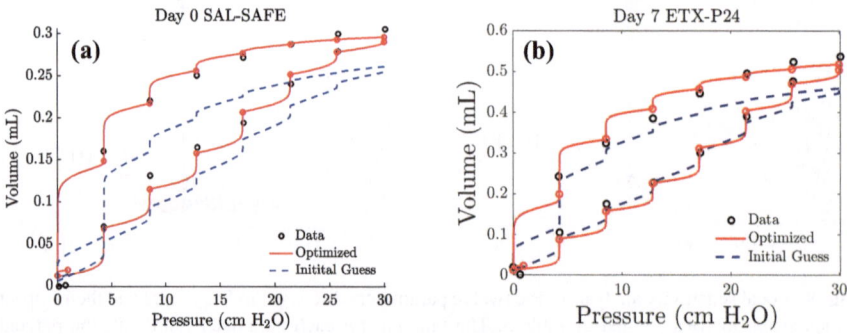

Fig. 10 Optimized pressure-volume curves (red solid curves), experimental quasi-static points (black circles), model simulated quasi-static points (red circles), and initial optimization guess (blue dashed curves) are plotted for a representative rat in (**a**) the D0 SAL-SAFE group and (**b**) the D7 ETX-P24 group. The optimized curves are close fits to the experimental data [2]

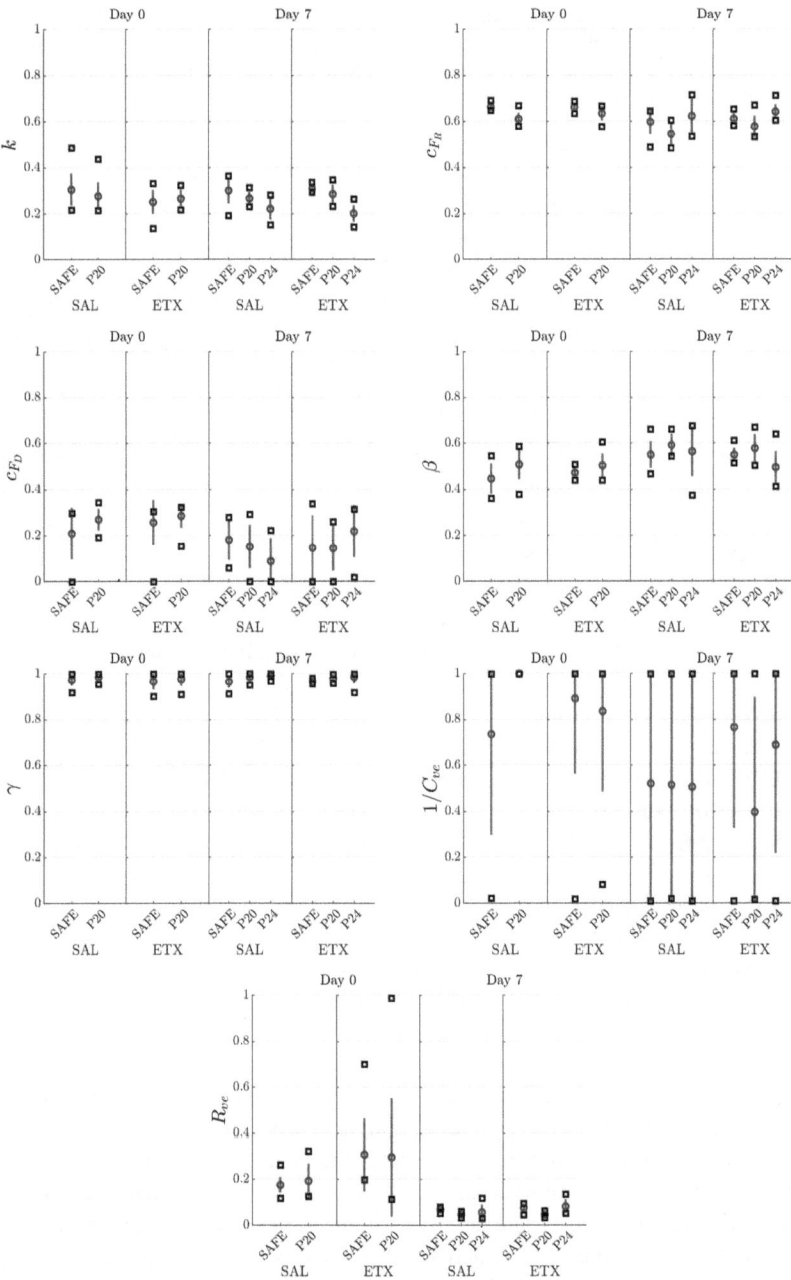

Fig. 11 Normalized optimal parameter values by rat group. We show mean values (blue circles), standard deviation range (red lines), and maximum and minimum values (black squares) for each group

Table 5 Optimal values were determined for parameters identified as sensitive by at least one metric (see Fig. 8 and Table 3). Optimal values were determined for each rat and then averaged across each group (n = number of rats in the group). Here, we report the mean μ optimal parameter values for each rat group as well as the variances σ^2

			Parameters (units)							
			k		c_{FR}		c_{FD}		β	
			$((cm\ H_2O)^{-1})$		$(cm\ H_2O)$		$(cm\ H_2O)$		(-)	
		n	μ	σ^2	μ	σ^2	μ	σ^2	μ	σ^2
D0	SAL-SAFE	14	0.18	0.001	16.58	0.08	4.18	5	0.45	0.004
	SAL-P20	12	0.17	0.001	15.27	0.44	5.39	0.87	0.51	0.004
	ETX-SAFE	9	0.15	0.0009	16.54	0.20	5.13	3.83	0.47	0.0006
	ETX-P20	10	0.16	0.0005	15.87	0.56	5.70	1.08	0.50	0.0026
D7	SAL-SAFE	8	0.18	0.0011	14.94	1.76	3.61	3.01	0.55	0.0032
	SAL-P20	8	0.16	0.0002	13.61	1.37	3.02	3.51	0.59	0.0023
	SAL-P24	6	0.13	0.0007	15.58	3.46	1.75	3.90	0.56	0.01
	ETX-SAFE	8	0.19	0.00007	15.27	0.38	2.93	7.65	0.55	0.0008
	ETX-P20	8	0.17	0.0005	14.44	1.32	2.89	3.82	0.58	0.0036
	ETX-P24	9	0.12	0.0004	16.05	0.61	4.37	5.10	0.49	0.0047

			Parameters (units)						
			γ		$1/C_{ve}$		R_{ve}		
			(-)		$(cm\ H_2O\ (mL)^{-1})$		$(cm\ H_2O \cdot s \cdot (mL)^{-1})$		
		n	μ	σ^2	μ	σ^2	μ	σ^2	
D0	SAL-SAFE	14	0.97	0.0005	367.27	47,673	17.52	10.8	
	SAL-P20	12	0.98	0.0002	500	0	19.26	53.35	
	ETX-SAFE	9	0.97	0.0012	445.47	26,760	30.61	250.13	
	ETX-P20	10	0.98	0.0008	417.74	30,587	29.46	652.86	
D7	SAL-SAFE	8	0.96	0.0006	259.94	65,988	6.94	0.90	
	SAL-P20	8	0.98	0.0002	257.14	67,417	4.74	0.84	
	SAL-P24	6	0.99	0.0002	252.50	73,508	5.67	10.50	
	ETX-SAFE	8	0.97	0.00004	382.01	47,883	7.49	2.08	
	ETX-P20	8	0.98	0.0001	198.14	62,539	4.61	0.82	
	ETX-P24	9	0.99	0.0007	343.73	55,132	8.16	8.74	

values (black squares) and one standard deviation above and below the mean (solid red line). Normalizing by the parameter constraint upper bound allows for easier comparison between groups and shows the degree of variability of optimized parameter values. Mean and variance for each optimized parameter per rat group are also reported in Table 5. The largest variation is observed for the optimal values for viscoelastic elastance $1/C_{ve}$ across all groups, with non-normalized variances on the order of 10^4. Large variation is also seen in viscoelastic resistance R_{ve} in the D0 ETX group. The mean optimized value for k for the D7 group decreases with increased ventilation for both the SAL and ETX groups. In contrast, c_{FD} decreases from P20 to P24 in the D7 SAL group but increases in the D7 ETX group.

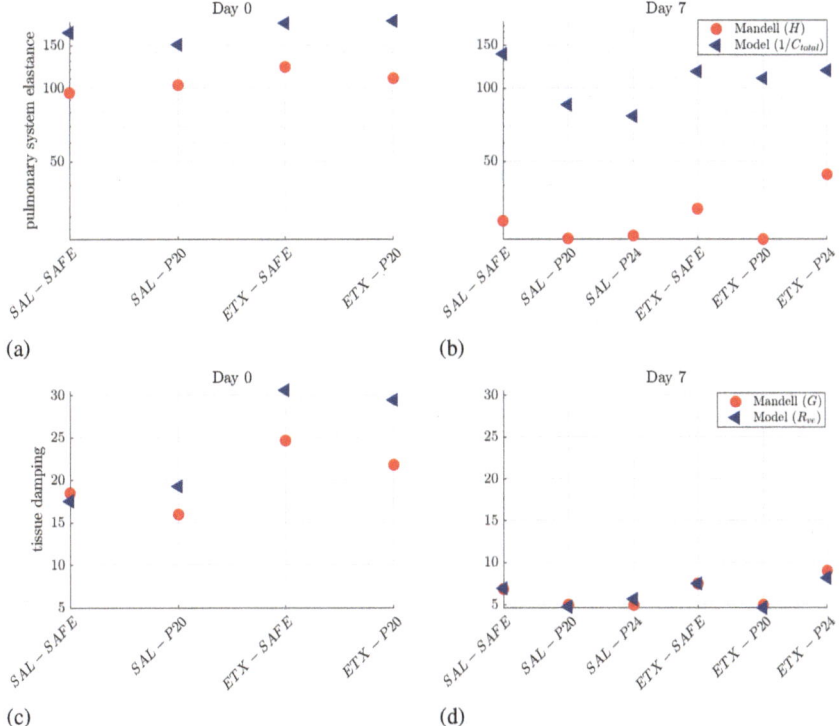

Fig. 12 Comparisons of optimized parameter values to estimated quantities from Mandell et al. [2]. (**a**)–(**b**) Comparing the averaged pulmonary system elastance from Eq. (35) in the current model ($1/C_{\text{total}}$) to the pulmonary system elastance from Mandell (H) for (**a**) day 0 and (**b**) day 7 rats. (**c**)–(**d**) Comparing the averaged viscoelastic resistance from the current model (R_{ve}) to the tissue damping from Mandell (G) for (**c**) day 0 and (**d**) day 7 rats. Simulated values are plotted as blue triangles, and experimental values are plotted as red circles

3.4.2 Biomechanical Metrics

Figure 12 shows comparisons to tissue damping coefficient and pulmonary system elastance, two biomechanical metrics obtained from Mandell et al. [2] who fit the constant phase model to forced oscillation technique data. We adopt their notation as G for tissue damping coefficient and H for pulmonary system elastance. Total pulmonary system elastance $H = 1/C_{\text{total}}$ was calculated in this study from lung compliance C_A and viscoelastic compliance C_{ve} in series by

$$\frac{1}{C_{\text{total}}} = \frac{1}{C_A} + \frac{1}{C_{ve}}, \tag{35}$$

(see also Fig. 2). Since C_A is a dynamic variable (cf. Sect. 2.2), the time-averaged lung compliance for each rat is computed via

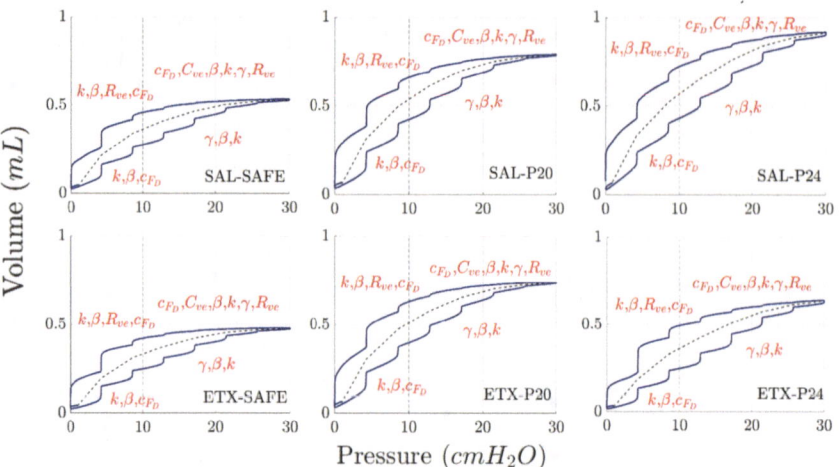

Fig. 13 Pressure-volume curves computed using averaged optimized parameter values for each group at day 7

$$C_{A_{\text{avg}}} = \frac{1}{T - t_0} \int_{t_0}^{T} C_A \, dt \qquad (36)$$

using optimal parameter values. Pulmonary system compliance/elastance is calculated from mean compliances for each group via Eq. (35). Viscoelastic resistance R_{ve} serves as a proxy for tissue damping, G. We observe in Fig. 12a,b that compared to the experimentally estimated H, average $1/C_{\text{total}}$ values are about 50% larger for D0 rats and approximately one order of magnitude larger for D7 rats. Figure 12c,d show closer agreement between R_{ve} and G, especially for the D7 rats.

3.4.3 Sensitivity Analyses

The sensitivity analysis (Sect. 3.2) addresses how model parameters affect the slopes of the PV curve. We considered recruitment and derecruitment slopes between pressures of 0 and 10 cm H_2O and between 10 and 30 cm H_2O (sl_R 0–10, sl_D 0–10, sl_D 10–30, and sl_D 10–30). PV curves simulated using averaged optimal parameter values for each D7 rat group are shown in Fig. 13. Each curve is divided into four regions bounded by the four slopes; divisions are shown by black dashed lines. The parameters that were determined to be sensitive by the Morris screening analysis for each slope of the curve are stated in red.

Lung elasticity constant k highly influences all four slopes, suggesting that small changes in k can significantly change the shape of the PV curve. The airway opening pressure c_{F_R} was not found to be highly influential for any of the PV slopes and thus is not represented in Fig. 13. In contrast, the airway closing pressure c_{F_D} was

Table 6 Relative change (%) from SAFE ventilation to VILI for each parameter in the D7 control group (SAL) and D7 endotoxin group (ETX). A red down arrow indicates a decrease from the baseline value, and a blue up arrow indicates an increase. Relative changes are calculated from average parameter values from optimizations taken over each group

| | | Parameter | | | | | | |
	Ventilation	k	c_{F_R}	c_{F_D}	β	γ	$1/C_{ve}$	R_{ve}
D7 SAL	P20	↓ 11.4	↓ 8.9	↓ 16.4	↑ 7.5	↑ 2	↓ 1.1	↓ 31.7
	P24	↓ 26.5	↑ 4.3	↓ 51.4	↑ 2.5	↑ 2.6	↓ 2.9	↓ 18.3
D7 ETX	P20	↓ 7.8	↓ 5.5	↓ 1.3	↑ 5.1	↑ 1.2	↓ 48.1	↓ 38.5
	P24	↓ 34.9	↑ 5.1	↑ 49.1	↓ 10	↑ 2	↓ 10	↑ 8.9

determined to be highly influential for the derecruitment slope between 10 and 30 cm H_2O and slightly less influential for the derecruitment slope between 0 and 10 cm H_2O. We find c_{F_D} to be higher in the D0 rat groups. For the D7 control group, c_{F_D} decreases as the ventilation pressure increases. However, for the D7 endotoxin group, it increases. The parameter β is the baseline fraction of lung recruited and was determined to be influential for all four PV slopes. In every rat group, we notice an increase in β from safe ventilation to P20 and P24 ventilation, except for the D7 endotoxin group where it decreases by 10% from safe to P24 ventilation. The viscoelastic elastance $1/C_{ve}$ was only found to be highly influential for the derecruitment slope between 10 and 30 cm H_2O. A lower viscoelastic elastance flattens the dynamics between the static points, while a higher value makes them more pronounced. Viscoelastic resistance, R_{ve}, is highly influential for the derecruitment slopes. The viscoelastic resistance is higher for the D0 rat pups and for the endotoxin rats in both age groups.

3.4.4   Comparisons of Saline Versus Endotoxin

The relative change, as a percent, of the D7 optimal parameter values for the P20 and P24 groups as compared to the SAFE ventilation groups are reported in Table 6. The two levels of injurious ventilation produce relative decreases in k with a larger change for P24 than P20. Similarly, γ increases with increasing ventilation for both the SAL and ETX groups, but the increases are small ($\sim$ 3%) and may be considered negligible. The P24 SAL group average c_{F_D} is roughly half the SAFE average; in contrast, the corresponding ETX value is a 50% increase. Although the changes are relatively small overall (up to 10%), the P24 SAL average β increases from the SAFE average whereas the same for ETX decreases. The parameter R_{ve} also shows differing responses under SAL versus ETX for P24 ventilation, as R_{ve} is decreased further from that of P20 for the SAL whereas the value increases beyond the SAFE average for ETX.

3.5   Image Analysis

We developed a customized image analysis procedure to analyze lung histology data in novel ways. Calculations of metrics for 424 histology images (207 for D0 and 217 for D7) of 47 individual rats (23 for D0 and 24 for D7) from [2] are presented below.

3.5.1   Distributions of Image Metrics

We performed model screening and, based on AICc, found the best statistical model. Lumen count per image depends significantly on the day and ventilation strategy ($p < 0.01$) but was not found to depend significantly on SAL versus ETX treatment ($p > 0.07$) (Fig. 14). The best model (by AICc) for lumen count N per standard-area image is

$$N = 48.8 + \left\{ \begin{array}{l} D0 \Rightarrow 0 \\ D7 \Rightarrow 22.3 \end{array} \right\} + \left\{ \begin{array}{l} NV \Rightarrow 6.0 \\ SAFE \Rightarrow 1.7 \\ P20 \Rightarrow -6.0 \\ P24 \Rightarrow -1.7 \end{array} \right\}. \tag{37}$$

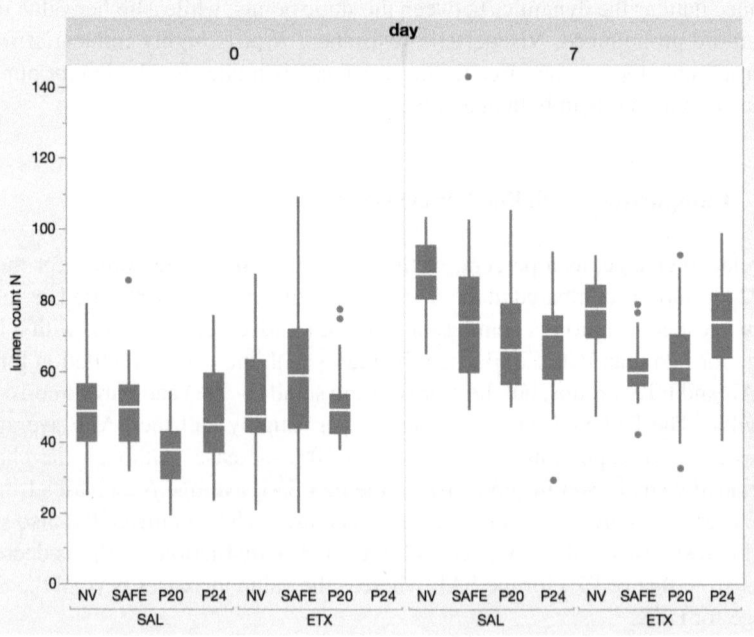

Fig. 14 Distribution of lumen counts categorized by experimental groups

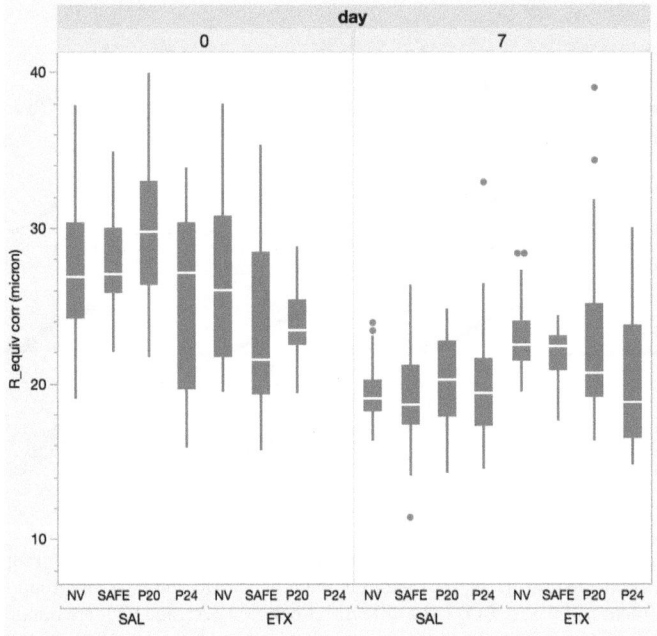

Fig. 15 Lumen equivalent radii R_{equiv} by experimental groups

The lumen equivalent radii R_{equiv} had a strong dependence on postnatal age, with lumens decreasing in size as the lungs developed (Fig. 15). We performed model screening and, based on AICc, found the best statistical model for lumen radius R_{equiv} to be (in microns)

$$R_{\mathrm{equiv}} = 25.4 + \begin{Bmatrix} \mathrm{D0} \Rightarrow 0 \\ \mathrm{D7} \Rightarrow -5.5 \end{Bmatrix} + \begin{Bmatrix} \mathrm{NV} \Rightarrow 1.2 \\ \mathrm{SAFE} \Rightarrow 0.4 \\ \mathrm{P20} \Rightarrow 1.7 \\ \mathrm{P24} \Rightarrow 0 \end{Bmatrix}. \tag{38}$$

For the area fraction, we calculated the proportion of each image ϕ that is tissue, and for each pruning width w, the proportion $\phi(t)$ of the w-pruned image that is tissue. For each image, tissue area fraction at each pruning width w is shown in Fig. 16 grouped by experimental day and treatment. Most of the tissue is thinner than 40 microns; only a few experimental categories show any amount of tissue thicker than 80 microns, notably D0 ETX P20. At D0, ETX-treated lungs show about 10% greater tissue fraction $\phi(0)$, corresponding to about a 30% increase in tissue area, when compared with the controls (SAL). At D7, the differences between SAL and ETX are minimal, but there is a substantial difference between the most aggressively ventilated lungs (P24) and the others.

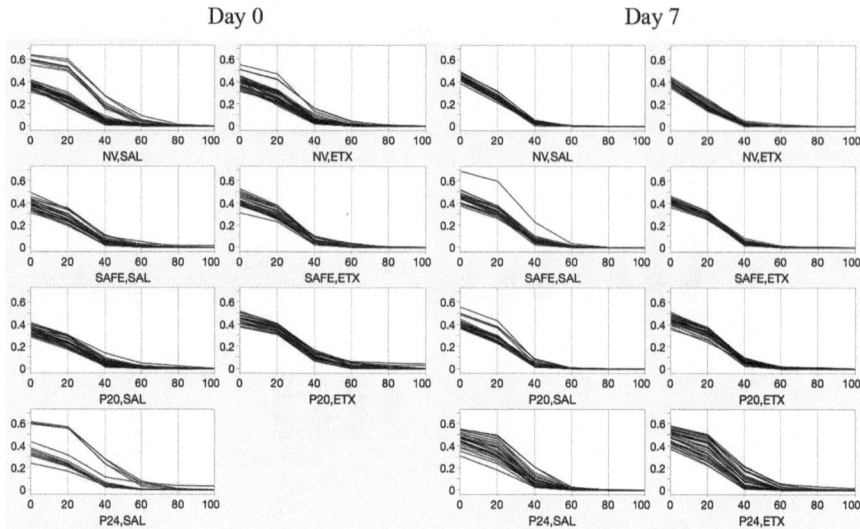

Fig. 16 Comparison of the tissue area fraction $\phi(w)$ across the 15 experimental groups for which histology was collected [2]. Here, w is width in microns. At D0, the ETX subgroup shows about 10% greater fraction of tissue. At D7, the differences between SAL and ETX are minimal, but there is a substantial difference between the most aggressively ventilated lungs (P24) and the others

Tissue length Λ (per standard image size) was confirmed to have the expected allometric relationship with lumen count N, with total length Λ proportional to $\sqrt{N}$ (Fig. 17). Comparison among all experimental groups revealed a significant dependence of tissue mean width w on all categories (Fig. 18). We performed model screening and, based on AICc, found the best model of mean tissue width (in microns) to be

$$
\bar{w} = 19.3 + \begin{Bmatrix} D0 \Rightarrow 0 \\ D7 \Rightarrow -1.7 \end{Bmatrix} + \begin{Bmatrix} NV \Rightarrow -1.3 \\ SAFE \Rightarrow -0.3 \\ P20 \Rightarrow 0.4 \\ P24 \Rightarrow 1.2 \end{Bmatrix} + \begin{Bmatrix} SAL \Rightarrow -0.2 \\ ETX \Rightarrow 0.2 \end{Bmatrix} \tag{39}
$$

with the SEM for each parameter ranging from 0.07–0.15 microns. Tissue is seen to thin by a mean of 1.7 microns from D0 to D7. Each pressure increase in the ventilation strategy correspondingly increases tissue width. Interestingly, the least significant variable was SAL versus ETX, which only made a difference at P20. Day and ventilation strategy were more significant.

The tortuosity metric from Eq. (31) had substantial variation between images, so even the best predictive models had a very small R^2. However, there were statistically significant trends (Fig. 19; $p < 0.01$ for SAL/ETX and P24 vs. other ventilation). Notably, there was an increase in tortuosity at the highest ventilation

Fig. 17 Total tissue length, as calculated by the length of centerlines per standard image, goes as $\sqrt{N}$, where N is the adjusted lumen count per image

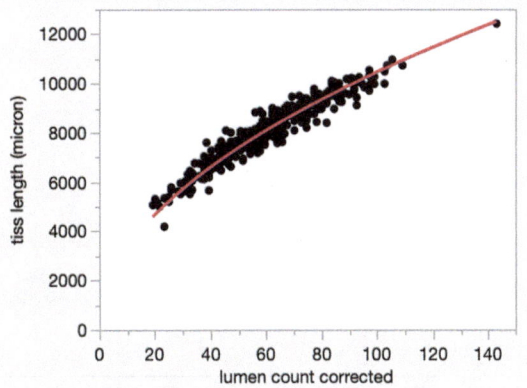

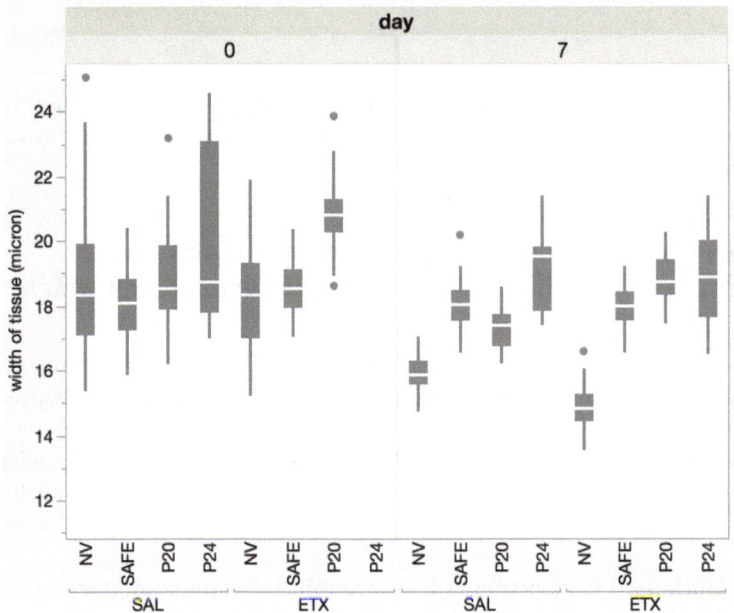

Fig. 18 Distribution of mean tissue width $\bar{w}$, categorized by experimental groups

pressures. Model screening determined the best fit to include linear terms for ventilation and SAL versus ETX, as well as an interaction term between those two variables. Postnatal day was not found to be significant.

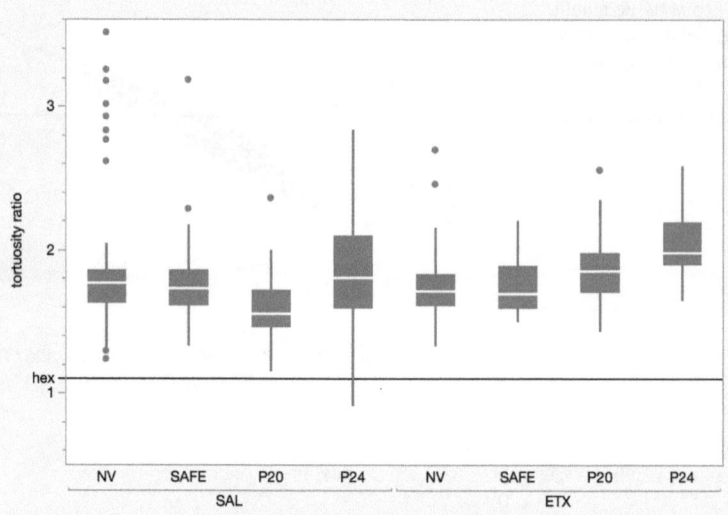

Fig. 19 Tortuosity ratio by experimental groups ($p < 0.01$ for SAL/ETX and P24 vs. other ventilation). Dependence on postnatal day was negligible. The reference line indicates the tortuosity ratio for standard hex packing

3.6 Correlations Between Optimization Metrics, Image Metrics, and Biomechanical and Inflammatory Markers

The results from our customized modeling and analysis procedures were examined via exploratory data analysis techniques to search for connections between the optimized model parameters, the imaging metrics, and biomarkers determined in [2]. Principal component analysis (PCA) provided a measure of broad relationships between variables, and cluster analysis indicated natural groupings of variables.

3.6.1 Optimized Parameters vs Biomechanical and Inflammatory Markers

PCA was applied to the mean optimized parameter values in the current study (Table 5) together with biomarkers reported by [2] for 10 available groups based on age (D0 or D7), exposure (SAL or ETX), and ventilation type (SAFE, P20, or P24). As seen in both the correlation heat map (Fig. 20) and factor map (Fig. 21), the parameter k, which characterizes the aggregate lung elasticity in the model, is negatively correlated with γ, the maximum recruitable lung fraction, and the inflammatory metrics IL-6, TNF-a, and CXCL2. Interestingly, k is also included in the sensitive Morris set $\boldsymbol{p}_{\mathrm{Mor}}$ (Eq. 33, Table 4). These generally represent the second principal component PC2 (23%). The parameter β describing baseline lung recruitment is strongly correlated with IC and biomechanical compliance

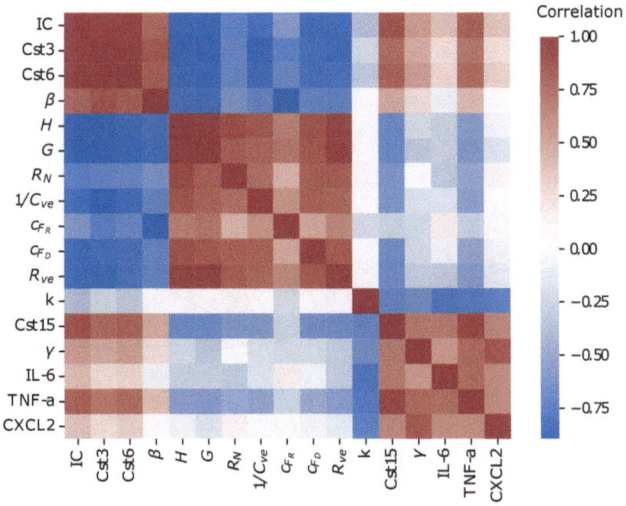

Fig. 20 Correlations of optimized parameter values with biomechanical and inflammatory markers from [2]. The parameter k, which characterizes aggregate lung elasticity in the model, is negatively correlated with inflammatory markers

biomarkers, Cst. Negatively correlated with β are the lung opening pressures c_F, viscoelastic parameters R_{ve} and $1/C_{ve}$, and biomechanical markers G and H. These generally comprise the first principal component PC1 (62%). Cst15 and TNF-a strongly correlate with each other but not as strongly with the other variables. Additionally, a cluster analysis revealed that k, γ, and Cst15 grouped together with inflammatory metrics IL-6, TNF-a, and CXCL2 in the same cluster, while the rest of the variables grouped into a second cluster. The cluster analysis results are not explicitly shown, as they are a subset of the analyses in Sect. 3.6.3.

3.6.2 Image Metrics vs Biomechanical and Inflammatory Markers

We calculated the mean values of image metrics, biomechanical variables, and chemical measures of inflammation for the 14 different groups for which there was data of all three types. The 14 groups are formed by dividing the available data according to the age of the rat (D0 or D7), its exposure to endotoxin (SAL or ETX), and the ventilation pressure that it received (NV, SAFE, P20, or P24) [2]. PCA was applied to the mean values calculated for each category, as seen schematically in the correlation heat map (Fig. 22) and factor map (Fig. 23).

The biomechanical measures of compliance, such as Cst, are strongly correlated among themselves and with inspiratory capacity IC. This is consistent with the findings in [2]. The measures of tissue stiffness, pulmonary system elastance H, tissue damping coefficient G, and central airway resistance R_N are all strongly correlated with each other and inversely correlated with measures of compliance

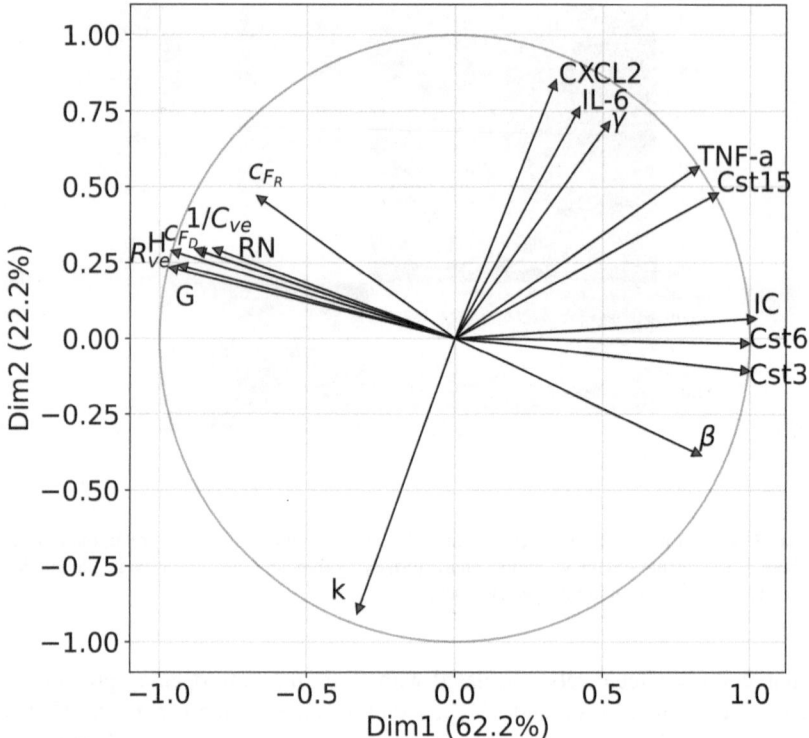

Fig. 21 Principal component analysis of optimized parameter values as seen in Table 5 with biomechanical and inflammatory markers from [2]

and IC. Among the image metrics, we observe a strong correlation between R_{equiv} and MLI measurements. Between the image metrics and biomechanical variables, biomechanical measures of tissue stiffness show a strong positive correlation with R_{equiv} and a strong negative correlation with the number of lumens. Therefore, the first principal component PC1, accounting for almost half (45%) of the variation (Fig. 23), can be described loosely as the compliance versus resistance axis. In terms of image metrics, this axis can also be described as many small lumens versus few large lumens.

In addition to these findings regarding the biomechanical data, we note a strong correlation between all the inflammatory markers. The biomechanical measures of compliance and IC are strongly correlated with TNF-a and less correlated with the other inflammatory markers. Each of the inflammatory markers has its highest correlation, among the measures of compliance, with Cst_{15}. Tortuosity $\tilde{T}$ is strongly correlated with tissue area fraction $\phi(0)$ and not strongly correlated with most other variables. The inflammatory variables are generally weakly correlated with the image metrics, with the following exceptions. IL-6 is moderately correlated with $\tilde{T}$, $\phi(0)$, and $\phi(20)$. $\tilde{T}$ is the best predictor for IL-6. IL-6 is moderately negatively

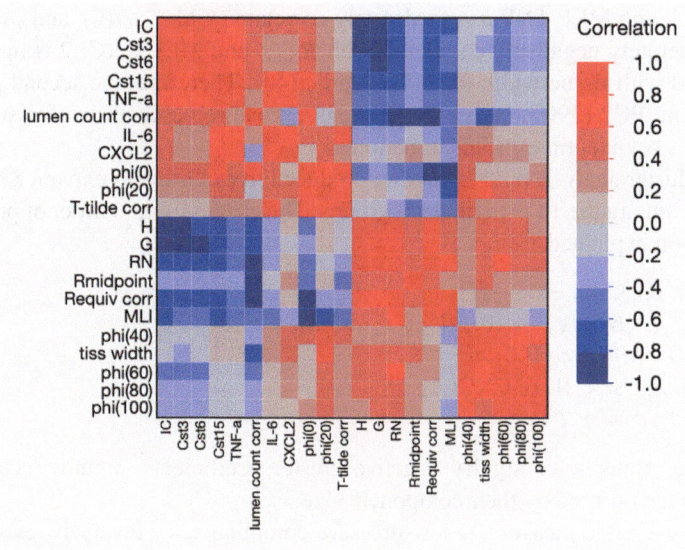

Fig. 22 Heat map of correlations for image, biomechanical, and inflammatory variables

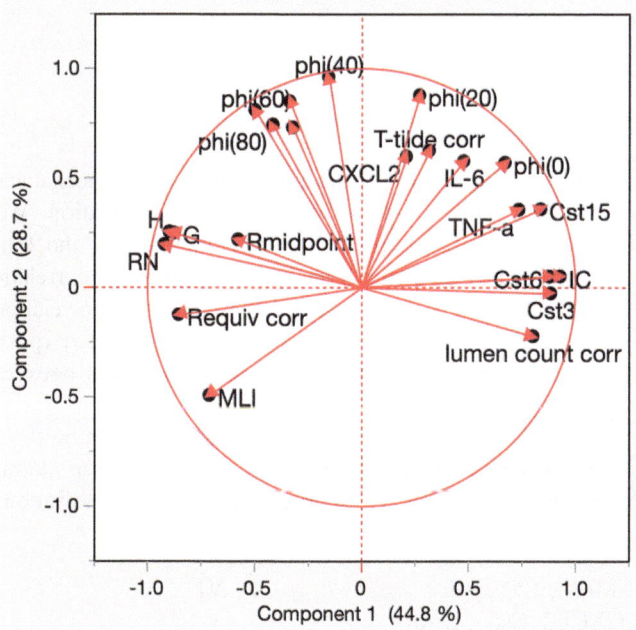

Fig. 23 Principal component analysis of image, biomechanical, and inflammatory metrics: first two principal components

correlated with MLI. TNF-a is moderately correlated with $\tilde{T}$, $\phi(0)$, and $\phi(20)$. It is also moderately negatively correlated with R_{equiv} and MLI. CXCL2 is moderately correlated with tissue width $\bar{w}$, $\phi(20)$, and $\phi(40)$. Therefore, the second principal component, PC2 (29%), is to a great extent a broad representation of tissue width, which is a key identifier of inflammation.

The cluster analysis on this data set produced five clusters to explain 82% of the variation among the 14 experimental groups. The clusters are, in order of proportion of variation explained,

C_1: [Cst$_3$, IC, Cst$_6$, $-G$, $-H$]
C_2: [R_{equiv}, $-N$, R_N, $-\phi(0)$, R_{midpoint}, MLI]
C_3: [$\phi(60)$, $\phi(80)$, $\phi(100)$, $\bar{w}$]
C_4: [TNF-a, Cst$_{15}$, IL-6, CXCL2]
C_5: [$\phi(20)$, $\phi(40)$, $\tilde{T}$]

Here, the minus signs signify negative cluster coefficients. Within each cluster, variables are ordered by their component size.

Cluster C_1 is a measure of low-pressure compliance, or inversely, elastance H and damping G. Notably, airway resistance R_N appears in cluster C_2, which is otherwise a measure of lumen size. Cluster C_3 represents tissue width. Cluster C_4 represents inflammatory markers and high-pressure compliance. Cluster C_5 represents tissue thinness and tortuosity.

3.6.3   All Variables

There were 10 groups for which we had image metrics, biomechanical markers, inflammatory markers, and P-V data for parameter optimization. We analyzed means for each of these 10 groups for metrics obtained by the four different approaches and applied PCA and cluster analysis. The strongest correlations (± 0.7) were between $1/C_{ve}$ and both R_{equiv} and R_{midpoint} (positive correlation), as well as with N (inverse correlation), which are strongly correlated (positively and negatively) with each other. The other strong correlation was between R_{ve} and $\phi(60)$.

We also performed a cluster analysis on this data set, spanning the four major approaches. Four clusters serve to explain 82% of the variation among these 10 experimental groups. The clusters are, in order of proportion of variation explained,

K_1: [H, G, $-$Cst$_3$, $-$Cst$_6$, R_{ve}, IC, $-\beta$, c_{FD}, $1/C_{ve}$, c_{FR}, R_N]
K_2: [$\phi(0)$, $-$MLI, $\phi(20)$, $-R_{\text{midpoint}}$, $-R_{\text{equiv}}$, $\tilde{T}$, N]
K_3: [$-k$, γ, CXCL2, Cst$_{15}$, IL-6]
K_4: [$\phi(60)$, $\phi(80)$, $\phi(100)$, $\bar{w}$, $\phi(40)$]

Here, the minus signs signify negative cluster coefficients.

Notably, cluster K_1 includes only biomechanical variables, but these span both the Mandell et al. [2] data and the optimized model parameters. Thus K_1 represents the measured and fitted biomechanics. Clusters K_2 and K_4 contain only metrics of histology. Of these, cluster K_4 is a measure of thickened regions of tissue, and cluster K_2 is a measure of thin, tortuous, and numerous septa between small lumens. Cluster K_3 contains inflammatory markers, the two optimized parameters k and γ, and C_{st6}.

4   Discussion

Mandell et al. [2] found that the rat lungs subjected to the most injurious ventilation pressure (P24) exhibited decreased compliance and increased stiffness compared to lower ventilation pressure (P20). Our model is able to replicate their findings and show the counter-intuitive pressure-volume curve trend for the highest ventilation (Fig. 13). The large variance in C_{ve} as seen in Table 5 and Fig. 11 suggests that the optimizer was not able to uniquely identify this parameter despite the sensitivity analysis findings. The structure of Eq. (6b) suggests that the viscoelastic parameters C_{ve} and R_{ve} have similar effects on the dynamics of P_{ve}, and the correlation heat map (Fig. 20) indicates a strong correlation with each other. While the PV curves are adequately replicated, this observation suggests that only one of the two viscoelastic parameters is observable.

4.1   Compartmental Model Analyses

The sensitivity analyses (Sects. 3.2–3.3) identified the key parameters of the fractional recruitment function k, γ, and β as sensitive, though they are correlated with different variables from the compartmental model and the data. The parameter k, which characterizes the aggregate lung elasticity, is a highly sensitive model parameter. A lower k value models a less compliant or stiffer lung, meaning higher pressure must be applied to expand and fill the lung for a given volume of air. While k does not strongly depend on the age of the rat, we observe k trending down as more aggressive ventilation is applied. We note that k and γ have no direct analog in the biomechanical or inflammatory markers from Mandell et al. [2]; however, they were anti-correlated (k) and correlated (γ) with inflammatory markers CXCL2 and IL-6. This suggests a strong connection between the level of inflammation and the ability of the lungs to inflate, as expected. The parameter β, the recruitment fraction at $P = 0$ cm H_2O, is anti-correlated with biomechanical markers H, G, and R_n from [2], suggesting that increases in these markers might oppose a healthy lung status. Of the Morris sensitive set $\boldsymbol{p}_{Mor}$ (Eq. 33), the mean opening airway pressures c_{F_R} and c_{F_D} were the least sensitive across the six metrics but were strongly correlated with optimized parameters and data markers representing biomechanics.

The relative changes in average optimized parameter values from SAFE ventilation to P20 and P24 for D7 rats (Table 6) are of particular interest because these may connect to the counter-intuitive nature of the PV curves from the observed data [2]. The direction of change differs noticeably between SAL and ETX groups for c_{F_D}, β, and R_{ve}. Generally, a lower mean opening pressure is a sign of a healthier respiratory system. Increased mechanical ventilation should aid in opening the lungs and reduce c_{F_D}. However, for the ETX group, when P24 ventilation has been applied, there is subsequently a much greater pressure required to open the lungs. This was speculated to occur due to alveolar flooding and increased elastance [2], which is consistent with our parameterization. The contrasting relative changes in β may not be surprising, as an unhealthy lung further stressed by a high level of injurious ventilation may not be able to recruit as large a fraction of alveoli at zero pressure. The inverse relationship between the relative changes in R_{ve} for P24 SAL and ETX may suggest that a stretching limit has been reached in the latter case. For the ETX group, the unhealthy and injured lung may be more resistant to expansion, and therefore a higher resistance is observed. Interestingly, the relative changes for the P20 SAL and ETX groups are all of the same sign, although they can differ significantly in magnitude. This may reinforce the findings of Mandell et al. [2], who observed such dynamics in their experiments.

4.2   Image Analysis

The customized image metrics from this study, in some cases, confirm other related measurements. The lumen metrics R_{equiv} and R_{mid} are seen to be highly correlated with MLI (Sect. 3.6.2), which was used in [2]. This makes intuitive sense and provides alternate methods for quantifying lumen size. Moreover, the image processing via segmentation that is summarized in R_{equiv} and R_{mid} provides a route to a more detailed analysis of lumen sizes and shapes, and their distribution, for future histological analysis.

Other image metrics in our analysis quantitatively reveal patterns in the experimental groups that may have only been reported qualitatively. For example, from our model fitting, we see a clear developmental signature in the increase in the lumen count N from D0 to D7 (Fig. 14 and Eq. 37) and the decrease in lumen radius R_{equiv} from D0 to D7 (Fig. 15, Eq. 38), which correspond to the process of secondary septation. We can also see quantitatively, from the model fit for N, that secondary septation is inhibited by aggressive ventilation; our statistical fit from Eq. (37) shows that P20 ventilation results in approximately 3 days' developmental delay relative to nonventilated pups. Our measurements of the distribution of tissue width for each treatment group (Fig. 18) show small but statistically significant differences in the quantitative histology by Eq. (39).

Similarly, we found that tortuosity shows modest increases with endotoxin treatment and/or higher-pressure ventilation, though, surprisingly, it is not seen to be dependent on the developmental day (Fig. 19). That our tortuosity metric does not correlate with developmental day, despite our expectation of increasing parenchymal surface area, suggests that either the differences are not significant at these specific stages, or that they are not visible at the length scale of the imaging or with the sample preparation techniques used. It may also suggest that changes in crimp (tortuosity) are histologically insignificant relative to the much larger changes observed in lumen count and size (Figs. 14 and 15).

The correlations (Sect. 3.6.2) between the image metrics and each other are somewhat expected, but the correlations (Figs. 22 and 23) between image metrics and biomechanical and inflammatory metrics from Mandell et al. [2] are to some extent surprising. The first principal component PC1 and the first two variable clusters C_1 and C_2 show a relationship between the biomechanical variables from [2] and lumen size (or number, which is inversely related to lumen size). Larger lumens, which might be expected to provide less fluid drag, and hence less resistance, were instead associated with greater airway resistance, damping, and elastance and lower compliance. Tortuosity, which would be expected to reflect the nonlinearity of the typical stress-strain curve, was not here found to be correlated with mechanical parameters. It is only one of several factors in tissue compliance, along with tissue width and material elasticity. It was, however, highly correlated with inflammatory marker IL-6.

4.3 Relationship Between Approaches

Examining the full set of metrics using PCA and cluster analysis was the first step toward generating hypotheses about possible relationships between compartmental model parameters and lung tissue histology metrics. Viscoelastic elastance (reciprocal of compliance) was strongly correlated with lumen metrics; however, the variance was large enough to suspect that the values for compliance were not uniquely identified. The other strong correlation was between viscoelastic resistance and tissue area fraction with a pruning diameter of 60 microns. This significance is unclear; however, it is notable that the largest correlations were with the viscoelastic parameters.

The cluster analysis identified groupings that may hold significance. Cluster K_1 grouped all the biomechanical metrics together regardless if derived from the data or from the compartmental model. Some of the related correlations as discussed earlier suggest that the model parameters could act as surrogates for the experimentally derived biomechanical markers. An additional biomechanical marker C_{st15}, the compliance of the PV curve during derecruitment at a pressure of 15 cm H_2O, was grouped with the inflammatory markers in a separate cluster K_3 along with k and γ. The compliances C_{st3} and C_{st6} represent compliance at low pressures of 3 and 6 cm H_2O, whereas γ impacts the curve at high pressures and k impacts the entire pressure range. This may indicate that inflammation has a greater effect on PV dynamics at higher pressures.

The separate correlation and cluster analyses were done because of different numbers of experimental groups for the P-V fitting than for the images. The results of the separate analyses are consistent, except that cluster C_2 reveals a relationship between the image metrics and one of the biomechanical variables, R_N.

4.4 Extensions and Clinical Implications

A goal of this work was to apply mathematical techniques to a neonatal rat model of chorioamnionitis and VILI to better understand the mechanisms of breathing and quantify differences between healthy and diseased groups in a challenging population. To this end, our approaches focus on what is feasible given the available experimental data [2]. This allows for several extensions given additional measurement types. Here, we detail potential future steps and subsequent clinical implications.

The image analysis metrics could be inputs into an augmentation of our compartment model. However, since these metrics cannot be obtained from human subjects except postmortem, the clinical applications of such a pipeline remain unclear. At present, we envision that our model optimization could be applied to recorded pressure-volume data from human patients, and then lung histology relationships could be inferred based on our identified correlations with model parameter values. For example, if an optimal parameter value that is positively correlated with inflammatory markers is high, this suggests that a scan of the patient's lung might show inflammation.

Our results suggest trends in safe versus injurious ventilation between healthy and unhealthy lungs that could be of clinical interest, but additional work is needed to verify these hypotheses, including a validation with a significantly larger data set. Analysis of our optimal parameters identified that for the ETX group with P24 ventilation, a much greater pressure is needed to open the lungs than for a healthy rat and that a stretching limit may be reached. Together these confirm the need for caution during ventilation of neonates that have mothers with histories of chorioamnionitis or other infections during pregnancy in order to prevent BPD and other respiratory conditions.

4.5 Limitations

The Morris screening and coarse local sensitivity analysis are both conducted using scalar model outputs, whereas a classical gradient-based local sensitivity analysis would calculate a sensitivity index across the full time course of data. Given the sizeable number of rats and associated data sets, the latter analysis was out of the scope of this study to perform on each rat pup. Thus, it is possible that parameters affected the scalar outputs differently than the quasi-static points of the pressure-volume loop, and the related optimizations in Sect. 3.4 might be based on an incomplete understanding of parameter sensitivities. In future work, an optimization algorithm could be formulated in which the objective function is weighted based on the WOB or v_{max}. Further, the Morris screening uses the mean ranking in Eq. (21) as a sensitivity threshold [31, 32]; we note that other options are available, such as using 5% of the maximum rank. Using this alternative method on our Morris rankings results in eleven out of thirteen parameters deemed sensitive, rather than the eight that we report in Sect. 3.2. The subjectivity of this choice allows for other interpretations of relative parameter importance; Colebank and Chesler [24] state the need for a consistent selection method.

It is expected that inflammation from VILI or infections stiffens lung tissue by increasing resistance and decreasing compliance. Indeed, we saw increases in the viscoelastic resistance parameter R_{ve} between most saline and endotoxin groups, but only one difference was statistically significant (Day 0, SAFE ventilation, $p < 0.04$, two-sample t-test with unequal variances). As previously mentioned, C_{ve} and R_{ve} may not be uniquely identifiable by our optimization; therefore, an important next step is to independently measure or estimate one quantity and re-run the optimizer.

We developed a new correction method for counting objects (lumens) that are clipped by the image edge. Our correction was, for simplicity, based on an assumption of monodispersity (equal sizes). It improves the accuracy of lumen count and other metrics based on lumen count, even though the lung lumens are quite polydisperse. A more detailed correction method might use a kind of bootstrapping to estimate lumen size distribution and, therefore, the size distribution of clipped lumens. Our image analysis protocols did not make a distinction between alveoli, alveolar ducts, bronchioli, and blood vessels, on the assumption that these other structures comprise a relatively small proportion of each image and can be neglected. Our quantitative analysis could potentially be improved by performing additional segmentation on the images to identify these structures and consider them separately. For example, by not separately segmenting the blood vessels, they contribute to the quantification of the non-vascular tissue and may skew the results.

The biggest challenges with quantitative image analysis reside in the image acquisition. Our image metrics were defined in a relatively straightforward fashion but can be thwarted by fields of view that encompass too few alveoli. Our estimates of the corrections for lumen counts, areas, etc. assumed regular, convex, and uniform lumen shapes and sizes. The actual lumens in lung slices are far from regular, convex, and uniform. If lumens are of even moderate size relative to the image size (as in Fig. 4a), most possibly even all—will be clipped by the edge of the image. Lumens clipped by the edge may appear small and distinct but may actually be fingers of the same larger lumen. These considerations complicate the estimation of true lumen counts and sizes. An alternative approach would ignore all lumens clipped by the image edge and only measure lumens with complete edges. However, this approach is again complicated by the presence of lumens that are moderate in size relative to the image, which will skew the statistics.

Ideally, each image would be large relative to the lumens it contains, but that is not always possible, either due to imaging constraints or due to the particular slice or lung itself. An additional factor out of our control is sample preparation.

The tissue in this study was fixed at 20 cm H_2O and, therefore, shows a tortuosity characteristic of a specific portion of the breathing cycle. Different inflation states at fixation would certainly be expected to alter most of our metrics, including lumen count, tissue width, and tortuosity.

4.6 Conclusions

We applied parameter estimation to a compartment model of pressure-volume lung dynamics and created novel image analysis metrics in an attempt to better understand the mechanisms of stiffening and inflammation and affected locations within the pulmonary structure. Importantly, our optimizations identified key parameter differences between healthy and unhealthy groups in data from a neonatal rat model from Mandell et al. [2] that may suggest the mechanisms of VILI in infected respiratory systems. Further, combined analyses of the two strategies identified correlations between inflammatory markers and model parameters with no analog in the data, suggesting that mathematical approaches provide an important path toward understanding VILI and infection.

Acknowledgments The work described herein was initiated during the Collaborative Workshop for Women in Mathematical Biology funded and hosted by UnitedHealth Group Optum of Minnetonka, MN and supported by University of Minnesota's Institute for Mathematics and its Applications in June 2022. Additionally, the authors and editors thank the anonymous peer reviewers for their feedback, which strengthened this work.

The authors would also like to thank Professor Bradford Smith and Dr. Erica Mandell (University of Colorado - Denver) for generously providing both the data analyzed in this study and useful medical insights.

References

1. J.Y. Islam, R.L. Keller, J.L. Aschner, T.V. Hartert, P.E. Moore, Am. J. Respir. Crit. Care Med. **192**(2), 134 (2015)
2. E.W. Mandell, C. Mattson, G. Seedorf, S. Ryan, T. Gonzalez, A. Wallbank, E.M. Bye, S.H. Abman, B.J. Smith, Front. Physiol. **11**, 614283 (2021). https://doi.org/10.3389/fphys.2020. 614283
3. D.S. Guzick, K. Winn, J. Obstet. Gynaecol. **65**(1), 11 (1985)
4. H. Parameswaran, A. Majumdar, S. Ito, A.M. Alencar, B. Suki, J. Appl. Physiol. **100**(1), 186 (2006). https://doi.org/10.1152/japplphysiol.00424.2005
5. C.C.W. Hsia, D.M. Hyde, M. Ochs, E.R. Weibel, Am. J. Respir. Crit. Care Med. **181**(4), 394 (2010)
6. L.E. Mount, Physiol. J. **127**(1), 157 (1955)
7. A.B. Otis, C.B. McKerrow, R.A. Bartlett, J. Mead, M.B. McIlroy, N.J. Selverstone, E.P. Radford, J. Appl. Physiol. **8**(4), 427 (1956)
8. R.W. Jodat, J.D. Horgan, R.L. Lange, Biophys. J. **6**(6), 773 (1966)
9. K.R. Lutchen, F.P. Primiano, G.M. Saidel, IEEE Trans. Biomed. Eng. **29**(9), 629 (1982)
10. R.K. Lambert, J. Appl. Physiol. **68**(6), 2550 (1990)

11. C.H. Liu, S.C. Niranjan, J.W. Clark, K.Y. San, J.B. Swischenburger, A. Bidani, J. Appl. Physiol. **84**(4), 1447 (1998)
12. V. Le Rolle, N. Samson, J.P. Praud, A.I. Hernandez, Acta Biotheor. **91**(1), 91 (2013)
13. M. Airen, H.B. Panitch, NeoReviews **5**(5), c194 (2004)
14. A.R. Carvalho, W.A. Zin, Biophys. Rev. **3**, 71 (2011)
15. A.R. Carvalho, W.A. Zin, Biomed. Eng. Online **11**(38), 1 (2012)
16. A. Hantos, B. Daroczy, B. Suki, S. Nagy, J.J. Fredberg, J. Appl. Physiol. **72**(1), 168 (1992)
17. J.H.T. Bates, C.G. Irvin, J. Appl. Physiol. **94**, 1297 (2003)
18. L. Ellwein Fix, J. Khoury, R.R. Moores, L. Linkous, M. Brandes, H. Rozycki, PLoS ONE **13**(6), 1 (2018). https://doi.org/10.1371/journal.pone.0198425
19. I. Bolle, G. Eder, S. Takenaka, K. Ganguly, A. Karrasch, C. Zeller, M. Neuner, W.G. Kreyling, A. Tsuda, H. Schul, J. Appl. Physiol. **104**, 1167 (2008)
20. E. Salazar, J.H. Knowles, J. Appl. Physiol. **19**(1), 97 (1964)
21. K.L. Hamlington, B.J. Smith, G.B. Allen, J.H.T. Bates, J. Appl. Physiol. **121**(1), 106 (2016)
22. G. Qian, A. Mahdi, Math. Biosci. **323**, 108306 (2020)
23. M.D. Morris, Technometrics **33**(2), 161 (1991)
24. M.J. Colebank, N.C. Chesler, PLOS Comput. Biol. **18**(9), e1010017 (2022)
25. M.T. Wentworth, R.C. Smith, H.T. Banks, SIAM-ASA J. Uncertain. Quantif. **4**(1), 266 (2016)
26. R.C. Smith, *Uncertainty Quantification: Theory, Implementation, and Applications*, vol. 12 (SIAM, Philadelphia, 2013)
27. F. Campolongo, J. Cariboni, A. Saltelli, Environ. Model Softw. **22**(10), 1509 (2007)
28. S.R. Lubkin, S.E. Funk, E.H. Sage, J. Theor. Med. **6**(3), 173 (2005)
29. S. Lê, J. Josse, F. Husson, J. Stat. Softw. **25**(1), 1 (2008). https://doi.org/10.18637/jss.v025.i01
30. R Core Team, *R: A Language and Environment for Statistical Computing*. R Foundation for Statistical Computing, Vienna, Austria (2013). http://www.R-project.org/
31. N. van Osta, A. Lyon, F. Kirkels, T. Koopsen, T. van Loon, M.J. Cramer, A.J. Teske, T. Delhaas, W. Huberts, J. Lumens, Philos. Trans. Royal Soc. A **378**(2173), 20190347 (2020)
32. W.P. Donders, W. Huberts, F.N. van de Vosse, T. Delhaas, Int. J. Numer. Method Biomed. Eng. **31**(10) (2015)

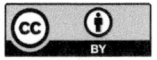

Estimation of Time-Dependent Transmission Rate for COVID-19 SVIRD Model Using Predictor–Corrector Algorithm

Ruiyan Luo, Alejandra D. Herrera-Reyes, Yena Kim, Susan Rogowski, Diana White, and Alexandra Smirnova

1 Introduction

Compartmental disease models, which track the progression of individuals between different disease stages and risk levels, remain at the kernel of epidemic theory [1]. A simple example of a compartmental framework is the Susceptible–Infected–Recovered (SIR) model proposed in [2]. This model has been extended to include other states, such as the Susceptible–Infectious–Recovered–Deceased (SIRD) [3] and the Susceptible–Infectious–Recovered–Vaccinated (SIRV) models

R. Luo
Department of Population Health Sciences, School of Public Health, Georgia State University, Atlanta, GA, USA
e-mail: rluo@gsu.edu

A. D. Herrera-Reyes
Department of Mathematical Science, George Mason University, Fairfax, VA, USA
e-mail: aherre6@gmu.edu

Y. Kim
Hawaii Pacific University, Honolulu, HI, USA
e-mail: yekim@hpu.edu

S. Rogowski
Department of Mathematics, Florida State University, Tallahassee, FL, USA
e-mail: srogowski@fsu.edu

D. White
Department of Mathematics & Statistics, Clarkson University, Potsdam, NY, USA
e-mail: dtwhite@clarkson.edu

A. Smirnova (✉)
Department of Mathematics & Statistics, Georgia State University, Atlanta, GA, USA
e-mail: asmirnova@gsu.edu

© The Author(s) 2024   213
A. N. Ford Versypt et al. (eds.), *Mathematical Modeling for Women's Health*,
The IMA Volumes in Mathematics and its Applications 166,
https://doi.org/10.1007/978-3-031-58516-6_7

[4]. Recently, generalizations of SIR models have been implemented to study the spread of COVID-19 with the adherence and non-adherence of social behavior protocols such as masking, social distancing, and the enforcement of closures and lockdowns [5–9]. Earlier models described the spread of the disease in uncontrolled systems and in the presence of different mitigation strategies such as social distancing and lockdown restrictions.

Since the development and widespread distribution of vaccines, incorporation of vaccination into such models has been an important development [10, 11]. However, few models have accounted for differing disease transmission within vaccinated and unvaccinated individuals. Here, we propose a new compartmental model of COVID-19 transmission that takes into consideration some of these important dynamics by including the vaccination status of both susceptible and infected humans. We also include the possibility of losing immunity and becoming reinfected within both vaccinated and unvaccinated populations. Thus, our new model incorporates important disease dynamics that have not been covered by previous COVID-19 models. Additionally, the proposed model can easily be adjusted to other seasonal outbreaks. With new variants of COVID-19 and other viruses occurring regularly, along with fluctuations of vaccine efficacy among these variants, this new model will help to understand past and current disease dynamics and make predictions about future cases.

Another important novel feature of our compartmental model is the use of a time-dependent transmission rate. Oftentimes, the transmission rate of a disease is the most challenging parameter to estimate [12]. The emerging new variants of COVID-19 make stable estimation of disease transmission even more complicated. To simplify this, many previous COVID-19 models incorporated constant transmission rates found in the literature. To better assess the efficiency of control and prevention and to account for new COVID-19 strains, in our proposed model, we introduce a time-dependent transmission rate for vaccinated and unvaccinated individuals. This rate is reconstructed from noise-contaminated data on new incidence cases and daily deaths by solving a parameter estimation inverse problem.

A commonly used method for estimating parameters of ordinary differential equations (ODEs) from noisy data is nonlinear least squares (NLS), where model predictions for an invading pathogen are fitted to reported incidence cases and daily new deaths [13–16]. In the NLS, a numerical method, such as Runge–Kutta or similar, is used to approximate the solution of a given ODE system using a trial set of values for parameters and initial conditions. The fit value is then input into an optimization algorithm that updates parameter estimates. As a result, the NLS framework can be computationally expensive when noisy data is considered or a highly nonlinear model is being used to describe a complex biological process. In [17, 18], a two-stage approach for this method was proposed, which first fit a smooth curve to given noisy data and then estimated the unknown parameters in the ODE system. Ramsay et al. [19] expanded on this method by proposing to alternate the two procedures and by imposing a smoothness penalty on curve fitting. To that end, Ramsay et al. developed a novel profiling estimation procedure where the data fitting and the fidelity to the ODE were combined into a penalized log-likelihood criterion, which provided the statistical inference for the ODE parameters. For other prior work on alternating minimization, also known as (block) coordinate descent, one may consult [20–25] and the references therein.

A more general nonlinear constrained minimization problem was studied in [26], where parameter estimation was carried out in a predictor–corrector manner. In the predictor–corrector algorithm of [26], one updates the epidemiological parameters by a regularized second-order method while freezing the state variables, and then the state variables are modified while the system (epidemiological) parameters are fixed. These updates are iterated until convergence. Here, we propose a new predictor–corrector algorithm that extends the earlier version in [26] to the case of parameter-dependent nonlinear observation operators. The new algorithm success-fully mitigates the associated computational costs and incorporates an extra layer of stability in the optimization process. In what follows, the proposed version of the predictor–corrector algorithm is used to get stable estimates of a time-dependent transmission rate and effective reproduction number from our new compartmental model, which is applied to the study of COVID-19 dynamics in a post-vaccination stage.

The chapter is organized as follows. In Sect. 2, we introduce our Susceptible–Vaccinated–Infectious–Recovered–Deceased (SVIRD) model. In Sect. 3, we describe the new computational algorithm for estimating disease parameters in the proposed epidemic model. In Sects. 4 and 5, the method is evaluated on synthetic and real data sets, respectively. Possible directions of future work are outlined in Sect. 6.

2 Mathematical Model: SVIRD

Prior studies have underscored the importance of stable parameter estimation related to infectious disease transmission models based on ordinary or partial differential equations [27–29]. Lack of stable parameter estimation, which is evident when parameter estimates are associated with large uncertainties, may be attributed to the model structure or to the lack of information in a given data set, which could be linked to the number of observations and to the spatial granularity of the data [28].

Within epidemiology, stable estimation of the effective reproduction number, $\mathcal{R}_e(t)$, and its underlying transmission rate, $\beta(t)$, is particularly important [30–32]. Unlike other system parameters, i.e., incubation and recovery rates, the effective reproduction number and the transmission rate of the disease are directly influenced by mitigation measures. Therefore, it is critical to develop both suitable epidemic models and regularized computational methods to reliably quantify disease-specific parameters, especially in the face of noise-contaminated data and substantial uncertainty in approximate solutions.

In this chapter, to model the COVID-19 dynamics and estimate the effective reproduction number, $\mathcal{R}_e(t)$, and its underlying transmission rate, $\beta(t)$, we propose the following system of ODEs:

$$\frac{dS}{dt} = -\beta(t)\frac{S(t)}{N - D(t)}(I_s(t) + I_v(t)) - pS(t) + \delta_r R(t) + \delta_v V(t) \qquad (1)$$

$$\frac{dV}{dt} = pS(t) - (1 - \alpha)\beta(t)\frac{V(t)}{N - D(t)}(I_s(t) + I_v(t)) - \delta_v V(t) \qquad (2)$$

$$\frac{dI_s}{dt} = \beta(t)\frac{S(t)}{N - D(t)}(I_s(t) + I_v(t)) - (\gamma_{s,r} + \gamma_{s,d})I_s(t) \qquad (3)$$

$$\frac{dI_v}{dt} = (1 - \alpha)\beta(t)\frac{V(t)}{N - D(t)}(I_s(t) + I_v(t)) - (\gamma_{v,r} + \gamma_{v,d})I_v(t) \qquad (4)$$

$$\frac{dR}{dt} = \gamma_{s,r}I_s(t) + \gamma_{v,r}I_v(t) - \delta_r R(t) \qquad (5)$$

$$\frac{dD}{dt} = \gamma_{s,d}I_s(t) + \gamma_{v,d}I_v(t). \qquad (6)$$

The system defined by Eqs. (1)–(6) includes susceptible unvaccinated (S), susceptible vaccinated (V), infected vaccinated (I_v), infected unvaccinated (I_s), recovered (R), and deceased (D) compartments. With N denoting the population size at the beginning time point of the study period, we use $N - D(t)$ as the total population size at time t. This is based on the assumption that the population increase (due to birth or immigration) and population decrease (due to reasons other than COVID-19) balance out, and the change in population size is just due to COVID-19 death. The diagram of the SVIRD model in Eqs. (1)–(6) is given in Fig. 1, which illustrates the transition of individuals between various disease compartments. Susceptible humans become vaccinated at a rate of p. Both vaccinated and unvaccinated individuals can be infected. The disease transmission rate, $\beta(t)$, for susceptible

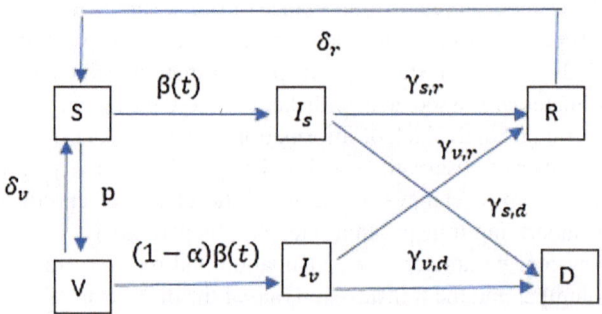

Fig. 1 Diagram of the SVIRD model used. Susceptible individuals get vaccinated at a rate p and become infected at a time-dependent transmission rate $\beta(t)$. A constant parameter, $0 < \alpha < 1$, is a measure of vaccine efficacy. The lower values correspond to less efficacy, and $(1 - \alpha)\beta(t)$ is the rate of disease transmission for vaccinated individuals. Both infected unvaccinated and vaccinated can recover at rates $\gamma_{s,r}$ and $\gamma_{v,r}$ and die at rates $\gamma_{s,d}$ and $\gamma_{v,d}$, respectively. Loss of immunity is accounted for by considering movement back to the susceptible class from the vaccinated and recovered classes at rates δ_v and δ_r

individuals is assumed to be time-dependent. We assume that vaccinated individuals become infected at a slower rate, which is taken into account by the incorporation of a vaccine efficacy parameter, denoted by α; that is, vaccinated individuals become infected at a rate of $(1 - \alpha)\beta(t)$, where $0 < \alpha < 1$.

Motivated by the report that unvaccinated individuals are more likely to have severe symptoms from COVID-19 infections leading to a higher risk of hospitalization and death [33], we assume different death rates for vaccinated and unvaccinated individuals, denoted by $\gamma_{v,d}$ and $\gamma_{s,d}$, respectively. The severity in symptoms also leads to differing recovery rates for vaccinated and unvaccinated populations. The recovery rates for vaccinated and unvaccinated individuals are denoted by $\gamma_{v,r}$ and $\gamma_{s,r}$, respectively.

We further consider the case of possible reinfection due to the loss of immunity by vaccinated individuals at a rate of δ_v and recovered individuals at a rate of δ_r. We note from Eq. (1) that the rate of transmission depends only on the number of contacts between the living susceptible and infected individuals (described by the division by $N - D(t)$, the total living population at any instance in time).

The disease transmission rate, $\beta(t)$, is an important underlying factor for the effective reproduction number, $\mathcal{R}_e(t)$, which quantifies the number of secondary cases per primary case in a completely susceptible population during the entire course of the outbreak. Similar to the transmission rate, the effective reproduction number is significantly impacted by environmental conditions and the behavior of the population. A sustainable reduction of $\mathcal{R}_e(t)$ to a level less than 1 would indicate that mitigation measures are successful and that the disease is contained, because every infected person, on average, can only transmit the virus to less than one other human.

Using the next-generation matrix [34, 35], the effective reproduction number for compartmental model (Eqs. (1)–(6)) is estimated as

$$\mathcal{R}_e(t) = \frac{\beta(t)}{(\gamma_{s,r} + \gamma_{s,d})} \frac{S(t)}{N - D(t)} + \frac{(1 - \alpha)\beta(t)}{(\gamma_{v,r} + \gamma_{v,d})} \frac{V(t)}{N - D(t)}. \qquad (7)$$

From Eq. (7), we note that $\mathcal{R}_e(t)$ increases with increasing disease transmission $\beta(t)$, as well as increasing numbers of susceptible individuals (vaccinated and unvaccinated). In addition, $\mathcal{R}_e(t)$ decreases with increasing recovery rates. Next, in Sect. 3, we describe our predictor–corrector algorithm that will be used to reconstruct the disease transmission rate, $\beta(t)$, which allows us to provide an estimate for the effective reproduction number, $\mathcal{R}_e(t)$.

3   Methodology and Algorithm

Let C and $\mathcal{T}$ be incidence data on new COVID-19 confirmed cases and deaths, respectively, and n be the number of data points in each set. Naturally, we assume that both data sets are noise contaminated. According to our SVIRD model given by Eqs. (1)–(6), the daily number of new COVID-19 cases is

$$\mathbb{C}(t) := \beta(t)\frac{S(t)(I_s(t) + I_v(t))}{N - D(t)} + (1 - \alpha)\beta(t)\frac{V(t)(I_s(t) + I_v(t))}{N - D(t)}, \tag{8}$$

which we define as the rate of new infections into the system. On the other hand, by Eq. (6), the daily number of new deaths is

$$\mathbb{T}(t) := \gamma_{s,d}I_s(t) + \gamma_{v,d}I_v(t). \tag{9}$$

Assume that in a particular region, the values $a = t_1$ and $b = t_n$ are the first and the last days of the study period. We note that, fortunately, the number of deceased individuals is considerably smaller than infectious ones. So, we multiply daily new deaths, $\mathbb{T}$, by a positive scaling parameter, λ, to ensure that new deaths and new cases have the same order of magnitude. Let the data, d, for new cases and deaths, C and $\mathcal{T}$, be reported on days $t_1, t_2, ..., t_n$. That is,

$$d := [C(t_1), ..., C(t_n), \lambda\mathcal{T}(t_1), ..., \lambda\mathcal{T}(t_n)]^T. \tag{10}$$

Combining Eqs. (8) and (9), we now introduce the observation operator as

$$\mathcal{B} := [\mathbb{C}(t_1), ..., \mathbb{C}(t_n), \lambda\mathbb{T}(t_1), ..., \lambda\mathbb{T}(t_n)]^T. \tag{11}$$

Then our goal is to recover the unknown time-dependent transmission rate, $\beta(t)$, from the nonlinear constrained minimization problem:

$$\min_{\beta,S,V,I_s,I_v,D} f(\beta, S, V, I_s, I_v, D) \tag{12}$$

subject to system in Eqs. (1)–(6), where

$$f(\beta, S, V, I_s, I_v, D) := \|\mathcal{B} - d\|^2$$

$$= \sum_{i=1}^{n} \left\{ (\mathbb{C}(t_i) - C(t_i))^2 + \lambda^2 (\mathbb{T}(t_i) - \mathcal{T}(t_i))^2 \right\}. \tag{13}$$

To solve Eqs. (12) and (13) numerically, we discretize unobserved state variables, S, V, I_s, and I_v, and the time-varying transmission rate, $\beta(t)$, using basis expansions. The vector of expansion coefficients for the transmission rate, $\beta(t)$, is of primary interest. The vector of expansion coefficients for the state variables is of less practical importance, and it is primarily needed for the estimation of $\beta(t)$. For this reason, in statistics literature, the expansion coefficients for state variables are often referred to as nuisance parameters [19]. Upon discretization, we iteratively update both sets of unknown expansion coefficients using alternating minimization as described below.

In order to obtain the discrete approximation of $\beta(t)$, we consider a finite subset spanned by shifted Legendre polynomials of degree $0, 1, ..., m - 1$, which are orthogonal on the interval $[a, b]$ with respect to L_2 inner product, defined recursively as follows:

$$x = \frac{2t - a - b}{b - a}, \quad P_0(x) = 1, \quad P_1(x) = x, \quad t \in [a, b],$$

$$(j + 1)P_{j+1}(x) = (2j + 1)x P_j(x) - j P_{j-1}(x), \quad j = 1, 2, ..., m - 2.$$

This gives rise to the following finite-dimensional approximation of the transmission rate:

$$\bar{\beta}_i[\theta] = \sum_{j=0}^{m-1} \theta_{j+1} P_j(t_i), \quad i = 1, 2, ..., n. \tag{14}$$

Likewise, we express the state variables S, V, I_s, and I_v as

$$\bar{S}_i[u] = \sum_{j=0}^{l-1} u_{j+1} P_j(t_i), \quad \bar{V}_i[u] = \sum_{j=0}^{l-1} u_{l+j+1} P_j(t_i),$$

$$\bar{I}_{s,i}[u] = \sum_{j=0}^{l-1} u_{2l+j+1} P_j(t_i), \quad \bar{I}_{v,i}[u] = \sum_{j=0}^{l-1} u_{3l+j+1} P_j(t_i), \tag{15}$$

which generates discretized daily rates of incidence and death, $\bar{\mathbb{C}}_{d,i}[\theta, u]$ and $\bar{\mathbb{T}}_{d,i}[u]$, respectively, if one substitutes $\bar{\beta}_i[\theta]$ from Eq. (14) and $\bar{S}_i[u]$, $\bar{V}_i[u]$, $\bar{I}_{s,i}[u]$, and $\bar{I}_{v,i}[u]$ from Eq. (15) for $\beta(t_i)$, $S(t_i)$, $V(t_i)$, $I_s(t_i)$, and $I_v(t_i)$ in Eqs. (1)–(6) and Eqs. (8) and (9). The derivatives of S, V, I_s, and I_v get discretized by replacing $P_j(t_i)$ with $P'_j(t_i)$ in the identities above.

Next, we define vectors for the unknown parameters, θ and u, from the discrete approximation of the transmission rate, $\beta(t_i)$, in identity Eq. (14) and from the discrete approximation of the state variables, $S(t_i)$, $V(t_i)$, $I_s(t_i)$, and $I_v(t_i)$, $i = 1, 2, ..., n$, in Eq. (15) as

$$\theta := [\theta_1, ..., \theta_m]^T \text{ and } u := [u_1, ..., u_l, u_{l+1}, ..., u_{2l}, u_{2l+1}, ..., u_{3l}, u_{3l+1}, ..., u_{4l}]^T.$$

This enables us to introduce the observation operator, B:

$$B(\theta, u) := \left[\bar{\mathbb{C}}_{d,1}[\theta, u], ..., \bar{\mathbb{C}}_{d,n}[\theta, u], \lambda \bar{\mathbb{T}}_{d,1}[\theta, u], ..., \lambda \bar{\mathbb{T}}_{d,n}[\theta, u] \right]^T \tag{16}$$

and the operator G to account for the constraints

$$G_i(\theta, u) := \bar{S}'_i[u] + \bar{\beta}_i[\theta]\frac{\bar{S}_i[u](\bar{I}_{s,i}[u] + \bar{I}_{v,i}[u])}{N - \bar{D}_i[u]} + p\bar{S}_i[u] - \delta_r\bar{R}_i[u] - \delta_v\bar{V}_i[u]$$

$$G_{n+i}(\theta, u) := \bar{V}'_i[u] - p\bar{S}_i[u] + (1 - \alpha)\bar{\beta}_i[\theta]\frac{\bar{V}_i[u](\bar{I}_{s,i}[u] + \bar{I}_{v,i}[u])}{N - \bar{D}_i[u]} + \delta_v\bar{V}_i[u]$$

$$G_{2n+i}(\theta, u) := \bar{I}'_{s,i}[u] - \bar{\beta}_i[\theta]\frac{\bar{S}_i[u](\bar{I}_{s,i}[u] + \bar{I}_{v,i}[u])}{N - \bar{D}_i[u]} + (\gamma_{s,r} + \gamma_{s,d})\bar{I}_{s,i}[u]$$

$$G_{3n+i}(\theta, u) := \bar{I}'_{v,i}[u] - (1 - \alpha)\bar{\beta}_i[\theta]\frac{\bar{V}_i[u](\bar{I}_{s,i}[u] + \bar{I}_{v,i}[u])}{N - \bar{D}_i[u]} + (\gamma_{v,r} + \gamma_{v,d})\bar{I}_{v,i}[u]$$

for $i = 1, 2, ..., n$. Here $\bar{D}_i[u]$ is the reported cumulative number of deaths on day t_i and

$$\bar{R}_i[u] := N - (\bar{S}_i[u] + \bar{V}_i[u] + \bar{I}_{s,i}[u] + \bar{I}_{v,i}[u] + \bar{D}_i[u]). \tag{17}$$

We can now recast the constrained minimization problem as follows:

$$\text{minimize} \quad \|B(\theta, u) - d\|^2 \quad \text{with respect to } \theta \text{ and } u$$

$$\text{subject to} \quad G(\theta, u) = 0. \tag{18}$$

Note that the data-fitting operator, B, also depends on the input data, $\bar{D}$, the cumulative number of deceased individuals. However, the cumulative data, as opposed to daily number of cases and deaths on the right-hand side, are smooth, and the noise in cumulative data is consistent with discretization and modeling errors.

To reconstruct the transmission rate, $\beta(t)$, we employ a predictor–corrector algorithm, where one updates θ while freezing u, and then u is modified while θ is kept unchanged. The process is repeated until a desired tolerance level is achieved. More specifically, given $\begin{pmatrix} \theta_k \\ u_k \end{pmatrix}$, one transitions from θ_k to θ_{k+1} by applying one step of the iteratively regularized Gauss–Newton (IRGN) procedure:

$$\theta_{k+1} = \theta_k - [G'^*_\theta(\theta_k, u_k)G'_\theta(\theta_k, u_k) + B'^*_\theta(\theta_k, u_k)B'_\theta(\theta_k, u_k) + \tau_k I]^{-1}$$

$$\{G'^*_\theta(\theta_k, u_k)G(\theta_k, u_k) + B'^*_\theta(\theta_k, u_k)(B(\theta_k, u_k) - d) + \tau_k(\theta_k - \bar{\theta})\}, \tag{19}$$

where τ_k is the regularization parameter needed to incorporate stability in the optimization process and $\bar{\theta}$ is a prior value of θ. Then, given $\begin{pmatrix} \theta_{k+1} \\ u_k \end{pmatrix}$, one computes u_{k+1} using the classical Gauss–Newton scheme

$$u_{k+1} = u_k - [G'^*_u(\theta_{k+1}, u_k)G'_u(\theta_{k+1}, u_k) + B'^*(\theta_{k+1}, u_k)B'(\theta_{k+1}, u_k)]^{-1}$$

$$\{G'^*_u(\theta_{k+1}, u_k)G(\theta_{k+1}, u_k) + B'^*(\theta_{k+1}, u_k)(B(\theta_{k+1}, u_k) - d)\}. \tag{20}$$

A simpler version of this algorithm was introduced and analyzed in [26]. In [26], the data-fitting operator, B, does not depend on the system parameter, θ, and is a function of the state variable only, i.e., $B = B(u)$. The IRGN scheme in Eq. (19) originates from variational regularization in the form

$$\min_{\theta \in \mathbb{R}^m} \left\{ \frac{1}{2}||G(\theta, u_k)||^2 + \frac{1}{2}||B(\theta, u_k) - d||^2 + \frac{\tau_k}{2}||\theta - \bar{\theta}||^2 \right\}. \qquad (21)$$

The method in Eq. (20), on the other hand, is the classical Gauss–Newton algorithm applied to the nonlinear minimization problem

$$\min_{u \in \mathbb{R}^{4l}} \left\{ \frac{1}{2}||G(\theta_{k+1}, u)||^2 + \frac{1}{2}||B(\theta_{k+1}, u) - d||^2 \right\}. \qquad (22)$$

The Gauss–Newton procedure in Eq. (20) does not need to be regularized, since solving the ODE system of equations in Eqs. (1)–(6), with respect to S, V, I_s, I_v, R, and D, is a forward problem, which is not generally ill-posed. Thus, its discrete approximation is also stable (as our numerical experiments below confirm).

The algorithm in Eqs. (19) and (20) was coded in MATLAB, using the optimization and parallel toolboxes. The code, along with figures, simulated data, and parameter estimates, can be found in our GitHub repository: https://github.com/donajialej/WIMB2022team5.git.

For all numerical simulations (with synthetic and real data), the unobserved state variables, S, V, I_s, and I_v, are normalized; that is, in place of S, V, I_s, and I_v, we reconstruct the expansion coefficients for S/N, V/N, I_s/N, and I_v/N, where N is the total population of the region.

To select the number of basis functions for $\beta(t)$ and for the unobserved state variables (m and n, respectively), we start with $m = n = 5$ and keep increasing them until the reconstructed functions, $\beta(t)$, $S(t)$, $V(t)$, $I_s(t)$, and $I_v(t)$, no longer visibly change.

An important part of parameter estimation is the choice of λ in Eqs. (10)–(11), which ensures that the two data sets—reported daily new cases and deaths—are well-balanced. In all our experiments, the value of λ is equal to 1000. For $\lambda = 1$, the misfit in daily new deaths is perceived as part of noise in incidence data, and the process is less sensitive to daily new deaths as compared to new incidence cases.

4 Numerical Experiments with Synthetic Data

In this section, we test our proposed predictor–corrector algorithm (Eqs. (19)–(20)) using two synthetic data sets for incidence cases and deaths. The first synthetic data set was generated using the transmission rate $\beta(t)$ shown in Fig. 2, which represents a case when initial success in disease prevention is followed by some setbacks causing the transmission rate to fluctuate. Specifically, this transmission rate was

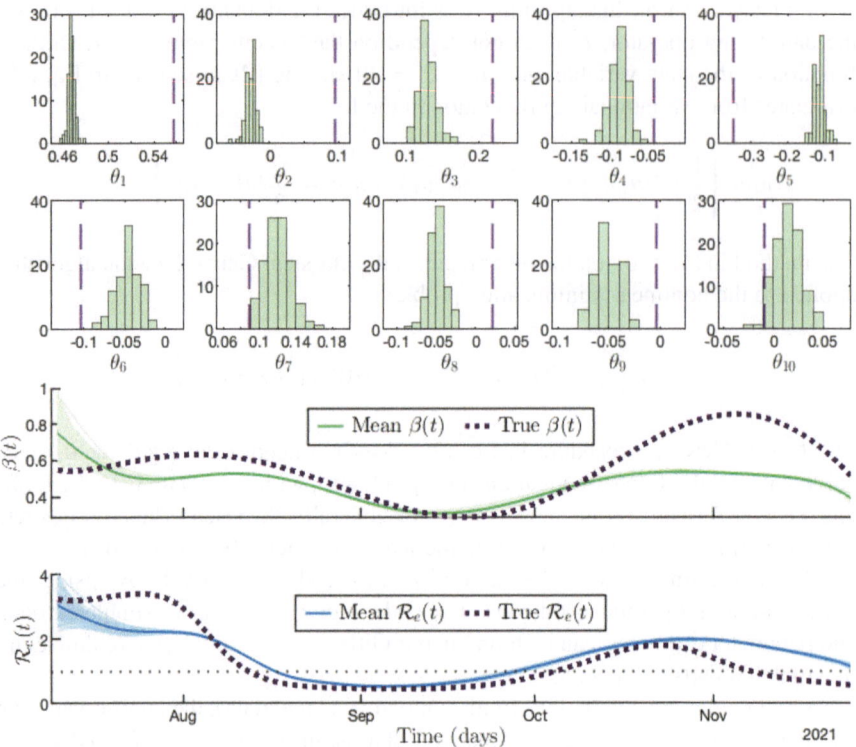

Fig. 2 Reconstruction of disease transmission $\beta(t)$ (along with coefficients) and the effective reproduction number $\mathcal{R}_e(t)$ for Scenario 1 (non-effective mitigation) from synthetic noisy data on new daily cases and deaths in Fig. 3. Simulations are carried out with 10 basis functions for the transmission rate $\beta(t)$ and 40 basis functions for each unobserved state variable, S, V, I_s, and I_v, i.e., 160 basis functions for all state variables combined. The regularization sequence is $\tau_k = 10^{10}/(k+1)^{15}$, and the iterations are stopped when $k = 43$. This stopping time is determined by the goodness of fit to both data sets

chosen to model a "non-effective mitigation" scenario where $\mathcal{R}_e(t)$ remains above 1 for multiple time periods showing that the disease persists and spreads quickly. This is illustrated in the graph of $\mathcal{R}_e(t)$ in Fig. 2. The second synthetic data set was generated using the transmission rate shown in Fig. 4 and represents an "effective mitigation" scenario where the disease transmission rate is reduced during the study period and where $\mathcal{R}_e(t)$ stays below 1 more consistently.

In what follows, we evaluate the performance of the proposed method in reconstructing the unknown time-dependent transmission rate, $\beta(t)$, given synthetic daily rates of incidence cases and new deaths over a certain period of time. Two model transmission rates, described above, were selected (see Figs. 2 and 4). Each model transmission rate was used to solve the forward problem, i.e., the system of ODEs (Eqs. (1)–(6)), and to generate clean data on incidence cases, $C(t)$, and

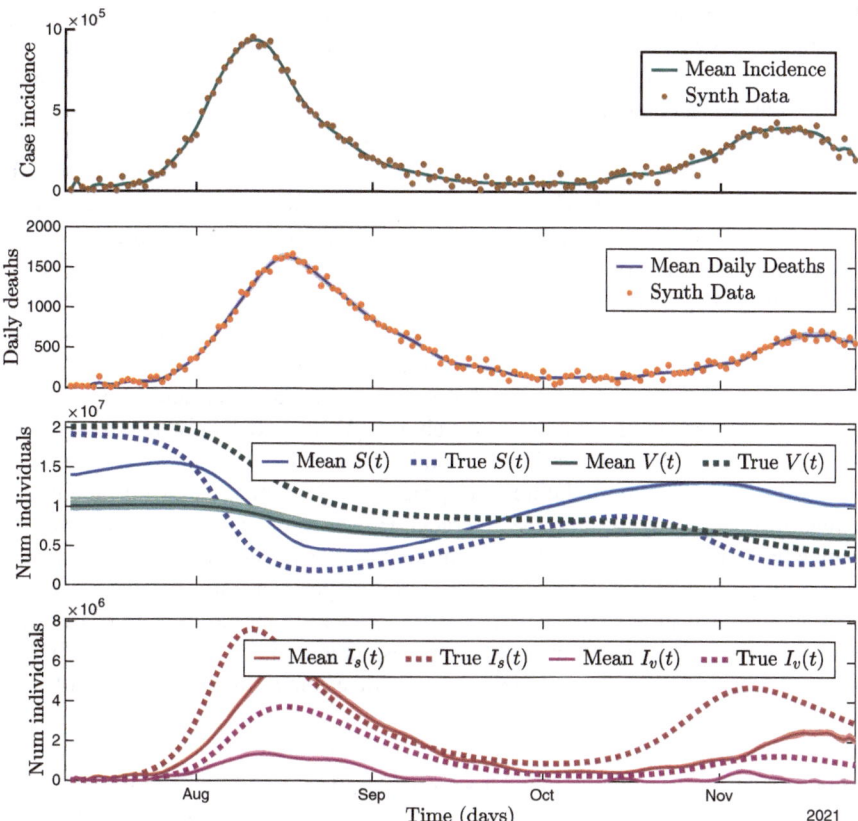

Fig. 3 Synthetic study of Scenario 1: non-effective mitigation. Top to bottom: synthetic (Synth) data (dots) and model fit (solid line) for daily new cases and daily new deaths; true synthetic values (dash line) and model reconstructions (solid line) for $S(t)$ (blue), $V(t)$ (green), $I_s(t)$ (red), and $I_v(t)$ (pink). There are 100 bootstrap model reconstructions, and the mean of them is a darker line of the color corresponding to each compartment

daily new deaths, $\mathcal{T}(t)$, on a given time interval $[t_1, t_n]$ according to expressions, Eqs. (8) and (9), respectively. Then, random Gaussian noise (with 0 mean and a rather aggressive standard deviation) was added to epidemic data in order to mimic noise-contaminated data in a real-life setting, as shown in the top panels of Figs. 3 and 5. Since real incidence cases and deaths are known to be positive, uniform noise was added if the incidence became negative at any point.

Given "real" data for incidence cases and daily new deaths, we employed the regularized algorithm (Eqs. (19) and (20)) to simultaneously reconstruct the unknown transmission rate, $\beta(t)$, and the state variables, S, V, I_s, and I_v, with discrete approximation given by Eqs. (14) and (15). In order to quantify uncertainty in the extracted transmission rate, we refit the model (using parallel programming via the *parfor* function in MATLAB) to $M = 100$ additional data sets for incidence cases and daily deaths assuming Poisson error structure. The resulting M best-fit parameter sets are used to build the histogram for each Legendre coefficient, θ_j, $j = 1, 2, ..., m$, representing the frequency distribution of the reconstructed values.

To ensure an unbiased choice of the initial guess for $\beta(t)$, we take $[\beta_0, 0, ..., 0]^T$ to serve as initial approximation for $[\theta_1, \theta_2, ..., \theta_m]^T$ at every bootstrap iteration, where $0.1 < \beta_0 < 1$. To find initial approximations for u, we solve the system of ODEs (Eqs. (1)–(6)) with $\beta(t) = \beta_0$ one time before the start of the iterative process and then evaluate Legendre expansion coefficients for the computed S, V, I_s, and I_v to form the initial vector $u :=$ $[u_1, ..., u_l, u_{l+1}, ..., u_{2l}, u_{2l+1}, ..., u_{3l}, u_{3l+1}, ..., u_{4l}]^T$.

For the non-effective mitigation scenario (Scenario 1) with transmission rate $\beta(t)$ shown in Fig. 2, the fitting procedure is initiated with $\beta_0 = 0.5$ and is carried out using $m = 10$ basis functions for the transmission rate, $\beta(t)$, and $n = 40$ basis functions for each unobserved state variable, S, V, I_s, and I_v, giving a total of 160 basis functions for all state variables combined.

With no regularization, the iterative process to estimate the transmission rate in Scenario 1 (Fig. 2) turns out to be divergent. However, the process can be stabilized with a broad range of initial values, τ_0, as long as they are consistent with the rate of decay of the regularization sequence, τ_k. In our experiment, we selected $\tau_0 = 10^{10}$ and the regularization sequence, $\tau_k = 10^{10}/(k + 1)^{15}$, the fastest rate of decrease that gives rise to a convergent iterative process. Iterations of Eqs. (19) and (20) are stopped when $k = 43$. This stopping time is determined by the goodness of fit to both data sets C and $\mathcal{T}$.

For the effective mitigation case (Scenario 2), where the transmission rate $\beta(t)$ is presented in Fig. 4, the parameter estimation process is initiated with $\beta_0 = 0.3$. As before, the reconstruction is done with $m = 10$, $n = 40$, and $\tau_0 = 10^{10}$, and the regularization sequence is driven to zero at the rate $10^{10}/(k + 1)^{15}$. But in this scenario, the iterative process is terminated when $k = 19$.

Figures 2 and 4 illustrate the connection between exact and reconstructed effective reproduction numbers, $\mathcal{R}_e(t)$, for the two scenarios with different model transmission rates. As stated in Sect. 2, $\mathcal{R}_e(t) > 1$ describes time periods for which the disease persists and spreads quickly, and $\mathcal{R}_e(t) < 1$ describes time periods for which the disease is contained (i.e., the disease is spreading slowly, eventually dying out). In the non-effective mitigation scenario described in Fig. 2, we see two approximately month-long windows for which the disease persists, highlighting that after the first push to decrease transmission ($\mathcal{R}_e(t)$ falls to less than 1 in mid-August), mitigation strategies are not successful at keeping the transmission rate low enough, and a second wave begins in early October. For the effective mitigation scenario, described in Fig. 4, we see that although the effective reproduction rate $\mathcal{R}_e(t)$ is greater than 1 for an extended initial period of time, once it drops below 1 (close to September) it stays below 1.

The top panels of Figs. 3 and 5 show how the bundles of incidence curves for daily new cases and deaths corresponding to the reconstructed transmission rates, $\beta(t)$, are compared to the noisy synthetic data used for data fitting.

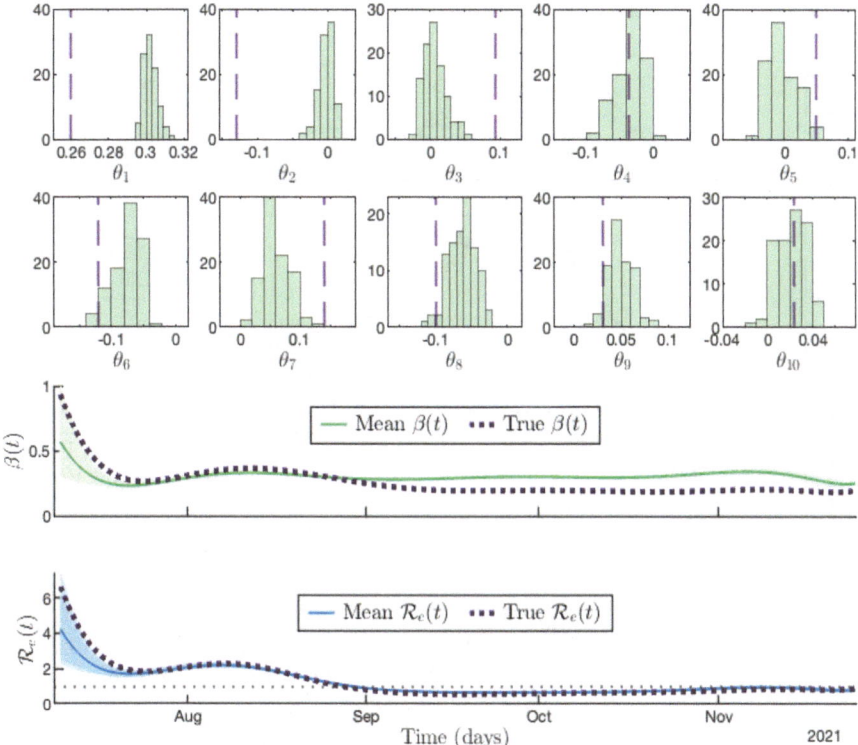

Fig. 4 Reconstruction of disease transmission $\beta(t)$ (along with coefficients) and the effective reproduction number $\mathcal{R}_e(t)$ in Scenario 2 (effective mitigation) from synthetic data on new daily cases and deaths in Fig. 5. Simulations are carried out with 10 basis functions for the transmission rate $\beta(t)$ and 40 basis functions for each unobserved state variable, S, V, I_s, and I_v, i.e., 160 basis functions for all state variables combined. The regularization sequence is $\tau_k = 10^{10}/(k+1)^{15}$, and the iterations are stopped when $k = 19$. This stopping time is determined by the goodness of fit to both data sets

Reconstructed $S(t)$, $V(t)$, $I_s(t)$, and $I_v(t)$ from these two scenarios can be viewed in the lower panels of Figs. 3 and 5, respectively. While there are inevitable errors due to noise contamination in both data sets and due to accuracy loss stemming from regularization, Figs. 2, 3, 4, and 5 illustrate numerical experiments for synthetic data where the uncertainty is very low and the reconstruction of all unknown parameters is very stable. Yet, as evident from Figs. 3 and 5, it is harder to reconstruct the dynamics of the vaccinated population compared to the susceptible one since vaccinated individuals are less likely to contribute to new incidence cases (and especially deaths).

When comparing the time series for the reconstructed state variables between our two scenarios in the lower panels of Figs. 3 and 5, the progression of the disease follows the trend of the disease transmission rates. In particular, two infection peaks

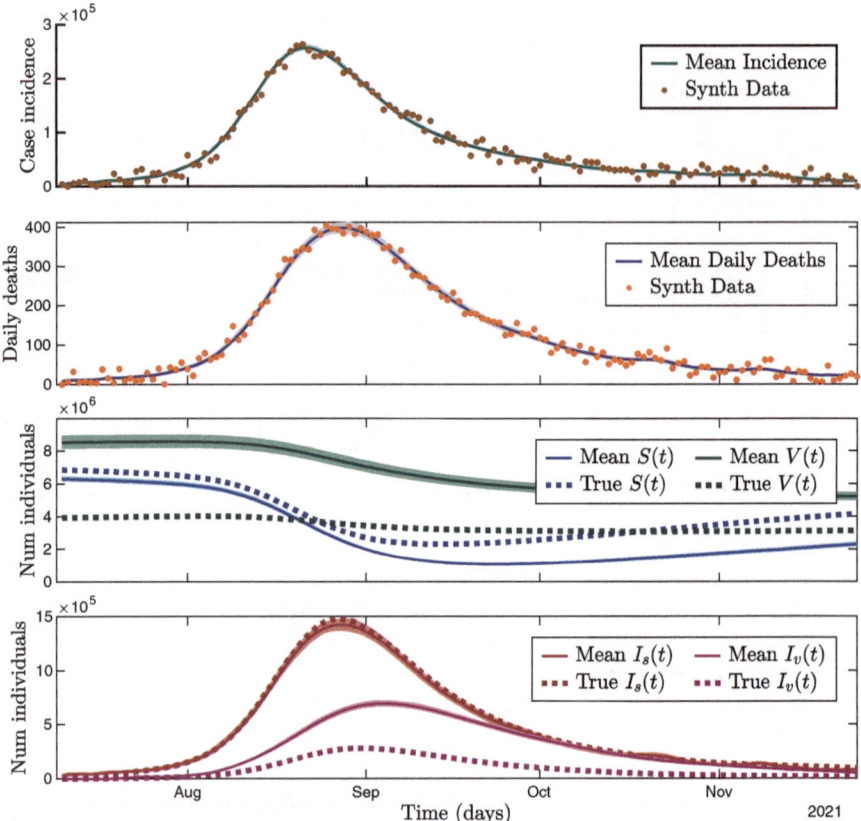

Fig. 5 Synthetic study of Scenario 2: effective mitigation. Top to bottom: synthetic (Synth) data (dots) and model fit (solid line) for daily new cases and daily new deaths; true synthetic values (dash line) and model reconstructions (solid line) for $S(t)$ (blue), $V(t)$ (green), $I_s(t)$ (red), and $I_v(t)$ (pink). There are 100 bootstrap model reconstructions, and the mean of them is a darker line of the color corresponding to each compartment

are in the lower panel of Fig. 3, which follow the peaks in the transmission rate and effective reproduction number curves in Fig. 2. A similar trend for a single infected peak is in the lower panel of Fig. 5, which follows the peaks in the transmission rate and effective reproduction number curves in Fig. 4. We also note that in the non-effective mitigation scenario (Fig. 3) the initial population is assumed to be $N = 39,237,836$ and for the effective mitigation scenario (Fig. 5) $N = 10,799,566$.

Our simulated data and the inversion results for both experiments with synthetic data largely depend on the values of pre-estimated parameters, p, α, $\gamma_{s,r}$, $\gamma_{v,r}$, $\gamma_{s,d}$, $\gamma_{v,d}$, δ_v, and δ_r, and the initial values for S, V, I_s, and I_v. In both scenarios, we simulated for 140 days with the parameters as those from the real epidemic listed in Table 2. For initial values of S, V, I_s, and I_v, see the lower panels of Figs. 3 and 5.

5   Simulations with Real Data for COVID-19 Pandemic

In this section, we apply our SVIRD model (Eqs. (1)–(6)) and regularized computational algorithm (Eqs. (19) and (20)) to real data on incidence cases and new daily deaths for the second wave of COVID-19 in the United States in 2021, when the Delta variant was one of the more widely spread strains [36]. Most states experienced this second wave during an approximately 4-month period between July 9 and November 25, 2021, while vaccines were distributed to the US general population starting in early 2021. So we can study the progression of the pandemic under the effect of vaccination. For our experiments, we choose data sets for two states, Georgia and California, as both have different population sizes (Georgia is much smaller with approximately 11 million people versus the nearly 40 million living in California), had different proportions of vaccinated individuals between July 9 and November 25, 2021, and had different COVID-19 protocols. In particular, California had more vaccinated people at the onset and at the end of this time window [36], and California had stricter masking protocols; masks were required indoors in most places during this time period, whereas they were only recommended in the state of Georgia. The model variables and initial conditions corresponding to the population sizes in Georgia and California at the onset of the second wave are given in Table 1. Initial conditions were found using Census and CDC data [36–39]. Here, $I(0) = I_s(0) + I_v(0)$ is the number of cases within the most recent week of the onset of the second wave, as most people with COVID-19 are no longer contagious 5 days after they first have symptoms and have been fever-free for at least 3 days.

Table 1 Initial conditions used in the SVIRD model for the Georgia and California data. Population size was based on the January 7, 2021 data from https://www.census.gov/quickfacts/GA and https://www.census.gov/quickfacts/CA

Variable	Meaning	
$S(t)$	Number of susceptible unvaccinated individuals	
$V(t)$	Number of susceptible vaccinated individuals	
$I_s(t)$	Number of infectious unvaccinated individuals	
$I_v(t)$	Number of infectious vaccinated individuals	
$R(t)$	Number of recovered individuals	
$D(t)$	Number of deceased individuals	
Initial condition	Georgia	California
$S(0)$	$10,799,566 - V(0) - I(0)$	$39,237,836 - V(0) - I(0)$
$V(0)$	$3,942,002$	$20,086,693$
$I_s(0)$	$3,580$	$25,039$
$I_v(0)$	$731 (= 3580 * 5116/25039)$	$5,116$
$R(0)$	0	0
$D(0)$	0	0

Table 2 Parameter values recorded for California and Georgia during the *second wave* of the pandemic, July 9–November 25, 2021 (approximately 4 months). The bars "–" in the last column mean that these values were calculated using $\gamma_{s,d}$, as described in the text

Parameter	Meaning	Value	Source
$\beta(t)$	Transmission rate		
p	Vaccination rate	0.00086 day^{-1}	[37–39]
α	Vaccine dose efficacy	0.8	[40, 43, 44]
$\gamma_{s,r}$	Recovery rate of unvaccinated	0.0995 day^{-1}	–
$\gamma_{v,r}$	Recovery rate for vaccinated	0.09996 day^{-1}	–
$\gamma_{s,d}$	Case-fatality for unvaccinated	0.00027 day^{-1}	[45]
$\gamma_{v,d}$	Case-fatality for vaccinated	0.000021 day^{-1}	–
δ_v	Loss of immunity for vaccinated	0 day^{-1}	[46]
δ_s	Loss of immunity for unvaccinated	0.011 day^{-1}	

System parameter values used for California and Georgia during the *second wave* of the pandemic are presented in Table 2. The rationale for the selection of these values is as follows:

- Vaccination rate p: Based on the CDC data [39], during the selected time window, the proportion of fully vaccinated people changed from 37.5% to 49.8% in Georgia and from 51.1% to 63.1% in California, both of which resulted in about 12% increase in vaccination. Dividing this by our 140-day window gives the approximate daily vaccination rate p of 0.00086 day^{-1}.
- Vaccine effectiveness α: We choose $\alpha = 0.8$ as the age-standardized crude vaccine effectiveness for infection was reported at 80% during July–November of 2021 [40].
- Death rate $\gamma_{s,d}$: We calculate $\gamma_{s,d} = 0.005/18.5 = 0.00027$ days^{-1} as the infectious fatality ratio IFR was reported as 0.5% from [41], and the median time from illness onset to death is 18.5 days (reported number for vaccinated vs unvaccinated [42]).
- Death rate $\gamma_{v,d}$: We take $\gamma_{v,d} = (0.005/12.7)/18.5 = 0.000021$ days^{-1} because during October–November, unvaccinated persons had 12.7 times the risks for COVID-19—associated death compared with those that were vaccinated without booster doses [33].
- Recovery rate $\gamma_{s,r}$: Assuming that individuals infected with COVID-19 either recover or die and using a recovery rate of 10 days, we conclude that the recovery rate for unvaccinated individuals is $\gamma_{s,r} = (1 - 0.005)/10 = 0.0995$ days^{-1}.
- Recovery rate $\gamma_{v,r}$: With a similar rationale as above, we estimate the recovery rate for vaccinated individuals as $\gamma_{v,r} = (1 - 0.005/12.7)/10 = 0.09996$ days^{-1}.
- Loss of immunity rate for recovered individuals δ_r: We set $\delta_s = 1/90 = 0.011$ days^{-1}.
- Loss of immunity rate for vaccinated individuals δ_v: We use $\delta_v = 0$ as the Moderna and Pfizer-BioNTech vaccines offer immunity against COVID-19 for at least 6 months, and most people in the USA were fully vaccinated by the

end of April 2021 or later. Therefore, they still had immunity against COVID-19 during most of the study period.

In the case of real data, apart from the measurement errors, which were incorporated in our earlier experiments, we also encounter modeling errors, which make the process considerably more unstable. Thus, apart from the penalty term, $\frac{\tau_k}{2}||\theta - \bar{\theta}||^2$, the iterative scheme also needs to be regularized by discretization. For this reason, fewer basis functions are used for the state variables. Specifically, we take 6 basis functions for each unobserved state variable, S, V, I_s, and I_v, for the Georgia data, and 12 basis functions for each unobserved state variable for the California data. To further stabilize the process, we also introduce a smaller step size, $\zeta = 0.1$, as we update $S(t)$, $V(t)$, $I_s(t)$, and $I_v(t)$. This calls for more iterations needed to achieve the desirable data fit. The iterative process is terminated when $k = 130$ for the Georgia data with regularization sequence $\tau_k = 1/(k+1)^{10}$ and $k = 58$ for the California data with $\tau_k = 10^3/(k+1)^7$. Overall, the time until convergence remains the same as for the case of synthetic data since the increase in the number of iterations is balanced by the reduction in the number of basis functions.

Another important aspect is the reporting rate of new cases. While it is natural to assume that the reporting rate for deaths due to COVID-19 is high, the reporting rate for daily new COVID-19 cases is unlikely to be anywhere close to 100% considering the large number of mild and asymptomatic cases ("silent spreaders" [47]). Figures 6

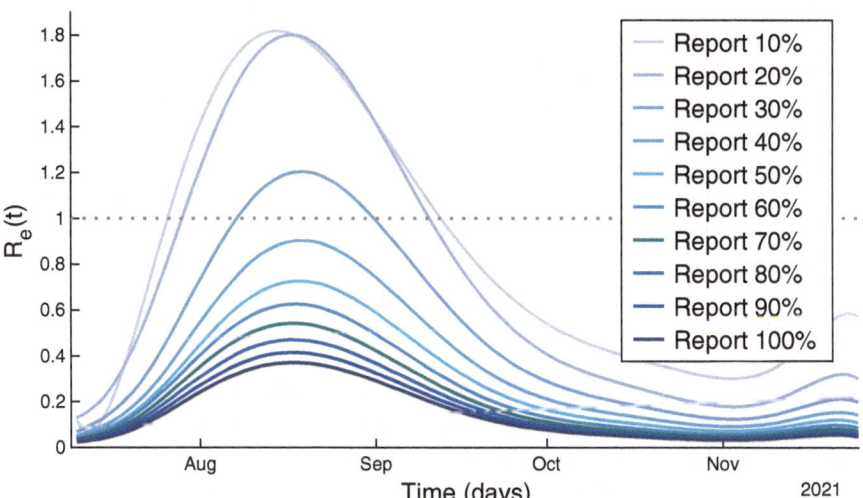

Fig. 6 Reconstructed effective reproduction numbers, $\mathcal{R}_e(t)$, for various assumed reporting rates in the state of Georgia. Simulations are carried out with 10 basis functions for the transmission rate, $\beta(t)$, and 6 basis functions for each unobserved state variable, S, V, I_s, and I_v, i.e., 24 basis functions for all state variables combined. The regularization sequence is $\tau_k = 1/(k+1)^{10}$, and the iterations are stopped when $k = 130$. This stopping time is determined by the goodness of fit to the Georgia data set

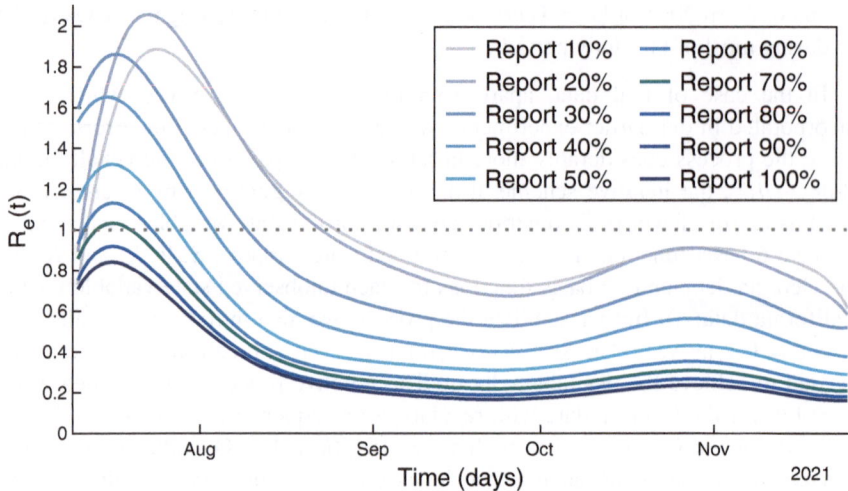

Fig. 7 Reconstructed effective reproduction numbers, $\mathcal{R}_e(t)$, for various assumed reporting rates in the state of California. Simulations are carried out with 10 basis functions for the transmission rate, $\beta(t)$, and 12 basis functions for each unobserved state variable, S, V, I_s, and I_v, i.e., 48 basis functions for all state variables combined. The regularization sequence is $\tau_k = 10^3/(k+1)^7$, and the iterations are stopped when $k = 58$. This stopping time is determined by the goodness of fit to the California data set

and 7 compare reconstructed time-dependent effective reproduction numbers, $\mathcal{R}_e(t)$, for various assumed reporting rates of daily new cases in Georgia and California, respectively (for both states, we fixed the reporting rate for daily new deaths due to COVID-19 at 90%). We know that at the onset of the Delta variant wave of the COVID-19 pandemic, the reproduction number must have been above 1 for some time. Thus, Fig. 6 suggests that the reporting rate of new COVID-19 incidence cases in the state of Georgia is 10–30%. For California, we see that the reporting rate is 10–60% as illustrated in Fig. 7. This is consistent with the estimation of COVID-19 incidence reporting rate carried out in [48]. In [48], the reporting rate was cast as one of the unknown parameters in the model and had to be reconstructed by the optimization algorithm. For the initial pre-vaccination stage of COVID-19 pandemic in the state of Georgia, the reporting rate for new incidence cases was estimated to be 0.23 (95% confidence interval (CI): [0.22,0.24]). For the reasons listed previously and as suggested by our numerical study, in simulations presented in Figs. 8, 9, 10, and 11, we assume a 90% reporting rate for new daily deaths due to COVID-19 and a 20% reporting rate for new incidence cases in the states of Georgia and California.

In Figs. 8 and 10, we show the transmission rate, $\beta(t)$, and the effective reproduction number, $\mathcal{R}_e(t)$, reconstructed from daily data on new cases and deaths for the states of Georgia and California, respectively, for the period from July 9 to November 25, 2021. The top panels of Figs. 9 and 11 show how incidence curves for daily new cases and deaths in the states of Georgia and California are compared to real data used for parameter estimation in the optimization process (Eqs. (19)

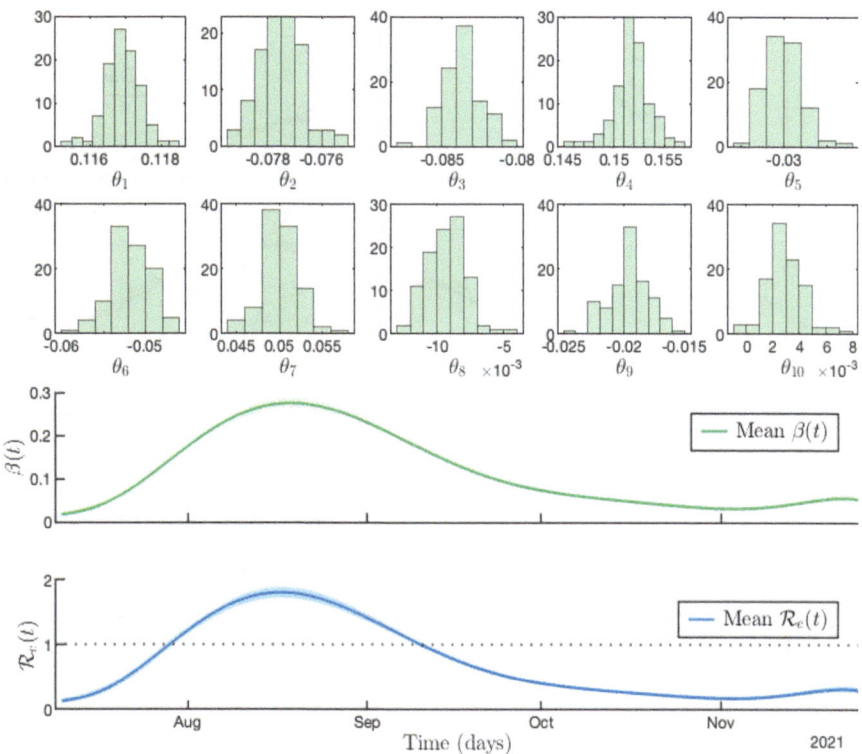

Fig. 8 Reconstruction of disease transmission $\beta(t)$ (along with coefficients) and the effective reproduction number $\mathcal{R}_e(t)$ for the state of Georgia

and (20)). Reconstructed $S(t)$, $V(t)$, $I_s(t)$, and $I_v(t)$ for the states of Georgia and California can be viewed in the lower panel of the same figures. One may notice that the California incidence data (top panel of Fig. 11) are more "spread out" than the Georgia incidence data (top panel of Fig. 9). This is because, for the Georgia data, a rolling 7-day average was recorded each week since in Georgia new cases were often not reported on the weekends when the Delta variant was dominant. So, the approximation of unobserved state variables for the state of California is more uncertain as compared to Georgia and to the sets of synthetic data.

The parameter estimation process is initiated with $\beta_0 = 0.5$ for both Georgia and California. The reconstruction is done with $m = 10$ in both cases (the number of basis functions for the transmission rate). For Georgia, the number of basis functions for each unobserved state variable is $n = 6$ (i.e., 24 basis functions for all state variables, S, V, I_s, and I_v, combined). The iterative process started with $\tau_0 = 1$. The regularization sequence is driven to zero at the rate $1/(k+1)^{10}$. Like in the case of Georgia, for the California data set the number of basis functions for S, V, I_s, and I_v is significantly reduced (from $n = 40$ to $n = 12$), as compared to

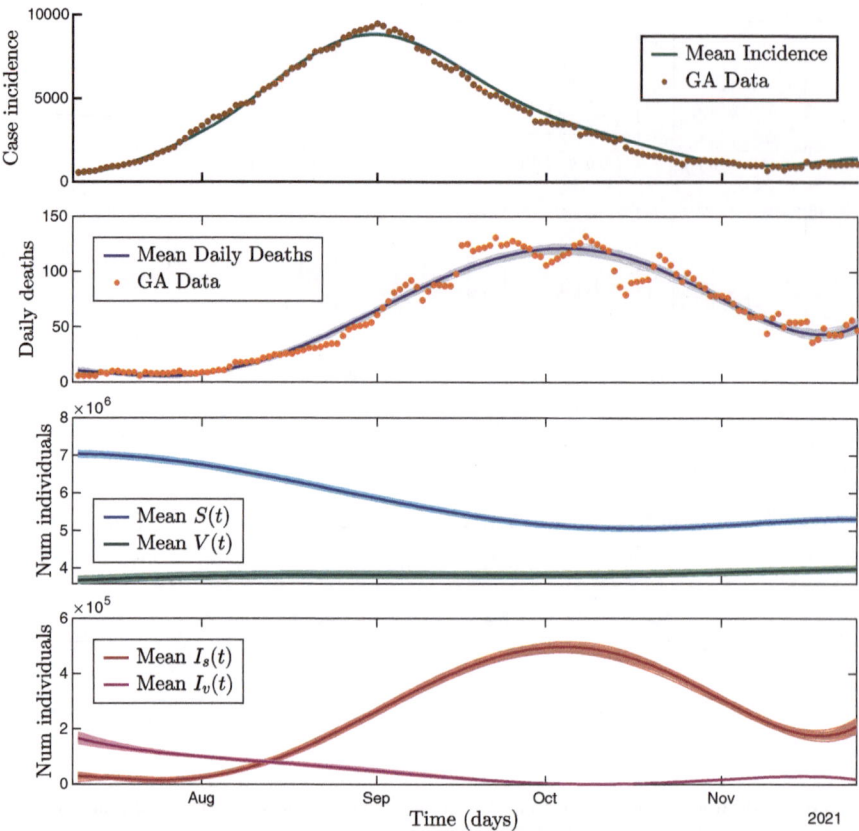

Fig. 9 State of Georgia (GA) case study. Top to bottom: state data (dots) and model fit (solid line) for daily new cases and daily new deaths; 100 bootstrap model reconstructions for $S(t)$ (blue), $V(t)$ (green), $I_s(t)$ (red), and $I_v(t)$ (pink). The mean of the bootstraps is a darker line of the color corresponding to each compartment

reconstructions with synthetic data in order to further stabilize predictor–corrector algorithm (Eqs. (19) and (20)) in the presence of modeling error.

By comparing Figs. 8 and 10, one can see that the start of the Delta variant wave in the state of California was more rapid as compared to Georgia, but it took longer for Georgia to get the virus under control (as compared to California). In California, the effective reproduction number, $\mathcal{R}_e(t)$, dropped under 1 around mid-August, while in Georgia $\mathcal{R}_e(t)$ remained greater than 1 until early September 2021. However, in California, the effective reproduction number almost bounced back to 1 in late October before going down again toward the end of the study period. In Georgia, on the other hand, $\mathcal{R}_e(t)$ remained very low after the end of September.

In the top panels of Figs. 9 and 11, we note the peak of around 9,000 new incidence cases in the state of Georgia in early September and the peak in mid-August of approximately 13,000 new incidence cases in the state of California.

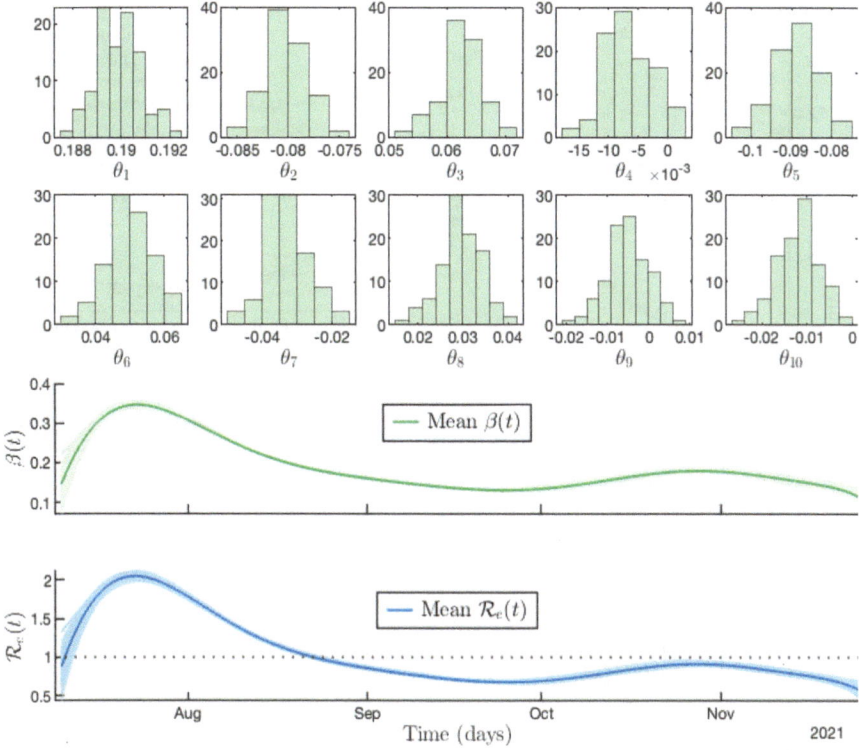

Fig. 10 Reconstruction of disease transmission $\beta(t)$ (along with coefficients) and the effective reproduction number $\mathcal{R}_e(t)$ for the state of California

In both states the daily reported new deaths are under 150 people. The peaks in deaths follow the peaks of incidence cases, in early October in Georgia and in early September in California. Reconstructed curves, $I_s(t)$ and $I_v(t)$, are consistent with the reported percentage of vaccinated individuals in the states of Georgia and California, respectively (Figs. 9 and 11).

6 Conclusion and Future Work

In this chapter, we propose a new dynamic model of COVID-19 transmission that takes into account the vaccination status of both susceptible and infected humans. It also includes a possible loss of immunity and reinfection within both vaccinated and unvaccinated populations. To estimate the unknown disease parameters, we develop a novel computational algorithm, which employs a parameter cascade approach. The proposed method is used to reconstruct time-dependent transition rates, $\beta(t)$,

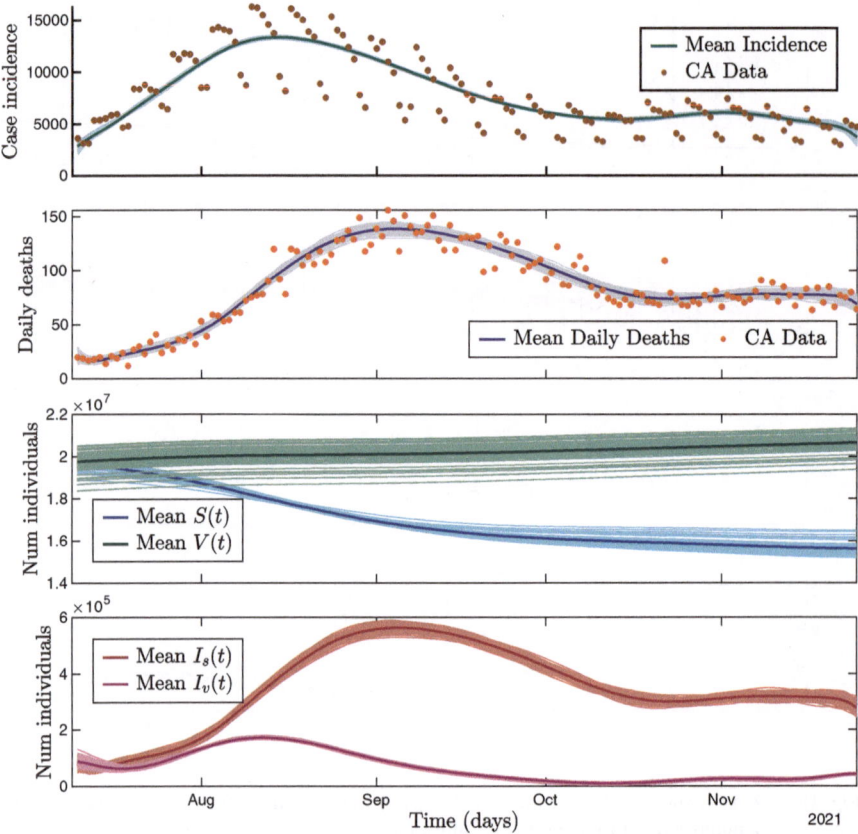

Fig. 11 State of California (CA) case study. Top to bottom: state data (dots) and model fit (solid line) for daily new cases and daily new deaths; 100 bootstrap model reconstructions for $S(t)$ (blue), $V(t)$ (green), $I_s(t)$ (red), and $I_v(t)$ (pink). The mean of the bootstraps is a darker line of the color corresponding to each compartment

and effective reproduction numbers, $\mathcal{R}_e(t)$, from synthetic and real data for the COVID-19 pandemic. Apart from COVID-19, the proposed compartmental model and iteratively regularized optimization method can be applied to the study of other infectious diseases.

In the course of our numerical study, the new optimization technique has emerged as a reliable alternative to more traditional trust-region and gradient-descent algorithms that are commonly used in parameter estimation. The efficiency of these algorithms is limited when a complex biological model (which may be a system of nonlinear ordinary or partial differential equations) constraining the underlying minimization problem does not have a closed-form solution and has to be solved numerically at every step of the iterative process. Our new method, on the other hand, does not require either exact or approximate solution to the constraining system.

In reconstructing time-dependent transmission rates, $\beta(t)$, in order to reduce the computational load and to improve the estimate efficiency, we pre-specified the values of other system parameters by conducting a thorough review of the literature. To assess the sensitivity of reconstructed transmission rates to slight variations in pre-estimated parameters, one can build a Bayesian model to assign priors to pre-specified parameters, and the posterior distributions of transmission rates will incorporate the uncertainty in these parameters. This is an important topic for future work. Note that for a simpler SIRD model corresponding to a pre-vaccination stage of the COVID-19 pandemic, the sensitivity analysis has been conducted in [48]. In [48], for every bootstrap iteration, the recovery rate, γ, and the fatality rate, ν, have been sampled from normal distributions, $N(0.20, 0.02)$ and $N(0.005, 0.001)$, respectively. The normal distribution, $N(0.20, 0.02)$, for the recovery rate, γ, reflected an average infectious period between 3 and 20 days, while the normal distribution $N(0.005, 0.001)$ for the fatality rate, ν, accounted for the variation of this parameter within different risk groups. The reconstructed values of $\beta(t)$ with normally distributed γ and ν were almost identical to those reconstructed with constant (mean) values of these pre-estimated parameters showing a very low sensitivity of $\beta(t)$ to inevitable variations in COVID-19 infectious periods and fatality rates.

With a considerable portion of mild and asymptomatic cases, the number of reported daily new cases is much lower than the actual value. In this chapter, we change the reporting rates of new incidence cases and investigate how different reporting rates affect the reconstruction of effective reproduction numbers, $\mathcal{R}_e(t)$, in our numerical simulations. Thus, another important direction of future research will be to modify our reconstruction process to include the estimation of the unknown percentages of new incidence cases along with the unknown time-dependent transmission rate, $\beta(t)$, and other system parameters. The problem of the reporting rate can also be addressed by extending the model to include the compartment of asymptomatic spreaders.

We also plan to add line search routines and incorporate nonnegativity constraints for unobserved state variables, S, V, I_s, and I_v, in iteratively regularized predictor–corrector algorithm (Eqs. (19) and (20)). This will allow further accuracy improvements and stability of the proposed optimization method.

Last but not least, the methodology must be extended to provide near real-time forecasting of future incidence cases and deaths (among vaccinated and unvaccinated individuals) from early data for an unfolding outbreak. This research is crucial for control and prevention, in particular, for the assessment of various vaccination strategies.

Acknowledgments The work described herein was initiated during the Collaborative Workshop for Women in Mathematical Biology funded and hosted by UnitedHealth Group Optum of Minnetonka, MN and supported by University of Minnesota's Institute for Mathematics and its Applications in June 2022. Additionally, the authors and editors thank the anonymous peer reviewers for their feedback, which strengthened this work.

AS was supported by the National Science Foundation (NSF) under Grant No. DMS-2011622.

References

1. A. Smirnova, G. Chowell, Infect. Dis. Modell. **2**(2), 268 (2017)
2. W. Kermack, A. McKendrick, Proc. R. Soc. London Ser. A **115**(772), 700 (1927)
3. N.T. Bailey, *The Mathematical Theory of Infectious Diseases and its Applications* (Charles Griffin & Company, High Wycombe, 1975)
4. R. Schlickeiser, M. Kroger, Physics **3**(2), 386 (2021)
5. Y. Mohamadou, A. Halidou, P. Kapen, Appl. Intell. **50**(11), 3913 (2020)
6. A. Mahajan, N. Sivadas, R. Solanki, Chaos Solitons Fractals **140**, 110156 (2020)
7. R. Singh, R. Adhikari, Preprint. arXiv:2003.12055 (2020)
8. T. Sardar, S. Nadim, S. Rana, J. Chattopadhyay, Chaos Solitons Fractals **139**, 110078 (2020)
9. M. Dalton, P. Dougall, F. Amoah-Darko, W. Annan, E. Asante-Asamani, S. Bailey, J. Greene, D. White, PLoS ONE **17**(11), e0274407 (2022). https://doi.org/10.1371/journal.pone.0274407
10. W. Wong, F. Juwono, T. Chua, Preprint. arXiv:2101.07494 (2021)
11. M. Angeli, G. Neofotistos, M. Mattheakis, E. Kaxiras, Chaos Solitons Fractals **154**, 111621 (2022)
12. C. Kirkeby, T. Halasa, M. Gussmann, Sci. Rep. **7**(9496) (2017)
13. D. Bates, D. Watts, *Nonlinear Regression Analysis and its Applications* (Wiley, Hoboken, 1988)
14. G. Seber, C. Wild, *Nonlinear Regression* (Wiley, Hoboken, New Jersey, 2003)
15. J. Cao, J. Huang, H. Wu, J. Comput. Gr. Stat. **21**(1), 42 (2012)
16. J. Ramsay, G. Hooker, *Dynamic Data Analysis* (Springer, New York, 2017)
17. J.M. Varah, SIAM J. Sci. Stat. Comput. **3**(1), 28 (1982)
18. H. Liang, H. Wu, J. Am. Stat. Assoc. **103**(484), 1570 (2008)
19. J. Ramsay, G. Hooker, D. Campbell, J. Cao, J. R. Stat. Soc. Ser. B **69**(5), 741 (2007)
20. H. Attouch, J. Bolte, P. Redont, A. Soubeyran, Math. Oper. Res. **35**(2), 438 (2010)
21. J. Bolte, S. Sabach, M. Teboulle, Math. Program. **146**, 459 (2014)
22. M. Hong, Z. Luo, M. Razaviyayn, SIAM J. Optim. **26**(1) (2016)
23. A. Patrascu, I. Necoara, J. Global. Optim. **61**(1), 19 (2015)
24. D. Driggs, J. Tang, J. Liang, M. Davies, C. Schönlieb, SIAM J. Imaging Sci. **14**(4) (2021)
25. D. Davis, M. Udell, B. Edmunds, *NIPS'16: Proceedings of the 30th International Conference on Neural Information Processing Systems* (2016), pp. 226–234
26. A. Smirnova, A. Bakushinsky, Inverse. Prob. **36**(12), 125015 (2020)
27. M. Eisenberg, S. Robertson, J. Tien, J. Theo. Biol. **324**, 84 (2013)
28. K. Roosa, G. Chowell, Theo. Biol. Med. Modell. **16**(1), 1 (2019)
29. N. Tuncer, H. Gulbudak, V. Cannataro, M. Martcheva, Bull. Math. Biol. **9**(78), 1796 (2016)
30. G. Giordano, F. Blanchini, R. Bruno, P. Colaneri, A.D. Filippo, A.D. Matteo, M. Colaneri, Nat. Med. **26**, 855 (2020)
31. K. Roosa, Y. Lee, R. Luo, A. Kirpich, R. Rothenberg, J.M. Hyman, P. Yan, G. Chowell, J. Clin. Med. **9**, 596 (2020)
32. R.N. Thompson, Epidemiological models are important tools for guiding COVID-19 interventions, BMC Med. **18**, 152 (2020). https://doi.org/10.1186/s12916-020-01628-4
33. A.G. Johnson, A.B. Amin, A.R. Ali, et al., COVID-19 incidence and death rates among unvaccinated and fully vaccinated adults with and without booster doses during periods of Delta and Omicron variant emergence - 25 U.S. Jurisdictions, April 4–December 25, 2021. Morb Mortal Wkly Rep (MMWR) **71**, 132–138 (2022). http://doi.org/10.15585/mmwr.mm7104e2
34. O. Diekmann, J. Heesterbeek, J. Metz, J. Math. Biol. **28**(4), 365 (1990)
35. P. van den Driessche, J. Watmough, Math. Biosci. **180**(1–2), 29 (2002)
36. Centers for Disease Control and Prevention. Trends in Number of COVID-19 Cases and Deaths in the US Reported to CDC, by State/Territory. https://covid.cdc.gov/covid-data-tracker/#trends_dailycases_select_06 (2022). Accessed 30 Sep 2022

37. America Counts Staff. California Remained Most Populous State but Growth Slowed Last Decade. https://www.census.gov/library/stories/state-by-state/california-population-change-between-census-decade.html. Accessed 30 Sep 2022

38. America Counts Staff. Georgia Among Top Five Population Gainers Last Decade. https://www.census.gov/library/stories/state-by-state/georgia-population-change-between-census-decade.html (2021). Accessed 30 Sep 2022

39. Centers for Disease Control and Prevention. Trends in Number of COVID-19 Vaccinations in the US. https://covid.cdc.gov/covid-data-tracker/#vaccination-trends (2022). Accessed 30 Sep 2022

40. Centers for Disease Control and Prevention. Morbidity and Mortality Weekly Report. https://www.cdc.gov/mmwr/volumes/71/wr/mm7104e2.htm (2022). Accessed 30 Sep 2022

41. M. O'Driscoll, G.R. dos Santos, L. Wang, D.A.T. Cummings, A.S. Azman, J. Paireau, A. Fontanet, S. Cauchemez, H. Salje, Nature **590**, 140–145 (21)

42. Drugs.com. How Do COVID-19 Symptoms Progress and What Causes Death? https://www.drugs.com/medical-answers/covid-19-symptoms-progress-death-3536264/ (2022). Accessed 30 Sep 2022

43. N. Lewis, L. Chambers, H. Chu, T. Fortnam, R.D. Vito, L. Gargano, P. Chan, J. McDonald, J. Hogan, JAMA Netw. Open, 1–11 (2022). https://doi.org/10.1001/jamanetworkopen.2022.23917

44. J. Lopez-Bernal, N. Andrews, C. Gower, E. Gallagher, R. Simmons, S. Thelwall, J. Stowe, E. Tessier, N. Groves, G. Dabrera, et al., N. Engl. J. Med. **385**(7), 585 (2021)

45. UCHealth. The Delta Variant of COVID-19. https://www.uchealth.org/services/infectious-diseases/coronavirus-covid-19/the-delta-variant-of-covid-19/ (2022). Accessed 30 Sep 2022

46. N. Doria-Rose, M.S. Suthar, M. Makowski, S. O'Connell, A.B. McDermott, B. Flach, J.E. Ledgerwood, J.R. Mascola, B.S. Graham, B.C. Lin, N. Engl. J. Med. **384**(23), 2259 (2021)

47. K. Hwang, C. Edholm, O. Saucedo, L. Allen, N. Shakiba, Bull. Math. Biol. **84**(9), 91 (2022)

48. A. Smirnova, B. Pidgeon, R. Luo, J. Inverse. Ill-Posed. Prob. **30**(6), 823 (2022). https://doi.org/10.1515/jiip-2021-0037

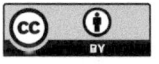

Index

© The Editor(s) (if applicable) and The Author(s) 2024
A. N. Ford Versypt et al. (eds.), *Mathematical Modeling for Women's Health*,
The IMA Volumes in Mathematics and its Applications 166,
https://doi.org/10.1007/978-3-031-58516-6